Optimised service centre location and allocated demand

Tripolis, in Arcadia, Greece

Coverage or p-centre location optimisation problem. See Section 7.4.2 for more details. Map produced using S-Distance software, © S A Sirigos

Contents

Foreword xiii

1 Introduction 1

1.1   Motivation and Media 1
   1.1.1   Guide overview 1
   1.1.2   Spatial analysis and GIS 1
1.2   Intended Audience and Companion Materials 5
   1.2.1   Scope 5
   1.2.2   GIS software 5
      1.2.2.1   Sample software products 6
      1.2.2.2   Software performance 7
   1.2.3   Suggested reading 11
1.3   Structure 13
1.4   Terminology and abbreviations 14
   1.4.1   Definitions 14
1.5   Common Measures and Notation 21
   1.5.1   Notation 21
   1.5.2   Statistical measures and related formulas 22
      1.5.2.1   Counts and specific values 23
      1.5.2.2   Measures of centrality 23
      1.5.2.3   Measures of spread 25
      1.5.2.4   Measures of distribution shape 27
      1.5.2.5   Measures of complexity and dimensionality 28
      1.5.2.6   Common distributions 29
      1.5.2.7   Data transforms and back transforms 29
      1.5.2.8   Selected functions 31
      1.5.2.9   Matrix expressions 32

2 Conceptual Frameworks for Spatial Analysis 33

2.1   The geospatial perspective 33
2.2   Basic Primitives 34
   2.2.1   Place 34
   2.2.2   Attributes 35
   2.2.3   Objects 36
   2.2.4   Maps 37
   2.2.5   Multiple properties of places 38
   2.2.6   Fields 38
   2.2.7   Networks 39
   2.2.8   Density estimation 39
   2.2.9   Detail, resolution, and scale 39
   2.2.10 Topology 40
2.3   Spatial Relationships 41
   2.3.1   Co-location 41
   2.3.2   Distance and direction 41
   2.3.3   Multidimensional scaling 42
   2.3.4   Spatial context 42

2.3.5	Neighbourhood	42
2.3.6	Spatial heterogeneity	43
2.3.7	Spatial dependence	43
2.3.8	Spatial sampling	44
2.3.9	Spatial interpolation	44
2.3.10	Smoothing and sharpening	45
2.4	**Spatial Statistics**	**47**
2.4.1	Spatial probability	47
2.4.2	Probability density	47
2.4.3	Uncertainty	47
2.4.4	Statistical inference	48
2.5	**Spatial Data Infrastructure**	**50**
2.5.1	Geoportals	50
2.5.2	Metadata	50
2.5.3	Interoperability	50
2.5.4	Conclusion	50

3 Historical and Methodological Context ... 51

3.1	**Historical context**	**51**
3.2	**Methodological context**	**54**
3.2.1	Spatial analysis as a process	54
3.2.2	Analytical methodologies	55
3.2.3	Spatial analysis and the PPDAC model	59
3.2.3.1	Problem: Framing the question	60
3.2.3.2	Plan: Formulating the approach	61
3.2.3.3	Data: Data acquisition	62
3.2.3.4	Analysis: Analytical methods and tools	63
3.2.3.5	Conclusions: Delivering the results	65
3.2.4	The changing context of PPDAC	65

4 Building Blocks of Spatial Analysis ... 69

4.1	**Spatial data models and methods**	**69**
4.2	**Geometric and Related Operations**	**71**
4.2.1	Length and area for vector datasets	71
4.2.2	Length and area for raster datasets	73
4.2.3	Surface area	74
4.2.3.1	Projected surfaces	74
4.2.3.2	Terrestrial (unprojected) surface area	76
4.2.4	Line Smoothing and point-weeding	77
4.2.5	Centroids and centres	79
4.2.5.1	Polygon centroids and centres	79
4.2.5.2	Point sets	82
4.2.5.3	Lines	84
4.2.6	Point (object) in polygon (PIP)	85
4.2.7	Polygon decomposition	86
4.2.8	Shape	87
4.2.9	Overlay and combination operations	89
4.2.10	Areal interpolation	92
4.2.11	Districting and re-districting	94
4.2.12	Classification and clustering	99
4.2.12.1	Univariate classification schemes	99

4.2.12.2 Multivariate classification and clustering 102
4.2.13 Boundaries and zone membership 106
4.2.13.1 Convex hulls.. 106
4.2.13.2 Non-convex hulls 107
4.2.13.3 Minimum Bounding Rectangles (MBRs) 108
4.2.13.4 Fuzzy boundaries 110
4.2.13.5 Breaklines and natural boundaries 112
4.2.14 Tessellations and triangulations... 113
4.2.14.1 Delaunay Triangulation 113
4.2.14.2 TINs — Triangulated irregular networks 115
4.2.14.3 Voronoi/Thiessen polygons 115
4.3   Queries and Computations **118**
4.3.1 Spatial selection and spatial queries 118
4.3.2 Simple calculations.. 119
4.3.3 Ratios, indices, normalisation and standardisation 122
4.3.4 Density, kernels and occupancy.. 126
4.3.4.1 Point density 126
4.3.4.2 Line and intersection densities 133
4.4   Distance Operations **135**
4.4.1 Metrics.. 137
4.4.1.1 Introduction 137
4.4.1.2 Terrestrial distances 138
4.4.1.3 Extended Euclidean and L_p-metric distances 139
4.4.2 Cost distance 142
4.4.2.1 Accumulated cost surfaces and least cost paths 142
4.4.2.2 Distance transforms.. 146
4.4.3 Network distance 150
4.4.4 Buffering 151
4.4.4.1 Vector buffering 151
4.4.4.2 Raster buffering 153
4.4.4.3 Hybrid buffering 153
4.4.5 Network buffering 153
4.4.6 Distance decay models 153
4.5   Directional Operations **157**
4.5.1 Directional analysis - overview.. 157
4.5.2 Directional analysis of linear datasets 157
4.5.3 Directional analysis of point datasets 162
4.5.4 Directional analysis of surfaces 164
4.6   Grid Operations and Map Algebra **165**
4.6.1 Operations on single and multiple grids 165
4.6.2 Linear spatial filtering 166
4.6.3 Non-linear spatial filtering 170
4.6.4 Erosion and dilation 170

5   Data Exploration and Spatial Statistics **173**
5.1   Statistical Methods and Spatial Data **173**
5.1.1 Descriptive statistics.. 174
5.1.2 Spatial sampling 175
5.1.2.1 Sampling frameworks 175
5.1.2.2 Declustering.. 179

5.2	**Exploratory Spatial Data Analysis**	181
5.2.1	EDA, ESDA and ESTDA	181
5.2.2	Outlier detection	181
5.2.2.1	Mapped histograms	181
5.2.2.2	Box plots	182
5.2.3	Cross tabulations and conditional choropleth plots	184
5.2.4	ESDA and mapped point data	187
5.2.5	Trend analysis of continuous data	188
5.2.6	Cluster hunting	188
5.3	**Grid-based Statistics**	190
5.3.1	Overview of grid-based statistics	190
5.3.2	Crosstabulated grid data	191
5.3.3	Quadrat analysis of grid datasets	193
5.3.4	Landscape Metrics	194
5.3.4.1	Non-spatial metrics	195
5.3.4.2	Spatial metrics	196
5.3.4.3	Landscape metrics – table of metrics	200
5.4	**Point Sets and Distance Statistics**	202
5.4.1	Basic distance-derived statistics	202
5.4.2	Nearest neighbour methods	202
5.4.3	Hot spot and cluster analysis	209
5.4.3.1	Hierarchical nearest neighbour clustering	210
5.4.3.2	K-means clustering	211
5.4.3.3	Kernel density clustering	211
5.5	**Spatial Autocorrelation**	213
5.5.1	Autocorrelation, time series and spatial analysis	213
5.5.2	Global spatial autocorrelation	214
5.5.2.1	Join counts	214
5.5.2.2	Moran I and Geary C	218
5.5.2.3	Weighting models and lags	225
5.5.3	Local indicators of spatial association (LISA)	226
5.5.4	Significance tests for autocorrelation indices	229
5.6	**Regression Methods**	231
5.6.1	Regression overview	231
5.6.2	Simple regression and trend surface modelling	236
5.6.3	Geographically Weighted Regression (GWR)	238
5.6.4	Spatial autoregressive and Bayesian modelling	242
5.6.4.1	Spatial autoregressive modelling	242
5.6.4.2	Conditional autoregressive and Bayesian modelling	244
5.6.5	Spatial filtering models	247
6	**Surface and Field Analysis**	**249**
6.1	**Modelling Surfaces**	249
6.1.1	Test datasets	249
6.1.2	Surfaces and fields	250
6.1.3	Raster models	251
6.1.4	Vector models	254
6.1.5	Mathematical models	255
6.1.6	Statistical and fractal models	257
6.2	**Surface Geometry**	259
6.2.1	Gradient, slope and aspect	259
6.2.1.1	Slope	259

6.2.1.2   Aspect 261
6.2.2   Profiles and curvature 264
    6.2.2.1   Profiles and cross-sections 264
    6.2.2.2   Curvature and morphometric analysis 265
    6.2.2.3   Profile curvature 267
    6.2.2.4   Plan curvature 268
    6.2.2.5   Tangential curvature 268
    6.2.2.6   Longitudinal and cross-sectional curvature 268
    6.2.2.7   Mean, maximum and minimum curvature 269
6.2.3   Directional derivatives 269
6.2.4   Paths on surfaces 269
6.2.5   Surface smoothing 271
6.2.6   Pit filling 272
6.2.7   Volumetric analysis 272
6.3   Visibility 274
6.3.1   Viewsheds and RF propagation 274
6.3.2   Line of sight 277
6.3.3   Isovist analysis 279
6.4   Watersheds and Drainage 280
6.4.1   Overview of watersheds and drainage 280
6.4.2   Drainage modelling 280
6.4.3   D-infinity model 281
6.4.4   Drainage modelling case study 282
    6.4.4.1   Flow accumulation 282
    6.4.4.2   Stream network construction 282
    6.4.4.3   Stream basin construction 283
6.5   Gridding, Interpolation and Contouring 284
6.5.1   Overview of gridding and interpolation 284
6.5.2   Gridding and interpolation methods 285
    6.5.2.1   Comparison of sample gridding and interpolation methods 285
    6.5.2.2   Contour plots of sample gridding and interpolation methods 288
6.5.3   Contouring 291
6.6   Deterministic Interpolation Methods 293
6.6.1   Inverse distance weighting (IDW) 294
6.6.2   Natural neighbour 296
6.6.3   Nearest-neighbour 298
6.6.4   Radial basis and spline functions 298
6.6.5   Modified Shepard 300
6.6.6   Triangulation with linear interpolation 300
6.6.7   Triangulation with spline-like interpolation 300
6.6.8   Rectangular or bi-linear interpolation 301
6.6.9   Profiling 301
6.6.10   Polynomial regression 302
6.6.11   Minimum curvature 302
6.6.12   Moving average 302
6.6.13   Local polynomial 302
6.6.14   Topogrid/Topo to raster 303
6.7   Geostatistical Interpolation Methods 304
6.7.1   Core concepts 304
    6.7.1.1   Geostatistics 304
    6.7.1.2   Geostatistical references 305

6.7.1.3   Semivariance 305
6.7.1.4   Sample size 306
6.7.1.5   Support.. 307
6.7.1.6   Declustering 307
6.7.1.7   Variogram.. 307
6.7.1.8   Stationarity... 307
6.7.1.9   Sill, range and nugget 308
6.7.1.10   Transformation 309
6.7.1.11   Anisotropy 310
6.7.1.12   Indicator semivariance 311
6.7.1.13   Cross-semivariance.. 311
6.7.1.14   Comments on geostatistical software packages............... 311
6.7.1.15   Semivariance modelling 312
6.7.1.16   Fractal analysis 316
6.7.1.17   Madograms and Rodograms 316
6.7.1.18   Periodograms and Fourier analysis 317
6.7.2   Kriging interpolation 317
6.7.2.1   Core process 317
6.7.2.2   Goodness of fit... 319
6.7.2.3   Simple Kriging... 320
6.7.2.4   Ordinary Kriging 320
6.7.2.5   Universal Kriging... 321
6.7.2.6   Median-Polishing and Kriging 321
6.7.2.7   Indicator Kriging... 321
6.7.2.8   Probability Kriging 321
6.7.2.9   Disjunctive Kriging 321
6.7.2.10   Stratified Kriging 322
6.7.2.11   Co-Kriging... 322
6.7.2.12   Factorial Kriging 322
6.7.2.13   Conditional simulation... 322

7   Network and location analysis 325
7.1   Introduction to network and location analysis............................. 325
7.1.1   Overview of network and location analysis............................ 325
7.1.2   Terminology 325
7.1.3   Source data 326
7.1.4   Algorithms and computational complexity theory 328
7.2   Key problems in network and location analysis............................ 330
7.2.1   Overview – network analysis problems 330
7.2.1.1   Key problems in network analysis 331
7.2.1.2   Network analysis and logistics software 334
7.2.2   Overview of location analysis problems...................................... 336
7.3   Network construction, optimal routes and optimal tours 338
7.3.1   Minimum spanning tree.. 338
7.3.2   Gabriel network.. 339
7.3.3   Steiner trees... 341
7.3.4   Shortest (network) path problems 342
7.3.4.1   Overview of shortest path problems 342
7.3.4.2   Dantzig algorithm 343
7.3.4.3   Dijkstra algorithm 344
7.3.4.4   A* algorithm 344
7.3.4.5   GIS implementations of SPA 344

    7.3.5   Tours, travelling salesman problems and vehicle routing........................... 346
        7.3.5.1   Capacitated vehicle routing 349
  7.4   **Location and service area problems** 351
    7.4.1   **Location problems**.. 351
    7.4.2   **Larger p-median and p-centre problems**... 354
        7.4.2.1   Simple heuristics 354
        7.4.2.2   Lagrangian relaxation.. 354
        7.4.2.3   Comparison of alternative p-median heuristics 357
    7.4.3   **Service areas** 359
        7.4.3.1   Travel time zones 360
  7.5   **Arc Routing** 362

Afterword 365

References 367

Web links 377
Principal software products cited 677
Associations and academic bodies..379
Online technical dictionaries/definitions 382
Spatial data, test data and spatial information sources...........................382
    Selected data sources.. 382
    Test datasets 383
Statistics links..383
Trade sites...384

Index 385
Index cross references..394

List of Figures

Figure 2-1 An example map showing points, lines, and areas appropriately symbolised 37
Figure 2-2 Topological relationships 40
Figure 2-3 Three alternative ways of defining neighbourhood, using simple GIS functions 43
Figure 2-4 Four distinct patterns of twelve points in a study area....................................... 46
Figure 2-5 The process of statistical inference... 48
Figure 3-1 The components of a GIS 51
Figure 3-2 Analytical process — after Mitchell (2005).. 55
Figure 3-3 Analytical process — after Draper et al. (2005) 56
Figure 3-4 PPDAC as an iterative process 58
Figure 4-1 Area calculation using Simpson's rule 71
Figure 4-2 Triangular approximation of surface area... 75
Figure 4-3 Surface model of DEM for OS TQ81NE tile 76
Figure 4-4 Smoothing techniques 78
Figure 4-5 Polygon centroid (M2) and alternative polygon centres 80
Figure 4-6 Centre and Centroid positioning... 81
Figure 4-7 Triangle centroid.. 81
Figure 4-8 Polygon centre selection... 82
Figure 4-9 Point set centres 83
Figure 4-10 Point in polygon — tests and special cases 85
Figure 4-11 Skeletonised convex polygon 87
Figure 4-12 GRASS overlay operations, v.overlay 90
Figure 4-13 Areal interpolation from census areas to a single grid cell 93
Figure 4-14 Proportionally assigned population values.. 93
Figure 4-15 Grouping data — Zone arrangement effects on voting results................................ 96
Figure 4-16 Creating postcode polygons... 98
Figure 4-17 Automated Zone Procedure (AZP) 98
Figure 4-18 AZP applied to part of Manchester, UK 99
Figure 4-19 Jenks Natural Breaks algorithm 102
Figure 4-20 Convex hull of sample point set.. 106
Figure 4-21 Interpolation within "centroid" MBR 109
Figure 4-22 Point locations inside and outside polygon 110
Figure 4-23 Sigmoidal fuzzy membership functions 111
Figure 4-24 Delaunay triangulation of spot height locations 114
Figure 4-25 Voronoi regions generated in ArcGIS and MATLab.. 116
Figure 4-26 Voronoi cells for a homogeneous grid 117
Figure 4-27 Cell-by-cell or Local operations 120
Figure 4-28 Map algebra: Index creation 121
Figure 4-29 Normalisation within ArcGIS 124
Figure 4-30 Quantile map of normalised SIDS data 125
Figure 4-31 Excess risk rate map for SIDS data 126
Figure 4-32 Point data... 127
Figure 4-33 Simple linear (box or uniform) kernel smoothing.. 127
Figure 4-34 Univariate Normal kernel smoothing and cumulative (Cum) densities................... 128
Figure 4-35 Alternative univariate kernel density functions 129
Figure 4-36 2D Normal Kernel 130
Figure 4-37 Normal kernel density map, lung cancer cases 130
Figure 4-38 Univariate kernel density functions, unit bandwidth 132
Figure 4-39 Alternative measures of terrain distance 135
Figure 4-40 Glasgow's Clockwork Orange Underground 138

Figure 4-41 Great circle and constant bearing paths, Boston to Bristol...............................139
Figure 4-42 p-metric circles 140
Figure 4-43 Cost distance model..142
Figure 4-44 Cost surface as grid 143
Figure 4-45 Grid resolution and cost distance 143
Figure 4-46 Accumulated cost surface and least cost paths...144
Figure 4-47 Alternative route selection by ACS 145
Figure 4-48 Steepest path vs tracked path 146
Figure 4-49 Urban traffic modelling..147
Figure 4-50 Notting Hill Carnival routes 148
Figure 4-51 Alternative routes selected by gradient constrained DT 148
Figure 4-52 Hellisheiði power plant pipeline route selection 149
Figure 4-53 Shortest and least time paths 150
Figure 4-54 Simple buffering 151
Figure 4-55 Manifold: Buffer operations..152
Figure 4-56 Manifold: Buffer and Common buffer 153
Figure 4-57 Distance decay models..156
Figure 4-58 Directional analysis of streams 160
Figure 4-59 Two-variable wind rose 161
Figure 4-60 Slope and aspect plot, Mt St Helens data, USA 162
Figure 4-61 Standard distance circle and ellipses 163
Figure 4-62 Correlated Random Walk simulation 164
Figure 4-63 Dilation and erosion operations 171
Figure 5-1 Grid generation examples 176
Figure 5-2 Grid sampling examples within hexagonal grid, 1 hectare area.............................177
Figure 5-3 Random point generation examples − ArcGIS 177
Figure 5-4 Random point sampling examples − ArcGIS 178
Figure 5-5 Histogram linkage 182
Figure 5-6 Simple box plot 183
Figure 5-7 Mapped box plot...184
Figure 5-8 Conditional Choropleth mapping 186
Figure 5-9 Exploratory analysis of radioactivity data..187
Figure 5-10 Trend analysis of radioactivity dataset 188
Figure 5-11 Quadrat counts...193
Figure 5-12 Nearest Neighbour distribution...203
Figure 5-13 Ripley's K function computation 206
Figure 5-14 Ripley K function, shown as transformed L function plot 207
Figure 5-15 Lung cancer incidence data..209
Figure 5-16 Lung cancer NNh clusters...211
Figure 5-17 KDE cancer incidence mapping 212
Figure 5-18 Join count patterns..215
Figure 5-19 Join count computation 216
Figure 5-20 Grouping and size effects 218
Figure 5-21 Irregular lattice dataset 219
Figure 5-22 Moran's I computation..221
Figure 5-23 Sample dataset and Moran I analysis 223
Figure 5-24 Moran I (co)variance cloud, lag 1 224
Figure 5-25 Local Moran's I computation 227
Figure 5-26 LISA map, Moran's I 228
Figure 5-27 Significance tests for revised sample dataset..229
Figure 5-28 Georgia educational attainment: GWR residuals map, Gaussian adaptive kernel.....241

Figure 6-1 East Sussex test surface, OS TQ81NE 249
Figure 6-2 Pentland Hills test surface, OS NT04.. 250
Figure 6-3 Linear regression surface fit to NT04 spot heights 251
Figure 6-4 Raster file neighbourhoods 252
Figure 6-5 Vector models of TQ81NE.. 255
Figure 6-6 First, second and third order mathematical surfaces 256
Figure 6-7 Random and fractal grids 257
Figure 6-8 Pseudo-random surfaces 258
Figure 6-9 D-infinity slope computation 260
Figure 6-10 Gradient and sampling resolution 261
Figure 6-11 Slope computation output... 261
Figure 6-12 Frequency distribution of aspect values 262
Figure 6-13 Aspect computation output 263
Figure 6-14 Profile of NS transect, TQ81NE 264
Figure 6-15 Multiple profile computation 265
Figure 6-16 Surface morphology 265
Figure 6-17 Path smoothing — vertical profile... 270
Figure 6-18 Grid smoothing 271
Figure 6-19 Viewshed computation... 275
Figure 6-20 3D Urban radio wave propagation modelling using Cellular Expert and ArcGIS 276
Figure 6-21 Radio frequency viewshed 277
Figure 6-22 Line of sight analysis 278
Figure 6-23 Viewsheds and lines of sight on a synthetic (Gaussian) surface........................ 278
Figure 6-24 Isovist analysis, Street network, central London 279
Figure 6-25 D-Infinity flow assignment 281
Figure 6-26 Flow direction and accumulation 282
Figure 6-27 Stream identification 283
Figure 6-28 Watersheds and basins.. 283
Figure 6-29 Contour plots for alternative interpolation methods — generated with Surfer 8 288
Figure 6-30 Linear interpolation of contours 291
Figure 6-31 Contour computation output.. 292
Figure 6-32 IDW as surface plot... 294
Figure 6-33 Contour plots for alternative IDW methods, OS NT04 295
Figure 6-34 Simple Voronoi polygon assignment 296
Figure 6-35 Contour plots for Natural Neighbour method, OS NT04 297
Figure 6-36 Clough-Tocher TIN interpolation 301
Figure 6-37 Sample variogram... 306
Figure 6-38 Sill, range and nugget... 308
Figure 6-39 Data transformation for Normality.. 309
Figure 6-40 Indicator variograms 310
Figure 6-41 Anisotropy 2D map, zinc data 310
Figure 6-42 Variogram models — graphs... 315
Figure 6-43 Fractal analysis of TQ81NE... 316
Figure 6-44 Ordinary Kriging of untransformed zinc dataset.. 320
Figure 6-45 Conditional simulation of untransformed zinc test dataset............................... 324
Figure 7-1 Minimum Spanning Tree... 338
Figure 7-2 Gabriel network construction 339
Figure 7-3 Relative neighbourhood network and related constructions 340
Figure 7-4 Steiner MST construction 342
Figure 7-5 Dantzig shortest path algorithm 343
Figure 7-6 Salt Lake City — Sample networking problems and solutions 345
Figure 7-7 MST, TSP and related problems.. 348

Figure 7-8 Heuristic solution and dual circuit TSP examples.. 349
Figure 7-9 Tanker delivery tours.. 350
Figure 7-10 Optimum facility location on a network — LOLA solution............................... 353
Figure 7-11 Comparison of heuristic p-median solutions, Tripolis, Greece 358
Figure 7-12 Facility location in Triplois, Greece, planar model 359
Figure 7-13 Service area definition.. 360
Figure 7-14 Travel time zones.. 361
Figure 7-15 Routing directions 362
Figure 7-16 Arc routing.. 363

List of Tables

Table 1.1 Sample software tools   8
Table 1.2 Selected terminology   15
Table 1.3 Notation and symbology   21
Table 1.4 Common formulas and statistical measures   22
Table 4.1 Geographic data models   69
Table 4.2 OGC OpenGIS Simple Features Specification — Methods   70
Table 4.3 Spatial overlay methods, Manifold GIS   91
Table 4.4 Regional employment data — grouping affects   95
Table 4.5 Selected univariate classification schemes   100
Table 4.6 Multivariate classification methods   105
Table 4.7 Selected MATLab/GRASS planar geometric analysis functions   117
Table 4.8 Widely used univariate kernel density functions   131
Table 4.9 Interpretation of p-values   141
Table 4.10 Linear spatial filters   169
Table 5.1 Implications of Data Models   174
Table 5.2 Voronoi-based ESDA   188
Table 5.3 Sample statistical tools for grid data — Idrisi   190
Table 5.4 Simple 2-way contingency table   191
Table 5.5 Landscape metrics, Fragstats   200
Table 5.6 NN Statistics and study area size   204
Table 5.7 Join count mean and variance results   217
Table 5.8 Selected regression analysis terminology   235
Table 5.9 Georgia dataset — global regression estimates and diagnostics   241
Table 5.10 Georgia dataset — SAR comparative regression estimates and diagnostics   244
Table 6.1 Morphometric features — a simplified classification   267
Table 6.2 Gridding and interpolation methods   286
Table 6.3 Variogram models (univariate, isotropic)   314
Table 7.1 Network analysis terminology   326
Table 7.2 Some key problems in network analysis   332
Table 7.3 Sample network analysis problem parameters   333
Table 7.4 Routing functionality in selected logistics software packages   335
Table 7.5 Taxonomy of location analysis problems   336

Foreword

This Guide started life as a document to accompany the spatial analysis module of the postgraduate MSc in Geographic Information Science at University College London delivered by the principal author, Dr Mike de Smith. As is often the case, from its conception in summer 2005 through to completion of the first draft in summer 2006, it developed a life of its own, growing into a substantial Guide designed for use by a wide audience. Once several of the chapters had been written — notably those covering the building blocks of spatial analysis and on surface analysis — the project was discussed with Professors Longley and Goodchild. They kindly agreed to contribute to the contents of the Guide itself. As such, this Guide may be seen as a companion to the pioneering book on *Geographic Systems and Science* (2nd edition, 2005) by Longley, Goodchild, Maguire and Rhind, particularly Chapters 14-16 of that work which deal with spatial analysis and modelling. Their participation has also facilitated links with broader "spatial literacy" and spatial analysis programmes, especially the materials provided at www.spatial-literacy.org, www.ncgia.ucsb.edu and www.csiss.org.

For our title we have adopted the now widely used term *geospatial analysis* which reflects the coming together of the broad field of spatial analysis with the latest generation of Geographic Information Systems (GIS) and related software. A unique aspect of this Guide is its independent evaluation of software, in particular the set of readily available tools and packages for conducting various forms of geospatial analysis. To our knowledge, there is no similarly extensive resource that is available in printed or electronic form and so we think there is a need for guidance on where to find and how to apply selected tools. Inevitably, some topics have been omitted, primarily where there is little or no readily available commercial or open source software to support particular analytical operations. Other topics, whilst included, have been covered relatively briefly and/or with limited examples, reflecting the inevitable constraints of time and the authors' limited access to some of the available software resources.

Every effort has been made to ensure the information provided is up-to-date, accurate, compact, comprehensive and representative —we do not claim it to be exhaustive. However, with fast-moving changes in the software industry and in the development of new techniques it would be impractical and uneconomic to publish the material in a conventional manner. Accordingly the Guide has been prepared without intermediary typesetting. This has enabled the time between producing the text and delivery in electronic (web) and printed formats to be greatly reduced, thereby reducing costs and enabling the work to be as current as possible. It will also enable the work to be updated on a regular basis (electronic formats) and on a more regular basis than might be otherwise possible for its printed version. Comments and suggestions regarding the scope, detailed content and associated materials (e.g. case studies) are welcome and may be made via the Guide web site, www.spatialanalysisonline.com.

Mike de Smith, Edinburgh ♦ Mike Goodchild, Santa Barbara ♦ Paul Longley, London

September 2006

1 Introduction

1.1 Motivation and Media

1.1.1 Guide overview

This Guide addresses the full spectrum of spatial analysis and associated modelling techniques that are provided within currently available and widely used geographic information systems (GIS) and associated software. Collectively such techniques and tools are often now described as *geospatial analysis*, although we use the more common form, spatial analysis, in most of our discussions. The objective is to be comprehensive both in terms of concepts and techniques (but not necessarily exhaustive), representative and independent in terms of software tools, and above all practical in terms of application and implementation. However, we believe that it is no longer appropriate to think of a standard, discipline specific textbook as capable of satisfying every kind of new user need. Accordingly, an innovative feature of our approach here is the range of formats, channels and releases that we propose to disseminate the material.

The interactive web version of this Guide may be accessed via the associated Internet site: www.spatialanalysisonline.com. The contents and sample sections of the PDF version may also be accessed from this site. In both cases the contents are regularly updated. Because print versions of the Guide may be made in colour or black and white/greyscale, references to colours in the text are augmented by greyscale equivalents. Readers are recommended to refer to the associated web site and/or the PDF version of the Guide for full colour details.

The Internet is now well established as society's principal mode of information exchange, and most aspiring GIS users are accustomed to searching for material that can easily be customised to specific needs. Our objective for such users is to provide an independent, reliable and authoritative first port of call for conceptual, technical, software and applications material that addresses the panoply of new user requirements.

Readers wishing to obtain a more in-depth understanding of the background to many of the topics covered in this Guide should review the "Suggested reading" topic (Section 1.2.3). Those seeking examples of software tools that might be used for geospatial analysis should refer to the "Sample software products" topic (Section 1.2.2.1) and related discussions throughout this Guide.

Applications are the driving force behind GIS, and many are illustrated in Longley, Goodchild, Maguire and Rhind (2005, Chapter 2, "A gallery of applications"). In a similar vein the web site provides companion material for this Guide focusing on applications. One of the first of these will be the London GIS Casebook — a series of sector-specific case studies drawing on recent work in and around London. These include details of a range of applications from the fields of: Health and Welfare; Emergency and Security Management; Environmental Engineering and Planning; Education; Enterprise Development; and Retailing, amongst others.

Throughout this document the spelling convention adopted is UK English.

1.1.2 Spatial analysis and GIS

Given the vast range of spatial analysis techniques that have been developed over the past half century many areas can only be covered to a limited depth, whilst others have been omitted because they are not implemented in current mainstream GIS products. This is a rapidly changing field and increasingly GIS packages are including analytical tools as standard built-in facilities or as optional *toolsets*, *add-ins* or *analysts*. In many instances such facilities are provided by the original software suppliers (commercial vendors or collaborative non-commercial development teams) whilst in other cases facilities have been developed and are provided by third parties. Finally, many products offer software development kits (SDKs), programming languages and language support, scripting facilities and/or special interfaces for developing one's own analytical tools or variants.

Commercial products rarely provide access to source code or full details of the algorithms employed. Typically they provide references to books and articles on which the procedure is based, coupled with online help and "white papers" describing their parameters and applications. This means that results produced using one GIS package on a given dataset can rarely be exactly matched to those produced using any other package or through hand-crafted coding. There are many reasons for these inconsistencies including: differences in the software architectures of the various packages and the algorithms used to implement individual methods; errors in the source materials or their interpretation; coding errors; inconsistencies arising out of the ways in which different GIS packages model, store and manipulate information; and differing treatments of special cases (e.g. missing values, boundaries, adjacency, obstacles, distance computations etc.).

Non-commercial packages sometimes provide source code and test data for some or all of the analytical functions provided, although it is important to understand that "non-commercial" often does not mean that users can download the full source code. Source code greatly aids understanding, reproducibility and further development. Such software will often also provide details of known bugs and restrictions associated with functions — although this information may also be provided with commercial products it is generally less transparent. In this respect non-commercial software may meet the requirements of scientific rigour more fully than many commercial offerings, but is often provided with limited documentation, training tools, cross-platform testing and/or technical support, and thus is generally more demanding on the users and system administrators. In many instances OpenSource and similar not-for-profit GIS software may also be less generic, focusing on a particular form of spatial representation (e.g. a grid or raster spatial model). Like some commercial software, it may also be designed with particular application areas in mind, such as addressing problems in hydrology or epidemiology.

The process of selecting software tools encourages us to ask: (i) "what is meant by geospatial analysis techniques?" and (ii) "what should we consider to be GIS software?" To some extent the answer to the second question is the simpler, if we are prepared to be guided by self-selection. For our purposes we will focus upon products that claim to be geographic information systems, supporting at least 2D mapping (display and output) of raster (grid based) and/or vector (point/line/polygon based) data, with a minimum of basic map manipulation facilities. We concentrate our review on a number of the products most widely used or with the most readily accessible analytical facilities. This leads us beyond the realm of pure GIS. For example: we use examples drawn from packages that do not directly provide mapping facilities (e.g. Crimestat) but which provide input and/or output in widely used GIS map-able formats; products that include some mapping facilities but whose primary purpose is spatial or spatio-temporal exploration and analysis (e.g. GeoDa, GS+, STARS); and products that are general or special purpose analytical engines incorporating mapping capabilities (e.g. MATLab with the Mapping Toolbox, WinBUGS with GeoBUGS — for more details on these and other example software tools, please see Table 1.1).

The more difficult of the two questions above is the first — what should be considered as "geospatial analysis"? In conceptual terms, the phrase identifies the sub-set of techniques that are applicable when, as a minimum, data can be referenced on a two-dimensional frame and relate to terrestrial activities. The results of geospatial analysis will change if the location or extent of the frame changes, or if objects are repositioned within it: if they do not, then "everywhere is nowhere", location is unimportant, and it is simpler and more appropriate to use conventional, *aspatial*, techniques.

Many GIS products apply the term (geo)spatial analysis in a very narrow context. In the case of vector-based GIS this typically means operations such as map overlay (combining two or more maps or map layers according to predefined rules), simple buffering (identifying regions of a map within a specified distance of one or more features, such as towns, roads or rivers) and similar basic operations. This reflects (and is reflected in) the use of the term *spatial analysis* within the Open Geospatial Consortium (OGC) "simple feature specifications" (see further Table 4.2). For raster-based GIS, widely used in the environmental sciences and remote sensing, this typically means a range of actions applied to the grid cells of one or more maps (or images) often involving filtering and/or algebraic operations (*map algebra*). These techniques involve processing one or more raster layers according to simple rules resulting in a new map layer, for example replacing each cell value with some combination of its neighbours' values, or computing the sum or difference of specific attribute values for each grid cell in two matching raster datasets. Descriptive statistics, such as cell counts, means, variances, maxima, minima, cumulative values, frequencies and a number of other measures and distance computations are also often included in this generic term "spatial analysis".

However, at this point only the most basic of facilities have been included, albeit those that may be the most frequently used by the greatest number of GIS professionals. To this initial set must be added a large variety of statistical techniques (descriptive, exploratory, and explanatory) that have been designed specifically for spatial and spatio-temporal data. Today such techniques are of

great importance in the social sciences, medicine and criminology, despite the fact that their origins may often be traced back to problems in the environmental and life sciences, in particular ecology, geology and epidemiology. It is also to be noted that spatial statistics is largely an observational science (like astronomy) rather than an experimental science (like agronomy or pharmaceutical research). This aspect of geospatial science has important implications for analysis, particularly the application of a range of statistical methods to spatial problems.

Limiting the definition of geospatial analysis to 2D mapping operations and spatial statistics remains too restrictive for our purposes. There are other very important areas to be considered. These include: surface analysis —in particular analysing the properties of physical surfaces, such as gradient, aspect and visibility, and analysing surface-like data "fields"; network analysis — examining the properties of natural and man-made networks in order to understand the behaviour of flows within and around such networks; and locational analysis. GIS-based network analysis may be used to address a wide range of practical problems such as route selection and facility location, and problems involving flows such as those found in hydrology. In many instances location problems relate to networks and as such are addressed with tools designed for this purpose, but in others existing networks may have little or no relevance or may be impractical to incorporate within the modelling process. Problems that are not specifically network constrained, such as new road or pipeline routing, regional warehouse location, mobile phone mast positioning or the selection of rural community health care sites, may be effectively analysed (at least initially) without reference to existing physical networks. Locational analysis "in the plane" is also applicable where suitable network datasets are not available, or are too large or expensive to be utilised, or where the location algorithm is very complex or involves the examination or simulation of a very large number of alternative configurations.

A further important aspect of geospatial analysis is visualisation — the creation and manipulation of images, maps, diagrams, charts, 3D views and their associated tabular datasets. GIS packages increasingly provide a range of such tools, providing static or rotating views, draping images over 2.5D surface representations, providing animations and fly-throughs, dynamic linking and brushing and spatio-temporal visualisations. This latter class of tools is the least developed, reflecting in part the limited range of suitable compatible datasets and the limited set of analytical methods available, although this picture is changing rapidly. All these facilities augment the core tools utilised in spatial analysis throughout the analytical process (exploration of data, identification of patterns and relationships, construction of models, and communication of results).

GIS software, especially in the commercial sphere, is driven primarily by demand and applicability, as manifest in willingness to pay. Hence, to an extent, the facilities available often reflect commercial and resourcing realities (including the development of improvements in processing and display hardware, and the ready availability of high quality datasets) rather than the status of development in geospatial science. Indeed, there may be many capabilities available in software packages that are provided simply because it is extremely easy for the designers and programmers to implement them, especially those employing object-oriented programming and data models. For example, a given operation may be provided for polygonal features in response to a well-understood application requirement, which is then easily enabled for other features (e.g. point sets, polylines) despite the fact that there may be no known or likely requirement for the facility. Despite this cautionary note, for specific well-defined or *core* problems, software developers will frequently utilise the most up-to-date research on algorithms in order to improve the quality (accuracy, optimality) and efficiency (speed, memory usage) of their products.

Furthermore, the quality, variety and efficiency of spatial analysis facilities provide an important discriminator between commercial offerings in an increasing competitive and open market for software. However, the ready availability of analysis tools does not imply that one product is necessarily better or more complete than another — it is the selection and application of *appropriate* tools in a manner that is *fit for purpose* that is important. Guidance documents exist in some disciplines that assist users in this process, e.g. Perry *et al.* (2002) dealing with ecological data analysis, and to a significant degree we hope that this Guide will assist users from many disciplines in the selection process.

1.2 Intended Audience and Companion Materials

1.2.1 Scope

This Guide has been designed to be accessible to a wide range of readers — from undergraduates and postgraduates studying GIS and spatial analysis, to GIS practitioners and professional analysts. It is intended to be much more than a cookbook of algorithms and techniques, and seeks to provide an explanation of the key techniques of spatial analysis using examples from widely available software packages. It stops short, however, of attempting a systematic evaluation of competing software products. It seeks to provide an explanation of the key techniques of geospatial analysis using examples drawn from widely available software packages. In this context a substantial range of application examples are provided, but any specific selection inevitably illustrates only a small subset of the huge range of facilities available. Wherever possible, examples have been drawn from non-academic sources, highlighting the growing understanding and acceptance of GIS technology in the commercial and government sectors.

The scope of this Guide incorporates the various spatial analysis topics included within the NCGIA Core Curriculum (Goodchild and Kemp, 1990) and as such may provide a useful accompaniment to GIS Analysis courses based closely or loosely on this programme. More recently (Summer 2006) the University Consortium for Geographic Information Science (UCGIS) has produced a comprehensive "Body of Knowledge" (BoK) document, which is available for download from their website (www.ucgis.com). This Guide covers materials that primarily relate to the BoK sections CF: Conceptual Foundations and DA: Data Analysis. In the general introduction to the DA knowledge area UCGIS summarises this component as follows:

"[Spatial] Data analysis seeks to understand both first-order (environmental) effects and second-order (interaction) effects. Approaches that are both data-driven (exploration of geospatial data) and model-driven (testing hypotheses and creating models) are included. Data driven techniques derive summary descriptions of data, evoke insights about characteristics of data, contribute to the development of research hypotheses, and lead to the derivation of analytical results. The goal of model-driven analysis is to create and test geospatial process models. In general, model-driven analysis is an advanced knowledge area where previous experience with exploratory spatial data analysis would constitute a desired prerequisite." (UCGIS, BoK, p51)

1.2.2 GIS software

GIS packages have a variety of origins, and a comment should be made at this juncture concerning image processing software. A number of GIS packages and related toolsets have particularly strong facilities for processing and analysing binary, greyscale and colour images. They may have been designed originally for the processing of remote sensed data, from satellite and aerial surveys, but have developed into much more sophisticated and complete GIS tools (e.g. MicroImage's TNT product set; Leica's ERDAS Imagine suite of products; RSI's ENVI software and associated packages such as RiverTools from RIVIX). Alternatively, image handling may have been deliberately included within the original design parameters for a generic GIS package (e.g. Manifold), or simply be toolsets for image processing that may be combined with mapping tools (e.g. the MATLab Image Processing Toolbox). Whatever their origins, the primary purpose of such tools has been the capture, manipulation and interpretation of image data, rather than spatial analysis per se, although the latter inevitably follows from the former. In this Guide we do not provide a separate chapter on image processing, despite its considerable importance in GIS, focusing instead on those areas where image processing tools and concepts are applied for spatial analysis. We have adopted a similar position with respect to other forms of data capture, such as field and geodetic survey systems and data cleansing software — although these incorporate analytical tools, their primary function remains the recording and georeferencing of datasets, rather than the analysis of such datasets once stored.

The GIS software and analysis tools that an individual, group or corporate body chooses to use will depend very much on the purposes to which they will be put. There is an enormous difference between the requirements of academic researchers and educators, and those with responsibility for planning and delivery emergency control systems or large scale physical infrastructure projects. The spectrum of products that may be described as a GIS includes (amongst others):

- highly specialised, sector specific packages: for example civil engineering design and costing systems; satellite image processing systems; and utility infrastructure management systems

- transportation and logistics management systems

- civil and military control room systems

- systems for visualising the built environment for architectural purposes, for public consultation or as part of simulated environments for interactive gaming

- land registration systems

- census data management systems

The list of software functions and applications is long and in many instances suppliers would not describe their offerings as a GIS. In many cases such systems fulfil specific operational needs, solving a well-defined subset of spatial problems and providing mapped output as an incidental but essential part of their operation. Many of the capabilities may be found in generic GIS products. In other instances a specialised package may utilise a GIS engine for the display and in some cases processing of spatial data (directly, or indirectly through interfacing or file input/output mechanisms). For this reason, and in order to draw a boundary around the present work, reference to application-specific GIS will be limited.

For most GIS professionals, spatial analysis and associated modelling is an infrequent activity. Even for those whose job focuses on analysis the range of techniques employed tends to be quite narrow and application focused. GIS consultants, researchers and academics on the other hand are continually exploring and developing analytical techniques. For this second group of users it is common to make use of a variety of tools, data and programming facilities. For the former and for consultants, especially in commercial environments, the imperatives of financial considerations, timeliness and corporate policy loom large, directing attention to: delivery of solutions within well-defined time and cost parameters; working within commercial constraints on the cost and availability of software, datasets and staffing; ensuring that solutions are fit for purpose/meet client and end-user expectations and agreed standards; and in some cases, meeting "political" expectations.

1.2.2.1 Sample software products

The principal products we have included in this edition of the Guide are summarised in Table 1.1. A number of these products are free whilst others are available (at least in some form) for a small fee for all or selected groups of users (e.g. GWR, MapCalc Learner). Others are licensed at varying per user prices, from a few hundred to over a thousand US$ per user. Our tests and examples have all been carried out using desktop/Windows versions of these software products, and different versions that support Unix-based operating systems and more sophisticated back-end database engines, have not been utilised. In the context of this Guide we do not believe these selections affect our discussions in any substantial manner, although such issues may have performance and systems architecture implications that are extremely important for many users.

1.2.2.2 Software performance

Suppliers should be able to provide advice on performance issues (e.g. see the ESRI web site, "Services" area for relevant documents relating to their products) and in some cases such information is provided within product Help files (e.g. see the Performance Tips section within the Manifold GIS help file). Some analytical tasks are very processor- and memory-hungry, particularly as the number of elements involved increases. For example, vector overlay and buffering is relatively fast with a few objects and layers, but slows appreciably as the number of elements involved increases. This increase is generally at least linear with the number of layers and features, but for some problems grows in a highly non-linear (i.e. geometric) manner. Many optimisation tasks, such as optimal routing through networks or trip distribution modelling, are known to be extremely hard or impossible to solve optimally and methods to achieve a best solution with a large dataset can take a considerable time to run (see Section 7.1.4 for a fuller discussion of this topic). Similar problems exist with the processing and display of raster files, especially large images or sets of images. Geocomputational methods, some of which are beginning to appear within GIS packages and related toolsets, are almost by definition computationally intensive. This certainly applies to large-scale (Monte Carlo) simulation models, cellular automata models and some raster-based optimisation techniques, especially where modelling extends into the time domain.

A frequent criticism of GIS software is that it is over-complicated, resource hungry and requires specialist expertise to understand and use. Such criticisms are often valid and for many problems it may prove simpler, faster

and more transparent to utilise specialised tools for the analytical work and draw on the strengths of GIS in data management and mapping to provide input/output and visualisation functionality. A related approach is to develop solutions using high-level programming facilities either: (i) within a GIS (e.g. macros, scripts, VBA, Python); (ii) within general purpose data processing toolsets (e.g. MATLab, Excel); or (iii) utilising mainstream programming languages (e.g. Java, C++). The advantage of this latter approach is control and transparency, the disadvantages are that software development is never trivial, is often subject to frustrating and unforeseen delays and errors, and generally requires ongoing maintenance.

At present there are no standardised tests for the quality, speed and accuracy of GIS procedures. It remains the buyer's and user's responsibility and duty to evaluate the software they wish to use for the specific task at hand, and by systematic controlled tests or by other means establish that the product and facility within that product they choose to use is truly fit for purpose — *caveat emptor*! Details of how to obtain these products are provided in Table 1.1 and via the Web links section of the Appendix. The list maintained on Wikipedia is also a useful source of information and links. A number of trade magazines and websites (such as Geoplace and Geocommunity) provide ad hoc reviews of GIS software offerings, especially new releases, although coverage of analytical functionality may be limited.

Table 1.1 Sample software tools

Product	Ver.	Type/key sectors (authors shown in brackets where applicable)	Free?
ArcGIS	8/9	General purpose, comprehensive, very extensive toolsets, vector focused with substantial raster support. Cross industry, Open Geospatial Consortium (OGC) compliant	
ANUDEM	5.2	M F Hutchinson's DEM grid generation program. Takes irregularly spaced points, contour lines and vector stream data as input and produces hydrologically consistent DEM output (implemented in ArcGIS)	
CCMaps	Oct04	Conditioned choropleth mapping. Java-based interactive mapping/visualisation tool, developed for health (cancer) studies and related analyses	✓
Concorde	1.1	High performance solver for symmetric TSP network problems	✓
CPLEX	10.0	Linear Programming (LP)/Mixed Integer Programming (MIP) solver. Part of the ILOG Optimisation suite (see below)	
Crimestat	II/III	Crime event analysis, vector (N Levine). See also crime analysis toolsets listed at http://www.iaca.net/Software.asp and http://www.ojp.usdoj.gov/nij/maps/software.html	✓
ENVI	4.2	"Environment for Visualising Images", provides powerful analysis of remote sensing image data, with support for vector format import and overlay (see also, RiverTools)	
Fragstats	3	Analysis of ecological raster data	✓
GAM	4.0	Geographic Analysis Machine — cluster hunting software	✓
GeoDa	095a	Exploratory spatial data analysis, vector (L Anselin)	✓
Geomedia	6.0	General purpose database-driven GIS suite	
GRASS	6.1	Geographic Resources Analysis Support System. OpenSource GIS with both raster and vector support. Earth sciences, government, academia	✓
GS+	5.3	Geostatistical analysis	
GWR	3	Geographically weighted regression (S Fotheringham, C Brunsdon, M Charlton)	
Hawth's Tools		ArcGIS extension for spatial analysis, especially ecological applications.	✓
Idrisi		Raster-based product, especially for environmental sciences. Remote sensing, land management	
ILOG Dispatcher	4	Vehicle routing, scheduling and dispatching (Logistics) suite. Part of the ILOG optimisation product set (see also, CPLEX)	

Product	Ver.	Type/key sectors (authors shown in brackets where applicable)	Free?
ISATIS	6	Geostatistical software for the Earth sciences	
Landserf	2.2	Surface analysis package, Java based, cross-platform (J Wood)	✓
LEDA	5.1	Algorithmic suite including extensive range of graph (network) analysis utilities. Provides the network analysis kernel for at least one GIS package provider	
LOLA	2.0	Locational analysis	✓
Manifold	6.0	General purpose, very extensive toolsets, vector focused with raster support. Cross industry, OGC compliant	
MapCalc	2.0	Raster-based mapping and analysis package, with very low cost variant for teaching use	
MapInfo	7.8	General purpose, vector focused with raster support. Cross industry/Marketing; HotSpot Detective (J Ratcliffe) for crime analysis	
MATLab	6.5	Matrix/mathematical package with optional mapping toolbox, image processing toolbox and statistics toolbox; free spatial statistics toolbox (L Pace)	
MFWorks	3.0	Raster-based analysis based on C Dana Tomlin's map algebra and developments thereof	
Oriana	2	Statistical analysis of circular datasets	
PCRaster		Raster-based analysis with strong hydrological modelling, many aspects of which are also derived from Tomlin's map algebra. Hydrology/Soil science	✓
RiverTools	3	Hydrological analysis package; written in IDL and compatible with ENVI	
Rookcase		Excel add-in for computing simple spatial autocorrelation (M Sawada)	✓
SANET	2	Spatial analysis on a network (A Okabe)	✓
S-Distance	0.7	Network and locational analysis (S Sirigos)	✓
SaTScan	7.01	Spatial, temporal and spatio-temporal analysis of geographic data. Particularly designed for disease pattern analysis and surveillance	✓
SITATION	5.5	Facility location software (M Daskin)	✓
SPLANCS	2.01	Spatial analysis of point patterns. (R-Plus version is free)	(✓)
STARS	0.8.2	Space-time analysis of regional systems. Some techniques mirror those in GeoDa (unrelated to the STARS logistics package)	✓

Product	Ver.	Type/key sectors (authors shown in brackets where applicable)	Free?
Surfer	8	Surface building and modelling package, very strong on gridding, geostatistics and visualisation. Earth sciences. Grapher 6 from same provider (Golden Software) also referred to in this Guide	
TAS	2.07	Terrain Analysis System — Compact, stand-alone program. provides wide range of terrain analysis/hydrological analysis functions and index computations	✓
TAUDEM	2.07	Terrain Analysis Using Digital Elevation Models — ArcGIS Add-in/toolbar. Provides wide range of terrain analysis/hydrological analysis functions and index computations	✓
Terraseer		Commercial space-time and statistical analysis packages (STIS, Clusterseer for cluster analysis, Boundaryseer for Boundary detection and analysis). Health	
TNTMips	7.1	Commercial generic cross-platform GIS developed from image processing background. Large analytics toolset. Free "Lite" version for non-commercial use	(✓)
TransCAD/ Maptitude	4.7	TransCAD is the transportation-focused implementation of the Maptitude package, with very strong network analysis and related facilities. Transport, Marketing	
Vincenty		Excel spreadsheet for computing ellipsoidal distances (download link)	✓
WinBUGS	1.4.1	WinBUGS with GeoBUGS is a statistical analysis package. Health	✓
Xpress-MP	17.01	General purpose modelling and optimisation suite (free student edition)	(✓)
ZDES	3b	Zone design system. University of Leeds, UK	✓

1.2.3 Suggested reading

There are numerous excellent modern books on GIS and spatial analysis, although few address software facilities and developments. Hypertext links are provided here, and throughout the text where they are cited, to the more recent publications and web resources listed.

As a background to this Guide any readers unfamiliar with GIS are encouraged to first tackle Longley *et al.* (2005), which we hope provides a comprehensive and highly accessible introduction to the subject as a whole. The GB Ordnance Survey's "GIS Files" document, downloadable from their website: www.ordnancesurvey.co.uk/oswebsite/gisfiles also provides an excellent brief introduction.

Some of the basic mathematics and statistics of relevance to GIS analysis is covered in Dale (2005) and Allan (2004). For detailed information on datums and map projections, see Iliffe (1999). A useful online resource for those involved in data analysis, particularly with a statistical content, is the e-Handbook of Statistical Methods produced by the US National Institute on Standards and Technology, NIST). The more informally produced set of articles on statistical topics provided under the Wikipedia umbrella are also an extremely useful resource. These works, and the mathematics reference site, Mathworld, are referred to (with hypertext links) at various points throughout this document. As a more specific source on geostatistics and associated software packages, the European Commission's AI-GEOSTATS website (www.ai-geostats.org) is highly recommended (recently updated). For those who find mathematics and statistics something of a mystery, de Smith (2006) and Bluman (2003) may provide useful starting points.

Undergraduates and MSc programme students will find Burrough and McDonnell (1998) provides excellent coverage of many aspects of geospatial analysis, especially from an environmental sciences perspective. Valuable

guidance on the relationship between spatial process and spatial modelling may be found in Cliff and Ord (1981) and Bailey and Gatrell (1995). The latter provides an excellent introduction to the application of statistical methods to spatial data analysis. O'Sullivan and Unwin (2003) is a more broad-ranging book covering the topic the authors describe as "Geographic Information Analysis". This work is best suited to advanced undergraduates and first year postgraduate students. In many respects a deeper and more challenging work is Haining's (2003) "Spatial Data Analysis – Theory and Practice". This book is strongly recommended as a companion to the present Guide, especially for postgraduate researchers and professional analysts involved in using GIS in conjunction with statistical analysis.

However, as with many other writers, Haining does not address the broader spectrum of geospatial analysis and associated modelling as we have defined it. For example, problems relating to networks and location are not covered and the literature relating to this area is scattered across many disciplines, being founded upon the mathematics of graph theory, with applications ranging from electronic circuit design to computer networking and from transport planning to the design of complex molecular structures. Useful recent volumes addressing this field include Miller and Shaw (2001) "Geographic Information Systems for Transportation" (especially Chapters 3, 5 and 6), and Rodrigue *et al.* (2006) "The geography of transport systems" (parts of this material are available at http://people.hofstra.edu/geotrans/).

As companion reading on these topics for the present Guide we suggest the two volumes from the Handbooks in Operations Research and Management Science series by Ball *et al.* (1995): "Network Models", and "Network Routing". These rather expensive volumes provide collections of reviews covering many classes of network problems, from the core optimisation problems of shortest paths and arc routing (e.g. street cleaning), to the complex problems of dynamic routing in variable networks, and a great deal more besides. This is challenging material for many and readers may prefer to seek out more approachable material, available in a number of other books and articles, e.g. Ahuja *et al.* (1993), Mark Daskin's excellent book "Network and Discrete Location" (1995) and the earlier seminal works by Haggett and Chorley (1969), and Scott (1977), together with the widely available online materials accessible via the Internet. A final recommendation here is Worboys and Duckham (2004) which addresses GIS from a computing perspective. This latter volume covers many topics, including the central issues of data modelling and data structures, key algorithms, system architectures and interfaces.

Many recent books described as covering (geo)spatial analysis are essentially edited collections of papers. As such most do not seek to provide comprehensive coverage of the field, but tend to cover information on recent developments, often with a specific application focus (e.g. health, transport, archaeology). The latter is particularly common where these works are selections from sector- or discipline-specific conference proceedings, whilst in other cases they are carefully chosen or specially written papers. Classic amongst these is Berry and Marble (1968) "Spatial Analysis: A reader in statistical geography". More recent examples include "Spatial Analysis" edited by Longley and Batty (1997), "GIS, Spatial Analysis and Modeling" edited by Maguire, Batty and Goodchild (2005), and the forthcoming "Handbook of Spatial Analysis" edited by Fotheringham (2007).

A second category of companion materials to the present work is the extensive product-specific documentation available from software suppliers. Some of the online help files and product manuals are excellent, as are associated example data files, tutorials, worked examples and white papers. In many instances we utilise these to illustrate the capabilities of specific pieces of software and to enable readers to replicate our results using readily available materials. In addition some suppliers, notably ESRI, have a substantial publishing operation, including more general

(i.e. not product specific) books of relevance to the present work. Amongst their publications we strongly recommend the "ESRI Guide to GIS Analysis Volume 1: Geographic patterns and relationships" (1999) by Andy Mitchell, which is full of valuable tips and examples. This is a basic introduction to GIS Analysis, which he defines in this context as "a process for looking at geographic patterns and relationships between features". Mitchell's recently published Volume 2 (July 2005) covers more advanced techniques of data analysis, notably some of the more accessible and widely supported methods of spatial statistics, and is equally highly recommended. A number of the topics covered in his Volume 2 are also considered in this Guide.

In parallel with the increasing range and sophistication of spatial analysis facilities to be found within GIS packages, there has been a major change in spatial analytical techniques. In large measure this has come about as a result of technological developments and the related availability of software tools and detailed publicly available datasets. One aspect of this has been noted already — the move towards network-based location modelling where in the past this would have been unfeasible. More general shifts can be seen in the move towards local rather than simply global analysis, for example in the field of exploratory data analysis, and in the development of a wide range of computationally intensive and simulation methods that address problems through micro-scale processes (geocomputational methods).

1.3 Structure

In order to cover such a wide range of topics, this Guide has been divided into a number of main sections or chapters. These are then further subdivided, in part to identify distinct topics as closely as possible, facilitating the creation of a web site from the text of the Guide. Hyperlinks embedded within the document enable users of the web version of this document to navigate around the Guide and to external sources of information, data, software, and reading materials.

Chapter 2 provides an introduction to spatial thinking, recently described by some as "spatial literacy", and addresses the central issues and problems associated with spatial data that need to be considered in any analytical exercise. This is followed in Chapter 3 by an examination of the historical and methodological background to GIS and spatial analysis. Subsequent chapters present the various analytical methods supported within widely available software tools. The majority of the methods described are implemented as standard facilities in modern commercial GIS packages such as ArcGIS, MapInfo, Manifold, TNTMips and Geomedia. Many are also provided in more specialised GIS products such as Idrisi, GRASS, Terraseer, ENVI and MapCalc. In addition we discuss a number of more specialised tools, designed to address the needs of specific sectors or technical problems that are otherwise not well-supported within the core GIS packages at present.

Throughout this Guide examples are drawn from and refer to specific products —these have been selected purely as examples and are not intended as recommendations. Extensive use has also been made of tabulated information — providing abbreviated summaries of techniques and formulas for reasons of both compactness and coverage. These tables are designed to provide a quick reference to the various topics covered and are, therefore, not intended as a substitute for fuller details on the various items covered.

This Guide does not currently cover advanced spatial statistics, spatial econometrics and spatio-temporal methods and models. Many of these topics tend to be more specialised, often having been developed for particular sector requirements, such as health research, socio-economic analysis or criminology. As such many of the tools described are not available in GIS packages, but are provided in more specialised software with limited map display capabilities and input/output of commonly used vector and raster spatial data formats. Increasingly, however, spatial statistics toolsets with a large prospective user base are beginning to appear in the mainstream commercial GIS products. Finally, we note that in this issue of the Guide geocomputational methods and new approaches to visualisation are also only briefly discussed.

Just as all datasets and software packages contain errors, known and unknown, so too do all books and websites, and the authors of this Guide expect that there will be errors despite our best efforts to remove these! Some may be genuine errors or misprints, whilst others may reflect our use of specific versions of software packages and their documentation. Inevitably with respect to the latter, new versions of the packages that we have used to illustrate this Guide will have appeared even before publication, so specific examples, illustrations and comments on scope or restrictions may have been superseded. In all cases the user should review the documentation provided with the software version they plan to use, check release notes for changes and known bugs, and look at any relevant online services (e.g. user/developer forums on the web) for additional materials and insights.

1.4 Terminology and abbreviations

GIS, like all disciplines, utilises a wide range of terms and abbreviations, many of which have well-understood and recognised meanings. For a large number of commonly used terms online dictionaries have been developed, for example those created by the Association for Geographic Information (AGI), the Open Geospatial Consortium (OGC) and by various suppliers. The latter includes many terms and definitions that are particular to specific products, but remain a valuable resource. The University of California maintains an online dictionary of abbreviations and acronyms used in GIS, cartography and remote sensing. Web site details for each of these are provided at the end of this Guide.

1.4.1 Definitions

Geospatial analysis utilises many of these terms, but many others are drawn from disciplines such as mathematics and statistics with the result that the same terms may mean entirely different things depending on their context and in many cases, on the software provider utilising them. In most cases terms used in this Guide are defined on the first occasion they are used, but a number warrant defining at this stage. Table 1.2 provides a selection of such terms, utilising definitions from widely recognised sources where available and appropriate.

Table 1.2 Selected terminology

Term	Definition
Adjacency	The sharing of a common side or boundary by two or more polygons (AGI). Note that adjacency may also apply to features that lie either side of a common boundary where these features are not necessarily polygons
Arc	Commonly used to refer to a straight line segment connecting two nodes or vertices of a polyline or polygon. Arcs may include segments or circles, spline functions or other forms of smooth curve. In connection with graphs and networks, arcs may be directed or undirected
Artefact	A result (observation or set of observations) that appears to show something unusual (e.g. a spike in the surface of a 3D plot) but which is of no significance. Artefacts may be generated by the way in which data have been collected, defined or re-computed (e.g. resolution changing), or as a result of a computational operation (e.g. rounding error or substantive software error). Linear artefacts are sometimes referred to as "ghost lines"
Aspect	The direction in which slope is maximised for a selected point on a surface (see also, Gradient and Slope)
Attribute	A data item associated with an individual object (record) in a spatial database. Attributes may be explicit, in which case they are typically stored as one or more fields in tables linked to a set of objects, or they may be implicit (sometimes referred to as *intrinsic*), being either stored but hidden or computed as and when required (e.g. polyline length, polygon centroid). Raster/grid datasets typically have a single explicit attribute (a value) associated with each cell, rather than an attribute table containing as many records as there are cells in the grid
Azimuth	The horizontal direction of a vector, measured clockwise in degrees of rotation from the positive Y-axis, for example, degrees on a compass (AGI)
Azimuthal Projection	A type of map projection constructed as if a plane were to be placed at a tangent to the Earth's surface and the area to be mapped were projected onto the plane. All points on this projection keep their true compass bearing (AGI)
(Spatial) Autocorrelation	The degree of relationship that exists between two or more (spatial) variables, such that when one changes, the other(s) also change. This change can either be in the same direction, which is a positive autocorrelation, or in the opposite direction, which is a negative autocorrelation (AGI). The term autocorrelation is usually applied to ordered datasets, such as those relating to time series or spatial data ordered by distance band. The existence of such a relationship suggests but does not definitely establish causality
Choropleth	A thematic map [i.e. a map showing a theme, such as soil types or rainfall levels] portraying properties of a surface using area symbols such as shading [or colour]. Area symbols on a choropleth map usually represent categorised classes of the mapped phenomenon (AGI)
Conflation	A term used to describe the process of combining (merging) information from two data sources into a single source, reconciling disparities where possible (e.g. by rubber-sheeting — see below). The term is distinct from *concatenation* which refers to combinations of data sources (e.g. by overlaying one upon another) but retaining access to their distinct components
Contiguity	The topological identification of adjacent polygons by recording the left and right polygons of each arc. Contiguity is not concerned with the exact locations

Term	Definition
	of polygons, only their relative positions. Contiguity data can be stored in a table, matrix or simply as [i.e. in] a list, that can be cross-referenced to the relevant co-ordinate data if required (AGI).
Curve	A one-dimensional geometric object stored as a sequence of points, with the subtype of curve specifying the form of interpolation between points. A curve is simple if it does not pass through the same point twice (OGC). A LineString (or polyline — see below) is a subtype of a curve
Datum	Strictly speaking, the singular of *data*. In GIS the word datum usually relates to a reference level (surface) applying on a nationally or internationally defined basis from which elevation is to be calculated. In the context of terrestrial geodesy datum is usually defined by a model of the Earth or section of the Earth, such as WGS84 (see below). The term is also used for horizontal referencing of measurements.
DEM	Digital elevation model (a DEM is a particular kind of DTM, see below)
DTM	Digital terrain model
EDM	Electronic distance measurement
EDA,ESDA	Exploratory data analysis/Exploratory spatial data analysis
Ellipsoid/Spheroid	An ellipse rotated about its minor axis determines a spheroid (sphere-like object), also known as an ellipsoid of revolution (see also, WGS84)
Feature	Frequently used within GIS referring to point, line (including polyline and mathematical functions defining arcs), polygon and sometimes text (annotation) objects (see also, vector)
Geoid	An imaginary shape for the Earth defined by mean sea level and its imagined continuation under the continents at the same level of gravitational potential (AGI)
Geodemographics	The analysis of people by where they live, in particular by type of neighbourhood. Such localised classifications have been shown to be powerful discriminators of consumer behaviour and related social and behavioural patterns
Geostatistics	Statistical methods developed for and applied to geographic data. These statistical methods are required because geographic data do not usually conform to the requirements of standard statistical procedures, due to spatial autocorrelation and other problems associated with spatial data (AGI). The term is widely used to refer to a family of tools used in connection with spatial interpolation (prediction) of (piecewise) continuous datasets and is widely applied in the environmental sciences. Spatial statistics is a term more commonly applied to the analysis of discrete objects (e.g. points, areas) and is particularly associated with the social and health sciences
GIS-T	GIS applied to transportation problems
GPS/DGPS	Global positioning system; Differential global positioning system — DGPS provides improved accuracy over standard GPS by the use of one or more fixed reference stations that provide corrections to GPS data

Term	Definition
Gradient	Used in spatial analysis with reference to surfaces (scalar fields) . Gradient is a vector field comprised of the *aspect* (direction of maximum slope) and *slope* computed in this direction (magnitude of rise over run) at each point of the surface. The magnitude of the gradient (the slope or inclination) is sometimes itself referred to as the gradient (see also, Slope and Aspect)
Graph	A collection of vertices and edges (links between vertices) constitutes a graph. The mathematical study of the properties of graphs and paths through graphs is known as graph theory
Heuristic	A term derived from the same Greek root as Eureka, heuristic refers to procedures for finding solutions to problems that may be difficult or impossible to solve by direct means. In the context of optimisation heuristic algorithms are systematic procedures that seek a good or near optimal solution to a well-defined problem, but not one that is necessarily optimal. They are often based on some form of intelligent trial and error or search procedure
iid	An abbreviation for "independently and identically distributed". Used in statistical analysis in connection with the distribution of errors or residuals
Invariance	In the context of GIS invariance refers to properties of features that remain unchanged under one or more transformations
Kernel	Literally, the core or central part of an item. Often used in computer science to refer to the central part of an operating system, the term kernel in geospatial analysis refers to methods (e.g. density modelling, local grid analysis) that involve calculations using a well-defined local neighbourhood (block of cells, radially symmetric function).
MBR/MER	Minimum bounding rectangle/Minimum enclosing (or envelope) rectangle (of a feature set)
Planar/non-planar/planar enforced	Literally, lying entirely within a plane surface. A polygon set is said to be planar enforced if every point in the set lies in exactly one polygon, or on the boundary between two or more polygons. See also, planar graph. A graph or network with edges crossing (e.g. bridges/underpasses) is non-planar
Planar graph	If a graph can be drawn in the plane (embedded) in such a way as to ensure edges only intersect at points that are vertices then the graph is described as planar
Pixel/image	Picture element — a single defined point of an image. Pixels have a "colour" attribute whose value will depend on the encoding method used. They are typically either binary (0/1 values), greyscale (effectively a colour mapping with values, typically in the integer range [0,255]), or colour with values from 0 upwards depending on the number of colours supported. Image files can be regarded as a particular form of raster or grid file
Polygon	A closed figure in the plane, typically comprised of an ordered set of connected vertices, $v_1, v_2, ... v_{n-1}, v_n = v_1$ where the connections (edges) are provided by straight line segments. If the sequence of edges is not self-crossing it is called a simple polygon. A point is inside a simple polygon if traversing the boundary in a clockwise direction the point is always on the right of the observer. If every pair of points inside a polygon can be joined by a straight line that also lies inside the polygon then the polygon is described as being convex (i.e. the interior is a connected point set). The OGC definition of a polygon is "a planar surface defined by 1 exterior boundary and 0 or more interior boundaries. Each interior boundary defines a hole in the polygon"

Term	Definition
Polyline	An ordered set of connected vertices, $v_1, v_2, ... v_{n-1}, v_n \neq v_1$ where the connections (edges) are provided by straight line segments. The vertex v_1 is referred to as the start of the polyline and v_n as the end of the polyline. The OGC specification uses the term LineString which it defines as: a curve with linear interpolation between points. Each consecutive pair of points defines a line segment
Raster/grid	A data model in which geographic features are represented using discrete cells, generally squares, arranged as a rectangular grid. A single grid is essentially the same as a two-dimensional matrix, but is typically referenced from the lower left corner rather than the norm for matrices, which are referenced from the upper left. Raster files may have one or more values (attributes or bands) associated with each cell position or pixel
Resampling	1. Procedures for (automatically) adjusting one or more raster datasets to ensure that the grid resolutions of all sets match when carrying out combination operations. Resampling is often performed to match the coarsest resolution of a set of input rasters
	2. The process of reducing image dataset size by representing a group of pixels with a single pixel. Thus, pixel count is lowered, individual pixel size is increased, and overall image geographic extent is retained. Resampled images are "coarse" and have less information than the images from which they are taken. Conversely, this process can also be executed in the reverse (AGI)
	3. In a statistical context the term resampling (or re-sampling) is sometimes used to describe the process of selecting a subset of the original data, such that the samples can reasonably be expected to be independent
Rubber sheeting	A procedure to adjust the co-ordinates all of the data points in a dataset to allow a more accurate match between known locations and a few data points within the dataset. Rubber sheeting ... preserves the interconnectivity or topology, between points and objects through stretching, shrinking or re-orienting their interconnecting lines (AGI)
Slope	The amount of *rise* of a surface (change in elevation) divided by the distance over which this rise is computed (the *run*), along a straight line transect in a specified direction. The run is usually defined as the *planar* distance, in which case the slope is the tan() function. Unless the surface is flat the slope at a given point on a surface will (typically) have a maximum value in a particular direction (depending on the surface and the way in which the calculations are carried out). This direction is known as the *aspect*. The *vector* consisting of the slope and aspect is the *gradient* of the surface at that point (see also, Gradient and Aspect)
Spheroid	A flattened (oblate) form of a sphere, or ellipse of revolution. The most widely used model of the Earth is that of a spheroid, although the detailed form is slightly different from a true spheroid
SQL/Structured Query Language	Within GIS software SQL extensions known as spatial queries are frequently implemented. These support queries that are based on spatial relationships rather than simply attribute values
Surface	A 2D geometric object. A simple surface consists of a single 'patch' that is associated with one exterior boundary and 0 or more interior boundaries. Simple surfaces in 3D are isomorphic to planar surfaces. Polyhedral surfaces are formed by 'stitching' together simple surfaces along their boundaries (OGC). Surfaces may be regarded as scalar fields, i.e. fields with a single value, e.g. elevation or

Term	Definition
	temperature, at every point
Tesseral/Tessellation	A gridded representation of a plane surface into disjoint polygons. These polygons are normally either square (raster), triangular (TIN — see below), or hexagonal. These models can be built into hierarchical structures, and have a range of algorithms available to navigate through them. A (regular or irregular) 2D tessellation involves the subdivision of a 2-dimensional plane into polygonal tiles (polyhedral blocks) that completely cover a plane (AGI). More generally the subdivision of the plane may be achieved using arcs that are not necessarily straight lines
TIN	Triangulated irregular network. A form of the tesseral model based on triangles. The vertices of the triangles form irregularly spaced nodes. Unlike the grid, the TIN allows dense information in complex areas, and sparse information in simpler or more homogeneous areas. The TIN dataset includes topological relationships between points and their neighbouring triangles. Each sample point has an X,Y co-ordinate and a surface, or Z-Value. These points are connected by edges to form a set of non-overlapping triangles used to represent the surface. TINs are also called irregular triangular mesh or irregular triangular surface model (AGI)
Topology	The relative location of geographic phenomena independent of their exact position. In digital data, topological relationships such as connectivity, adjacency and relative position are usually expressed as relationships between nodes, links and polygons. For example, the topology of a line includes its from- and to-nodes, and its left and right polygons (AGI). In mathematics, a property is said to be *topological* if it survives stretching and distorting of space
Transformation 1. Map	Map transformation: A computational process of converting an image or map from one coordinate system to another. Transformation ... typically involves rotation and scaling of grid cells, and thus requires resampling of values (AGI)
Transformation 2. Affine	Affine transformation: When a map is digitised, the X and Y coordinates are initially held in digitiser measurements. To make these X,Y pairs useful they must be converted to a real world coordinate system. The affine transformation is a combination of linear transformations that converts digitiser coordinates into Cartesian coordinates. The basic property of an affine transformation is that parallel lines remain parallel (AGI, with modifications). The principal affine transformations are contraction, expansion, dilation, reflection, rotation, shear and translation
Transformation 3. Data	Data transformation (see also, subsection 6.7.1.10): A mathematical procedure (usually a one-to-one mapping or function) applied to an initial dataset to produce a result dataset. An example might be the transformation of a set of sampled values $\{x_i\}$ using the log() function, to create the set $\{\log(x_i)\}$. Affine and map transformations are examples of mathematical transformations applied to coordinate datasets
Transformation 4. Back	Back transformation: If a set of sampled values $\{x_i\}$ has been transformed by a one-to-one mapping function $f()$ into the set $\{f(x_i)\}$, and $f()$ has a one-to-one inverse mapping function $f^{-1}()$, then the process of computing $f^{-1}\{f(x_i)\}=\{x_i\}$ is known as back transformation. Example $f()=\ln()$ and $f^{-1}=\exp()$
Vector	1. Within GIS the term vector refers to data that are comprised of lines or arcs, defined by beginning and end points, which meet at nodes. The locations of these nodes and the topological structure are usually stored explicitly. Features are defined by their boundaries only and curved lines are represented as a series

Term	Definition
	of connecting arcs. Vector storage involves the storage of explicit topology, which raises overheads, however it only stores those points which define a feature and all space outside these features is "non-existent" (AGI)
	2. In mathematics the term refers to a directed line, i.e. a line with a defined origin, direction and orientation. The same term is used to refer to a single column or row of a matrix, in which case it is denoted by a bold letter
WGS84	World Geodetic System, 1984 version. This models the Earth as a spheroid with major axis 6378.137 kms and flattening factor of 1:298.257, i.e. roughly 0.3% flatter at the poles than a perfect sphere. One of a number of such global models

Note: Where cited, references are drawn from the Association for Geographic Information (AGI), and the Open Geospatial Consortium (OGC) . Square bracketed text denotes insertion by the present authors into these definitions

1.5 Common Measures and Notation

1.5.1 Notation

Throughout this Guide a number of terms and associated formulas are used that are common to many analytical procedures. In this Section we provide a brief summary of those that fall into this category. Others, that are more specific to a particular field of analysis, are treated within the section to which they primarily apply. Many of the measures we list will be familiar to readers, since they originate from standard single variable (univariate) statistics. For brevity we provide details of these in tabular form. In order to clarify the expressions used here and elsewhere in the text, we use the notation shown in Table 1.3. *Italics* are used within the text and formulas to denote variables and parameters, as well as selected terms.

Table 1.3 Notation and symbology

$[a,b]$	A closed interval of the Real line, for example [0,1] means the set of all values between 0 and 1, including 0 and 1
(a,b)	An open interval of the Real line, for example (0,1) means the set of all values between 0 and 1, NOT including 0 and 1. This should not be confused with the notation for coordinate pairs, (x,y), or its use within bivariate functions such as f(x,y), or in connection with graph edges (see below) — the meaning should be clear from the context
(i,j)	In the context of graph theory, which forms the basis for network analysis, this pairwise notation is often used to define an edge connecting the two vertices i and j
(x,y)	A (spatial) data pair, usually representing a pair of coordinates in two dimensions. Terrestrial coordinates are typically Cartesian (i.e. in the plane, or *planar*) based on a pre-specified projection of the sphere, or Spherical (latitude, longitude). Spherical coordinates are often quoted in positive or negative degrees from the Equator and the Greenwich meridian, so may have the ranges [-90,+90] for latitude (north-south measurement) and [-180,180] for longitude (east-west measurement)
(x,y,z)	A (spatial) data triple, usually representing a pair of coordinates in two dimensions, plus a third coordinate (usually height or depth) or an attribute value, such as soil type or household income
$\{x_i\}$	A *set* of n values $x_1, x_2, x_3, \ldots x_n$, typically continuous ratio-scaled variables in the range $(-\infty,\infty)$ or $[0,\infty)$. The values may represent measurements or attributes of distinct objects, or values that represent a collection of objects (for example the population of a census tract)
$\{X_i\}$	An ordered set of n values $X_1, X_2, X_3, \ldots X_n$, such that $X_i <= X_{i+1}$ for all i
X,x	The use of bold symbols in expressions indicates matrices (upper case) and vectors (lower case)
$\{f_i\}$	A set of k frequencies ($k<=n$), derived from a dataset $\{x_i\}$. If $\{x_i\}$ contains discrete values, some of which occur multiple times, then $\{f_i\}$ represents the number of occurrences or the *count* of each distinct value. $\{f_i\}$ may also represent the number of occurrences of values that lie in a range or set of ranges, $\{r_i\}$. If a dataset contains n values, then the sum $\Sigma f_i=n$. The set $\{f_i\}$ can also be written f(x_i). If $\{f_i\}$ is regarded as a set of weights (for example attribute values) associated with the $\{x_i\}$, it may be written as the set $\{w_i\}$ or w(x_i)
$\{p_i\}$	A set of k probabilities ($k<=n$), estimated from a dataset or theoretically derived. With a finite set of values $\{x_i\}$, $p_i=f_i/n$. If $\{x_i\}$ represents a set of k classes or ranges then p_i is the probability of finding an occurrence in the ith class or range, i.e. the proportion of events or values occurring in that class or range. The sum $\Sigma p_i=1$. If a set of frequencies, $\{f_i\}$, have been standardised by dividing each value f_i by their sum, Σf_i, then $\{p_i\}\equiv\{f_i\}$
Σ	Summation symbol, e.g. $x_1+x_2+x_3+\ldots+x_n$. If no limits are shown the sum is assumed to apply to all subsequent elements, otherwise upper and/or lower limits for summation are provided

Π	Product symbol, e.g. $x_1 \cdot x_2 \cdot x_3 \cdot \ldots \cdot x_n$. If no limits are shown the product is assumed to apply to all subsequent elements, otherwise upper and/or lower limits for multiplication are provided
^	Used here in conjunction with Greek symbols (directly above) to indicate a value is an estimate of the true population value. Sometimes referred to as "hat"
~	Is distributed as, for example $y \sim N(0,1)$ means the variable y has a distribution that is Normal with a mean of 0 and standard deviation of 1
!	Factorial symbol. $z=x!$ means $z=x(x-1)(x-2)\ldots1$. $x \geq 0$. Usually applied to integer values of x. May be defined for fractional values of x using the Gamma function (Table 1.4, Section 1.5.2.8)
≡	'Equivalent to' symbol
∈	'Belongs to' symbol, e.g. $x \in [0,2]$ means that x belongs to/is drawn from the set of all values in the closed interval $[0,2]$; $x \in \{0,1\}$ means that x can take the values 0 and 1

1.5.2 Statistical measures and related formulas

Table 1.4 provides a list of common measures (univariate statistics) applied to datasets, and associated formulas for calculating the measure from a sample dataset in summation form (rather than integral form) where necessary. In some instances these formulas are adjusted to provide estimates of the population values rather than those obtained from the sample of data one is working on.

Many of the measures can be extended to two-dimensional forms in a very straightforward manner, and thus they provide the basis for numerous standard formulas in spatial statistics. For a number of univariate statistics (variance, skewness, kurtosis) we refer to the notion of (estimated) *moments* about the mean. These are computations of the form

$$\sum (x_i - \bar{x})^r, r = 1,2,3\ldots$$

When $r=1$ this summation will be 0, since this is just the difference of all values from the mean. For values of $r>1$ the expression provides measures that are useful for describing the shape (spread, skewness, peakedness) of a distribution, and simple variations on the formula are used to define the correlation between two or more datasets (the *product moment* correlation). The term *moment* in this context comes from physics, i.e. like momentum and moment of inertia,

and in a spatial (2D) context provides the basis for the definition of a centroid – the centre of mass or centre of gravity of an object, such as a polygon (see further, Section 4.2.5).

Table 1.4 Common formulas and statistical measures

This table of measures has been divided into 9 subsections for ease of use. Each is provided with its own subheading:

- Counts and specific values

- Measures of centrality

- Measures of spread

- Measures of distribution shape

- Measures of complexity and dimensionality

- Common distributions

- Data transforms and back transforms

- Selected functions

- Matrix expressions

1.5.2.1 Counts and specific values

Measure	Definition	Expression(s)
Count	The number of data values in a set	$Count(\{x_i\})=n$
Top m, Bottom m	The set of the largest (smallest) m values from a set. May be generated via an SQL command	$Top_m\{x_i\}=\{X_{n-m+1},...X_{n-1},X_n\};$ $Bot_m\{x_i\}=\{X_1,X_2,...\ X_m\};$
Variety	The number of distinct i.e. different data values in a set. Some packages refer to the variety as diversity, which should not be confused with information theoretic and other diversity measures	
Majority	The most common i.e. most frequent data values in a set. Similar to mode (see below), but often applied to raster datasets at the neighbourhood or zonal level. For general datasets the term should only be applied to cases where a given class is 50%+ of the total	
Minority	The least common i.e. least frequently occurring data values in a set. Often applied to raster datasets at the neighbourhood or zonal level	
Maximum, Max	The maximum value of a set of values. May not be unique	$Max\{x_i\}=X_n$
Minimum, Min	The minimum value of a set of values. May not be unique	$Min\{x_i\}=X_1$
Sum	The sum of a set of data values	$\displaystyle\sum_{i=1}^{n}x_i$

1.5.2.2 Measures of centrality

Measure	Definition	Expression(s)
Mean (arithmetic)	The arithmetic average of a set of data values (also known as the *sample mean* where the data are a sample from a larger population). Note that if the set $\{f_i\}$ are regarded as weights rather than frequencies the result is known as the *weighted mean*. Other mean values include the geometric and harmonic mean. The population mean is often denoted by the symbol μ. In many instances the sample mean is the best (unbiased) estimate of the population mean and is sometimes denoted by μ with a ^ symbol above it) or as a variable such as x with a bar above it.	$\displaystyle\bar{x}=\frac{1}{n}\sum_{i=1}^{n}x_i$ $\displaystyle\bar{x}=\sum_{i=1}^{n}f_ix_i\left/\sum_{i=1}^{n}f_i\right.$ $\displaystyle\bar{x}=\sum_{i=1}^{n}p_ix_i$

Measure	Definition	Expression(s)
Mean (harmonic)	The harmonic mean, H, is the mean of the reciprocals of the data values, which is then adjusted by taking the reciprocal of the result. The harmonic mean is less than or equal to the geometric mean, which is less than or equal to the arithmetic mean	$$H = \left(\frac{1}{n} \sum_{i=1}^{n} \frac{1}{x_i} \right)^{-1}$$
Mean (geometric)	The geometric mean, G, is the mean defined by taking the products of the data values and then adjusting the value by taking the nth root of the result. The geometric mean is greater than or equal to the harmonic mean and is less than or equal to the arithmetic mean	$$G = \left(\prod_{i=1}^{n} x_i \right)^{1/n} \text{, hence}$$ $$\log(G) = \frac{1}{n} \sum_{i=1}^{n} \log(x_i)$$
Mean (power)	The general (limit) expression for mean values. Values for p give the following means: p=1 arithmetic; p=2 root mean square; p=-1 harmonic. Limit values for p (i.e. as p tends to these values) give the following means: p=0 geometric; p=-∞ minimum; p=∞ maximum	$$M = \left(\frac{1}{n} \sum_{i=1}^{n} x_i^{p} \right)^{1/p}$$
Trim-mean, *TM*, *t*, Olympic mean	The mean value computed with a specified percentage (proportion), *t*/2, of values removed from each tail to eliminate the highest and lowest outliers and extreme values. For small samples a specific number of observations (e.g. 1) rather than a percentage, may be ignored. In general an equal number, *k*, of high and low values should be removed and the number of observations summed should equal *n*(1-*t*) expressed as an integer. This variant is sometimes described as the Olympic mean, as is used in scoring Olympic gymnastics for example	$$TM = \frac{1}{n(1-t)} \sum_{i=nt/2}^{n(1-t/2)} X_i \text{ , } t \in [0,1]$$
Mode	The most common or frequently occurring value in a set. Where a set has one dominant value or range of values it is said to be unimodal; if there are several commonly occurring values or ranges it is described as multi-modal. Note that arithmetic mean-mode≈3 (arithmetic mean-median) for many unimodal distributions	
Median, *Med*	The middle value in an ordered set of data if the set contains an odd number of values, or the average of the two middle values if the set contains an even number of values. For a continuous distribution the median is the 50% point (0.5) obtained from	$Med\{x_i\}=X_{(n+1)/2}$; *n* odd $Med\{x_i\}=iX_{n/2}+X_{n/2+1})/2$; *n* even

Measure	Definition	Expression(s)
	the cumulative distribution of the values or function	
Mid-range, *MR*	The middle value of the Range	$MR\{x_i\}=Range/2$
Root mean square (*RMS*)	The root of the mean of squared data values. Squaring removes negative values	$\sqrt{\dfrac{1}{n}\sum\limits_{i=1}^{n} x_i^2}$

1.5.2.3 Measures of spread

Measure	Definition	Expression(s)
Range	The difference between the maximum and minimum values of a set	$Range\{x_i\}=X_n\text{-}X_1$
Lower quartile (25%), *LQ*	In an ordered set, 25% of data items are less than or equal to the upper bound of this range. For a continuous distribution the LQ is the set of values from 0% to 25% (0.25) obtained from the cumulative distribution of the values or function. Treatment of cases where *n* is even and *n* is odd, and when *i* runs from 1 to n or 0 to *n* vary	$LQ=\{\,X_1, \ldots X_{(n+1)/4}\}$
Upper quartile (75%), *UQ*	In an ordered set 75% of data items are less than or equal to the upper bound of this range. For a continuous distribution the UQ is the set of values from 75% (0.75) to 100% obtained from the cumulative distribution of the values or function. Treatment of cases where *n* is even and *n* is odd, and when *i* runs from 1 to n or 0 to *n* vary	$UQ=\{X_{3(n+1)/4}, \ldots X_n\}$
Inter-quartile range, *IQR*	The difference between the lower and upper quartile values, hence showing the middle 50% of the distribution. The inter-quartile range can be obtained by taking the median of the dataset, then finding the median of the upper and lower halves of the set. The IQR is then the difference between these two secondary medians	$IQR=UQ\text{-}LQ$
Trim-range, *TR*, *t*	The range computed with a specified percentage (proportion), *t*/2, of the highest and lowest values removed to eliminate outliers and extreme values. For small samples a specific number of observations (e.g. 1) rather than a percentage, may be ignored. In general an equal number, *k*, of high and low values are removed If possible)	$TR_t=X_{n(1-t/2)}\text{-}X_{nt/2},\ t\in[0,1]$ $TR_{50\%}=IQR$

Measure	Definition	Expression(s)
Variance, Var, σ^2, s^2, μ_2	The average squared difference of values in a dataset from their population mean, μ, or from the sample mean (also known as the sample variance where the data are a sample from a larger population). Differences are squared to remove the effect of negative values (the summation would otherwise be 0). The third formula is the frequency form, where frequencies have been standardised, i.e. $\Sigma f_i=1$. Var is a function of the 2nd moment about the mean. The population variance is often denoted by the symbol μ_2 or σ^2. The estimated population variance is often denoted by s^2 or by σ^2 with a ^ symbol above it	$Var = \sigma^2 = \dfrac{1}{n}\sum_{i=1}^{n}(x_i - \mu)^2$ $Var = \dfrac{1}{n}\sum_{i=1}^{n}(x_i - \bar{x})^2$ $Var = \sum_{i=1}^{n}f_i(x_i - \bar{x})^2$ $Var = \dfrac{1}{n}\sum_{i=1}^{n}(x_i - \bar{x})(x_i - \bar{x})$ $s^2 = \hat{\sigma}^2 = \dfrac{1}{n-1}\sum_{i=1}^{n}(x_i - \bar{x})^2$
Standard deviation, SD, s or $RMSD$	The square root of the variance, hence it is the Root Mean Squared Deviation (RMSD). The population standard deviation is often denoted by the symbol σ. SD* shows the estimated population standard deviation (sometimes denoted by σ with a ^ symbol above it or by s)	$SD = \sqrt{Var} = \sigma$ $SD = \sqrt{\dfrac{1}{n}\sum_{i=1}^{n}(x_i - \bar{x})^2}$ $SD^* = \hat{\sigma} = \sqrt{\dfrac{1}{n-1}\sum_{i=1}^{n}(x_i - \bar{x})^2}$
Standard error of the mean, SE	The estimated standard deviation of the mean values of n samples from the same population. It is simply the sample standard deviation reduced by a factor equal to the square root of the number of samples, $n \geq 1$	$SE = \dfrac{SD}{\sqrt{n}}$
Root mean squared error, $RMSE$	The standard deviation of samples from a known set of true values, x_i^*. If x_i^* are estimated by the mean of sampled values RMSE is equivalent to RMSD	$RMSE = \sqrt{\dfrac{1}{n}\sum_{i=1}^{n}(x_i - x_i^*)^2}$
Mean deviation/error, MD or ME	The mean deviation of samples from the known set of true values, x_i^*	$MD = \dfrac{1}{n}\sum_{i=1}^{n}(x_i - x_i^*)$
Mean absolute deviation/error, MAD or MAE	The mean absolute deviation of samples from the known set of true values, x_i^*	$MAE = \dfrac{1}{n}\sum_{i=1}^{n}|x_i - x_i^*|$

Measure	Definition	Expression(s)
Covariance, Cov	Literally the pattern of common (or co-) variation observed in a collection of two (or more) datasets, or partitions of a single dataset. Note that if the two sets are the same the covariance is the same as the variance	$Cov(x,y) = \frac{1}{n}\sum_{i=1}^{n}(x_i - \bar{x})(y_i - \bar{y})$ $Cov(x,x) = Var(x)$
Correlation/ product moment or Pearson's correlation coefficient, r	A measure of the similarity between two (or more) datasets. The correlation coefficient is the ratio of the covariance to the product of the standard deviations. If the two datasets are the same or perfectly matched this will give a result=1	$r = Cov(x,y)/SD_x SD_y$ $r = \dfrac{\sum_{i=1}^{n}(x_i - \bar{x})(y_i - \bar{y})}{\sqrt{\sum_{i=1}^{n}(x_i - \bar{x})^2}\sqrt{\sum_{i=1}^{n}(y_i - \bar{y})^2}}$
Coefficient of variation, CV	The ratio of the standard deviation to the mean, sometime computed as a percentage. If this ratio is close to 1, and the distribution is strongly left skewed, it may suggest the underlying distribution is Exponential. Note, mean values close to 0 may produce unstable results	$CV = SD/\bar{x}$
Variance mean ratio, VMR	The ratio of the variance to the mean, sometime computed as a percentage. If this ratio is close to 1, and the distribution is unimodal and relates to count data, it may suggest the underlying distribution is Poisson. Note, mean values close to 0 may produce unstable results	$VMR = Var/\bar{x}$

1.5.2.4 Measures of distribution shape

Measure	Definition	Expression(s)
Skewness, α_3	If a frequency distribution is unimodal and symmetric about the mean it has a skewness of 0. Values greater than 0 suggest skewness of a unimodal distribution to the right, whilst values less than 0 indicate skewness to the left. A function of the 3rd moment about the mean (denoted by α_3 with a ^ symbol above it for the sample skewness)	$\alpha_3 = \frac{1}{n\sigma^3}\sum_{i=1}^{n}(x_i - \mu)^3$ $\alpha_3 = \frac{1}{n\hat{\sigma}^3}\sum_{i=1}^{n}(x_i - \bar{x})^3$ $\hat{\alpha}_3 = \frac{n}{(n-1)(n-2)\hat{\sigma}^3}\sum_{i=1}^{n}(x_i - \bar{x})^3$
Kurtosis, α_4	A measure of the peakedness of a frequency distribution. More pointy distributions tend to have high kurtosis values. A function of the 4th moment about the mean. It is customary to subtract 3	$\alpha_4 = \frac{1}{n\sigma^4}\sum_{i=1}^{n}(x_i - \mu)^4$

Measure	Definition	Expression(s)
	from the raw kurtosis value (which is the kurtosis of the Normal distribution) to give a figure relative to the Normal (denoted by α_4 with a ^ symbol above it for the sample kurtosis)	$\alpha_4 = \dfrac{1}{n\hat{\sigma}^4}\displaystyle\sum_{i=1}^{n}(x_i - \bar{x})^4$ $\hat{\alpha}_4 = \dfrac{a}{\hat{\sigma}^4}\displaystyle\sum_{i=1}^{n}(x_i - \bar{x})^4 - b$ where $a = \dfrac{n(n+1)}{(n-1)(n-2)(n-3)}$, $b = \dfrac{3(n-1)^2}{(n-2)(n-3)}$

1.5.2.5 Measures of complexity and dimensionality

Measure	Definition	Expression(s)
Information statistic (Entropy), I (Shannon's)	A measure of the amount of pattern, disorder or *information*, in a set $\{x_i\}$ where p_i is the proportion of events or values occurring in the ith class or range. Note that if p_i=0 then $p_i\log_2(p_i)$ is 0. I takes values in the range $[0,\log_2(k)]$. The lower value means all data falls into 1 category, whilst the upper means all data are evenly spread	$I = -\displaystyle\sum_{i=1}^{k} p_i \log_2(p_i)$
Information statistic (Diversity), Div	Shannon's entropy statistic standardised by the number of classes, k, to give a range of values from 0 to 1	$Div = \dfrac{-\displaystyle\sum_{i=1}^{k} p_i \log_2(p_i)}{\log_2(k)}$
Dimension (topological), D_T	Broadly, the number of (intrinsic) coordinates needed to refer to a single point anywhere on the object. The dimension of a point=0, a rectifiable line=1, a surface=2 and a solid=3. See text for fuller explanation. The value 2.5 (often denoted 2.5D) is used in GIS to denote a planar region over which a single-valued attribute has been defined at each point (e.g. height). In mathematics topological dimension is now equated to a definition similar to cover dimension (see below)	D_T=0,1,2,3,...
Dimension (capacity, cover or fractal), D_c	Let $N(h)$ represent the number of small elements of edge length h required to cover an object. For a line, length 1, each element has length $1/h$. For a plane surface each element (small square of side length $1/h$) has area $1/h^2$, and for a volume, each element is a cube with volume $1/h^3$. More generally $N(h)=1/h^D$,	$D_c = -\lim \dfrac{\ln N(h)}{\ln(h)}, n \to 0^+$ $D_c{\geq}0$

Measure	Definition	Expression(s)
	where D is the topological dimension, so $N(h)= h^{-D}$ and thus $\log(N(h))=-D\log(h)$ and so $D_c=-\log(N(h))/\log(h)$. D_c may be fractional, in which case the term *fractal* is used	

1.5.2.6 Common distributions

Measure	Definition	Expression(s)
Uniform (continuous)	All values in the range are equally likely. Mean=$a/2$, variance=$a^2/12$. Here we use $f(x)$ to denote the probability distribution associated with continuous valued variables x, also described as a *probability density function*	$f(x) = \dfrac{1}{a}; x \in [0,a]$
Binomial (discrete)	The terms of the Binomial give the probability of x successes out of n trials, for example 3 heads in 10 tosses of a coin, where p=probability of success and $q=1-p$=probability of failure. Mean, $m=np$, variance=npq. Here we use $p(x)$ to denote the probability distribution associated with discrete valued variables x	$p(x) = \dfrac{n!}{(n-x)!\,x!} p^x q^{1-x}$; x=1,2,...n
Poisson (discrete)	An approximation to the Binomial when p is very small and n is large (>100), but the mean $m=np$ is fixed and finite (usually not large). Mean=variance=m	$p(x) = \dfrac{m^x}{x!} e^{-m}$; x=1,2,...n
Normal (continuous)	The distribution of a measurement, x, that is subject to a large number of independent, random, additive errors. The Normal distribution may also be derived as an approximation to the Binomial when p is not small (e.g. $p \approx 1/2$) and n is large. If μ=mean and σ=standard deviation, we write $N(\mu,\sigma)$ as the Normal distribution with these parameters. The Normal- or z-transform $z=(x-\mu)/\sigma$ changes (normalises) the distribution so that it has a zero mean and unit variance, $N(0,1)$. The distribution of n mean values of independent random variables drawn from *any* underlying distribution is also Normal (*Central Limit Theorem*)	$f(z) = \dfrac{1}{\sqrt{2\pi}} e^{-z/2}; z \in [-\infty, \infty]$

1.5.2.7 Data transforms and back transforms

Measure	Definition	Expression(s)
Log	If the frequency distribution for a dataset is broadly unimodal and left-skewed, the natural log transform (logarithms base e)	z=ln(x) or

Measure	Definition	Expression(s)
	will adjust the pattern to make it more symmetric/similar to a Normal distribution. For variates whose values may range from 0 upwards a value of 1 is often added to the transform. Back transform with the exp() function	$z = \ln(x+1)$ nb: $\ln(x) = \log_e(x) = \log_{10}(x) * \log_{10}(e)$ $x = \exp(z)$ or $x = \exp(z) - 1$
Square root (Freeman-Tukey)	A transform that may adjust the dataset to make it more similar to a Normal distribution. For variates whose values may range from 0 upwards a value of 1 is often added to the transform. For $0 \le x \le 1$ (e.g. rate data) the combined form of the transform is often used, and is known as the Freeman-Tukey (FT) transform	$z = \sqrt{x}$, or $z = \sqrt{x+1}$, or $z = \sqrt{x} + \sqrt{x+1}$ (FT) $x = z^2$, or $x = z^2 - 1$
Logit	Often used to transform binary response data, such as survival/non-survival or present/absent, to provide a continuous value in the range $(-\infty, \infty)$, where p is the proportion of the sample that is 1 (or 0). The inverse or back-transform is shown as p in terms of z. This transform avoids concentration of values at the ends of the range. For samples where proportions p may take the values 0 or 1 a modified form of the transform may be used. This is typically achieved by adding $1/2n$ to the numerator and denominator, where n is the sample size. Often used to correct S-shaped (*logistic*) relationships between response and explanatory variables	$z = \ln\left(\dfrac{p}{1-p}\right), p \in [0,1]$ $p = \dfrac{e^z}{1 + e^z}$
Normal, z-transform	This transform normalises or standardises the distribution so that it has a zero mean and unit variance. If $\{x_i\}$ is a set of n sample *mean* values from *any* probability distribution with mean μ and variance σ^2 then the z-transform shown here as z_2 will be distributed $N(0,1)$ for large n (Central Limit Theorem). The divisor in this instance is the standard error. In both instances the standard deviation must be non-zero	$z_1 = \dfrac{(x - \mu)}{\sigma}$ $z_2 = \dfrac{(x - \mu)}{\sigma/\sqrt{n}}$
Box-Cox, power transforms	A family of transforms defined for positive data values only, that often can make datasets more Normal; k is a parameter. The inverse or back-transform is also shown as x in terms of z	$z = \dfrac{(x^k - 1)}{k}, k > 0, x > 0$ $x = (kz + 1)^{1/k}, k > 0$

Measure	Definition	Expression(s)
Angular transforms (Freeman-Tukey)	A transform for proportions, p, designed to spread the set of values near the end of the range. k is typically 0.5. Often used to correct S-shaped relationships between response and explanatory variables. If $p=x/n$ then the Freeman-Tukey (FT) version of this transform is the averaged version shown. This is a variance-stabilising transform	$z = \sin^{-1}\left(p^k\right),$ $p = \sin(z)^{1/k}$ $z = \sin^{-1}\left(\sqrt{\dfrac{x}{n+1}}\right) +$ $\sin^{-1}\left(\sqrt{\dfrac{x+1}{n+1}}\right)$ (FT)

1.5.2.8 Selected functions

Measure	Definition	Expression(s)
Bessel function of the first kind	Bessel functions occur as the solution to specific differential equations. They are described with reference to a parameter known as the order, shown as a subscript. For integer orders Bessel functions can be represented as an infinite series. Order 0 and Order 1 expansions are shown here. The graph of a Bessel function is similar to a dampening sine wave. Usage in spatial analysis arises in connection with directional statistics and spline curve fitting. See the Mathworld website entry for more details	$I_0(\kappa) = \sum_{i=0}^{\infty} \dfrac{(-1)^i (\kappa/2)^{2i}}{(i!)^2}$, and $I_1(\kappa) = \dfrac{\kappa}{2} \sum_{i=0}^{\infty} \dfrac{(-1)^i (\kappa/2)^{2i+1}}{i!(i+1)!}$
Exponential integral function, $E_1(x)$	A definite integral function. Used in association with spline curve fitting. See the Mathworld website entry for more details	$E_1(x) = \int_1^{\infty} \dfrac{e^{-tx}}{t}\, dt$
Gamma function, Γ	A widely used definite integral function. For integer values of x: $\Gamma(x)=(x-1)!$ and $\Gamma(x/2)=(x/2-1)!$ so $\Gamma(3/2)=(1/2)!/2=(\sqrt{\pi})/2$ See the Mathworld website entry for more details	$\Gamma(x) = \int_0^{\infty} x^{1/2} e^{-x}\, dx$ $\Gamma(1/2) = \sqrt{\pi}$

1.5.2.9 Matrix expressions

Measure	Definition	Expression(s)
Identity	A matrix with diagonal elements 1 and off-diagonal elements 0	$$I = \begin{bmatrix} 1 & 0 & 0 & 0 \\ 0 & 1 & 0 & 0 \\ .. & .. & .. & .. \\ 0 & 0 & 0 & 1 \end{bmatrix}$$
Determinant	Determinants are only defined for square matrices. Let **A** be an n by n matrix with elements $\{a_{ij}\}$. The matrix M_{ij} here is a subset of **A** known as the *minor*, formed by eliminating row i and column j from **A**. An n by n matrix, **A**, with Det=0 is described as *singular*, and such a matrix has no inverse. If Det(**A**) is very close to 0 it is described as *ill-conditioned*	$\mid A \mid$, **Det(A)** $\mid A \mid = \sum_{i}^{n} a_{ij} a^{ij}$, where $a^{ij} = \left(-1\right)^{i+j} M_{ij}$
Inverse	The matrix equivalent of division in conventional algebra. For a matrix, **A**, to be invertible its determinant must be non-zero, and ideally not very close to zero. A matrix that has an inverse is by definition non-singular. A symmetric real-valued matrix is *positive definite* if all its eigenvalues are positive, whereas a *positive semi-definite* matrix allows for some eigenvalues to be 0. A matrix, **A**, that is invertible satisfies the relation $AA^{-1}=I$	A^{-1}
Transpose	A matrix operation in which the rows and columns are transposed, i.e. in which elements a_{ij} are swapped with a_{ji} for all i,j. The inverse of a transposed matrix is the same as the transpose of the matrix inverse	A^{T} or A' $(A^{T})^{-1}=(A^{-1})^{T}$
Symmetric	A matrix in which element $a_{ij}=a_{ji}$ for all i,j	$A=A^{T}$
Trace	The sum of the diagonal elements of a matrix, a_{ii} — the sum of the eigenvalues of a matrix equals its trace	**Tr(A)**
Eigenvalue, Eigenvector	If **A** is a real-valued k by k square matrix and **x** is a non-zero real-valued vector, then a scalar λ that satisfies the equation shown in the adjacent column is known as an eigenvalue of **A** and **x** is an eigenvector of **A**. There are k eigenvalues of **A**, each with a corresponding eigenvector. The matrix **A** can be decomposed into three parts, as shown, where **E** is a matrix of its eigenvectors and **D** is a diagonal matrix of its eigenvalues	$(A-\lambda I)x=0$ $A=EDE^{-1}$ (diagonalisation)

2 Conceptual Frameworks for Spatial Analysis

2.1 The geospatial perspective

Geospatial analysis provides a distinct perspective on the world, a unique lens through which to examine events, patterns, and processes that operate on or near the surface of our planet. It makes sense, then, to introduce the main elements of this perspective, the conceptual framework that provides the background to spatial analysis, as a preliminary to the main body of this Guide's material. This chapter provides that introduction. It is divided into four main sections. The first describes the basic components of this view of the world — the classes of things that a spatial analyst recognises in the world, and the beginnings of a system of organisation of geographic knowledge. The second section describes some of the structures that are built with these basic components and the relationships between them that interest geographers and others. The third section introduces the concepts of spatial statistics, including probability, that provide perhaps the most sophisticated elements of the conceptual framework. Finally, the fourth section discusses some of the basic components of the Internet-based infrastructure that increasingly provides the essential facilities for spatial analysis.

The domain of geospatial analysis is the surface of the Earth, extending upwards in the analysis of topography and the atmosphere, and downwards in the analysis of groundwater and geology. In scale it extends from the most local, when archaeologists record the locations of pieces of pottery to the nearest centimetre or property boundaries are surveyed to the nearest millimetre, to the global, in the analysis of sea surface temperatures or global warming. In time it extends backwards from the present into the analysis of historical population migrations, the discovery of patterns in archaeological sites, or the detailed mapping of the movement of continents, and into the future in attempts to predict the tracks of hurricanes, the melting of the Greenland ice-cap, or the likely growth of urban areas. Methods of spatial analysis are robust, and capable of operating over a range of spatial and temporal scales.

Ultimately, geospatial analysis concerns *what* happens *where*, and makes use of geographic information that links features and phenomena on the Earth's surface to their locations. This sounds very simple and straightforward, and it is not so much the basic information as the structures and arguments that can be built on it that provide the richness of spatial analysis. In principle there is no limit to the complexity of spatial analytic techniques that might find some application in the world, and might be used to tease out interesting insights and support practical actions and decisions. In reality, some techniques are simpler, more useful, or more insightful than others, and the contents of this Guide reflect that reality. This chapter is about the underlying concepts that are employed, whether it be in simple, intuitive techniques or in advanced, complex mathematical ones.

Spatial analysis exists at the interface between the human and the computer, and both play important roles. The concepts that humans use to understand, navigate, and exploit the world around them are mirrored in the concepts of spatial analysis. So the discussion that follows will often appear to be following parallel tracks — the track of human intuition on the one hand, with all its vagueness and informality, and the track of the formal, precise world of spatial analysis on the other. The relationship between these two tracks forms one of the recurring themes of this Guide.

2.2 Basic Primitives

2.2.1 Place

At the centre of all spatial analysis is the concept of *place*. The Earth's surface comprises some 500,000,000 sq km, so there would be room to pack half a billion industrial sites of 1 sq km each (assuming that nothing else required space, and that the two-thirds of the Earth's surface that is covered by water was as acceptable as the one-third that is land); and 500 trillion sites of 1 sq m each (roughly the space occupied by a sleeping human). People identify with places of various sizes and shapes, from the room to the parcel of land, to the neighbourhood, the city, the county, the state or province, or the nation-state. Places may overlap, as when a watershed spans the boundary of two counties, and places may be nested hierarchically, as when counties combine to form a state or province.

Places often have names, and people use these to talk about and distinguish between places. Some names are official, having been recognised by national or state agencies charged with bringing order to geographic names. In the U.S., for example, the Board on Geographic Names exists to ensure that all agencies of the federal government use the same name in referring to a place, and to ensure as far as possible that duplicate names are removed from the landscape. A list of officially sanctioned names is termed a *gazetteer*, though that word has come to be used for any list of geographic names.

Places change continually, as people move, climate changes, cities expand, and a myriad of social and physical processes affect virtually every spot on the Earth's surface. For some purposes it is sufficient to treat places as if they were static, especially if the processes that affect them are comparatively slow to operate. It is difficult, for example, to come up with instances of the need to modify maps as continents move and mountains grow or shrink in response to earthquakes and erosion.

On the other hand it would be foolish to ignore the rapid changes that occur in the social and economic makeup of cities, or the constant movement that characterises modern life. Throughout this Guide, it will be important to distinguish between these two cases, and to judge whether time is or is not important.

People associate a vast amount of information with places. Three Mile Island, Sellafield, and Chernobyl are associated with nuclear reactors and accidents, while Tahiti and Waikiki conjure images of (perhaps somewhat faded) tropical paradise. One of the roles of places and their names is to link together what is known in useful ways. So for example the statements "I am going to London next week" and "There's always something going on in London" imply that I will be having an exciting time next week. But while "London" plays a useful role, it is nevertheless vague, since it might refer to the area administered by the Greater London Authority, the area inside the M25 motorway, or something even less precise and determined by the context in which the name is used. Science clearly needs something better, if information is to be linked exactly to places, and if places are to be matched, measured, and subjected to the rigors of spatial analysis.

The basis of rigorous and precise definition of place is a *coordinate system*, a set of measurements that allows place to be specified unambiguously and in a way that is meaningful to everyone. The Meridian Convention of 1884 established the Greenwich Observatory in London as the basis of longitude, replacing a confusing multitude of earlier systems. Today, the World Geodetic System of 1984 and subsequent adjustments provide a highly accurate pair of coordinates for every location on the Earth's surface (and incidentally place the line of zero longitude about 100m east of the Greenwich Observatory). Elevation continues to be problematic, however, since countries and even agencies within countries insist on their own definitions of what marks zero elevation, or exactly how to define "sea level". Many other coordinate systems are in use, but most are easily converted to and from latitude/longitude. Today it is possible to measure location directly, using the Global Positioning System (GPS) or its Russian counterpart GLONASS (and soon its European counterpart Galileo).

Spatial analysis is most often applied in a two-dimensional space. But applications that extend above or below the surface of the Earth must often be handled as three-dimensional. Time sometimes adds a fourth dimension, particularly in studies that examine the dynamic nature of phenomena.

2.2.2 Attributes

Attribute has become the preferred term for any recorded characteristic or property of a place (see Table 1.2 for a more formal definition). A place's name is an obvious example of an attribute, but a vast array of other options has proven useful for various purposes. Some are measured, including elevation, temperature, or rainfall. Others are the result of classification, including soil type, land use or land cover type, or rock type. Government agencies provide a host of attributes in the form of statistics, for places ranging in size from countries all the way down to neighbourhoods and streets. The characteristics that people assign rightly or mistakenly to places, such as "expensive", "exciting", "smelly", or "dangerous" are also examples of attributes. Attributes can be more than simple values or terms, and today it is possible to construct information systems that contain entire collections of images as attributes of hotels, or recordings of birdsong as attributes of natural areas. But while these are certainly feasible, they are beyond the bounds of most techniques of spatial analysis.

Many terms have been adopted to describe attributes. From the perspective of spatial analysis the most useful divides attributes as follows:

- **Nominal.** An attribute is nominal if it successfully distinguishes between locations, but without any implied ranking

or potential for arithmetic. For example, a telephone number can be a useful attribute of a place, but the number itself generally has no numeric meaning. It would make no sense to add or divide telephone numbers, and there is no sense in which the number 9680244 is more or better than the number 8938049. Likewise, assigning arbitrary numerical values to classes of land type, e.g. 1=arable, 2=woodland, 3=marsh, 4=other is simply a convenient form of naming (the values are nominal)

- **Ordinal.** An attribute is ordinal if it implies a ranking, in the sense that Class 1 may be better than Class 2, but as with nominal attributes no arithmetic operations make sense, and there is no implication that Class 3 is worse than Class 2 by the precise amount by which Class 2 is worse than Class 1. An example of an ordinal scale might be preferred locations for residences — an individual may prefer some areas of a city to others, but such differences between areas may be barely noticeable or quite profound

- **Interval.** The remaining three types of attributes are all quantitative, representing various types of measurements. Attributes are interval if differences make sense, as they do for example with measurements of temperature on the Celsius or Fahrenheit scales, or for measurements of elevation above sea level

- **Ratio.** Attributes are ratio if it makes sense to divide one measurement by another. For example, it makes sense to say that one person weighs twice as much as another person, but it makes no sense to say that a temperature of 20 Celsius is twice as warm as a temperature of 10 Celsius, because while weight has an absolute zero Celsius temperature does not (but on an absolute scale of temperature, such as the Kelvin scale, 200 degrees can indeed be said to be twice as warm as 100 degrees). It follows that negative values cannot exist on a ratio scale

- **Cyclic.** Finally, it is not uncommon to encounter measurements of attributes that represent directions or cyclic phenomena, and to encounter the awkward property that two distinct points on the scale can be equal — for example, 0 and 360 degrees are equal. Directional data are cyclic, as are calendar dates. Arithmetic operations are problematic with cyclic data, and special techniques are needed, such as the techniques used to overcome the Y2K problem, when the year after (19)99 was (20)00. For example, it makes no sense to average 1° and 359° to get 180°, since the average of two directions close to north clearly is not south. Mardia and Jupp (1999) provide a comprehensive review of the analysis of directional or cyclic data

While this terminology of measurement types is standard, spatial analysts find that another distinction is particularly important. This is the distinction between attributes that are termed *spatially intensive* and *spatially extensive*. Spatially extensive attributes include total population, measures of a place's area or perimeter length, and total income — they are true only of the place as a whole. Spatially intensive attributes include population density, average income, and percent unemployed, and if the place is homogeneous they will be true of any part of the place as well as of the whole. For many purposes it is necessary to keep spatially intensive and spatially extensive attributes apart, because they respond very differently when places are merged or split, and when many types of spatial analysis are conducted.

2.2.3 Objects

The places discussed in Section 2.2.1 vary enormously in size and shape. Weather observations are obtained from stations that may occupy only a few square meters of the Earth's surface (from instruments that occupy only a small fraction of the station's area),

whereas statistics published for Russia are based on a land area of more than 17 million sq km. In spatial analysis it is customary to refer to places as *objects*. In studies of roads or rivers the objects of interest are long and thin, and will often be represented as lines of zero width. In studies of climate the objects of interest may be weather stations of minimal extent, and will often be represented as points. On the other hand many studies of social or economic patterns may need to consider the two-dimensional extent of places, which will therefore be represented as areas, and in some studies where elevations or depths are important it may be appropriate to represent places as volumes. To a spatial statistician, these points, lines, areas, or volumes are known as the attributes' *spatial support*

Each of these four classes of objects has its own techniques of representation in digital systems. The software for capturing and storing spatial data, analysing and visualising them, and reporting the results of analysis must recognise and handle each of these classes. But digital systems must ultimately represent everything in a language of just two characters, 0 and 1 or "off" and "on", and special techniques are required to represent complex objects in this way. In practice, points, lines, and areas are most often represented in the following standard forms:

- **Points** as pairs of coordinates, in latitude/longitude or some other standard system

- **Lines** as ordered sequences of points connected by straight lines

- **Areas** as ordered rings of points, also connected by straight lines to form polygons. In some cases areas may contain holes, and may include separate islands, such as in representing the State of Michigan with its separate Upper Peninsula, or the State of Georgia with its offshore islands. This use of polygons to represent areas is so pervasive that many spatial analysts refer to all areas as polygons, whether or not their edges are actually straight

Lines represented in this way are often termed *polylines*, by analogy to polygons (see Table 1.2 for a more formal definition). Three-dimensional volumes are represented in several different ways, and as yet no one method has become widely adopted as a standard. The related term *edge* is used in several ways within GIS. These include: to denote the border of polygonal regions; to identify the individual links connecting nodes or vertices in a network; and as a general term relating to the distinct or indistinct boundary of areas or zones. In many parts of spatial analysis the related term, *edge effect* is applied. This refers to possible bias in the analysis which arises specifically due to proximity of features to one or more edges. For example, in point pattern analysis computation of distances to the nearest neighbouring point, or calculation of the density of points per unit area, may both be subject to edge effects.

Figure 2-1 shows a simple example of points, lines, and areas, as represented in a typical map display. The hospital, boat ramp, and swimming area will be stored in the database as points with associated attributes, and symbolised for display. The roads will be stored as polylines, and the road type symbols (U.S. Highway, Interstate Highway) generated from the attributes when each object is displayed. The lake will be stored as two polygons with appropriate attributes. Note how the lake consists of two geometrically disconnected pieces, linked in the database to a single set of attributes — objects in a GIS may consist of multiple parts, as long as each part is of the same type.

It can be expensive and time-consuming to create the polygon representations of complex area objects, and so analysts often resort to simpler approaches, such as choosing a single *representative point*. But while this may be satisfactory for some purposes, there are obvious problems with representing the entirety of a large country such as Russia as a single point. For example, the distance from Canada to the U.S. computed between representative points in this way would be very misleading, given that they share a very long common boundary.

Figure 2-1 An example map showing points, lines, and areas appropriately symbolised

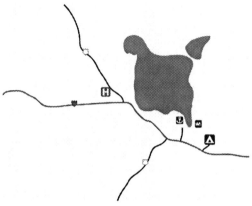

See text for detailed explanation

2.2.4   Maps

Historically, maps have been the primary means to store and communicate spatial data. Objects and their attributes can be readily depicted, and the human eye can quickly discern patterns and anomalies in a well-designed map. Points can be shown as symbols of various kinds, depicting anything from a windmill to a church; lines can be symbolised to distinguish between major roads, minor roads, and rivers; and areas can be symbolised with colour, shading, or annotation.

Maps have traditionally existed on paper, as individual sheets or bound into atlases (a term that originated with Mercator, who produced one of the first atlases in the late 16th century). The advent of digital computers has broadened the concept of a map substantially, however. Maps can now take the form of images displayed on the screens of computers or even mobile phones. They can be dynamic, showing the Earth spinning on its axis or tracking the movement of migrating birds. Their designs can now go far beyond what was traditionally possible when maps had to be drawn by hand, incorporating a far greater range of colour and texture, and even integrating sound.

2.2.5 Multiple properties of places

Dividing objects into points, lines, areas, and volumes is only the crudest of the many ways in which humans classify the things they find surrounding them in their environment. A topographic map may show areas of woodland, rivers, roads, contours, windmills, and several hundred other kinds of objects, each of them classifiable as points, lines, areas, or volumes. Objects may overlap — a river may pass through a wood, a road may cross a river, and all three may lie inside the area shown as "California". Spatial analysts think of multiple classes of objects as forming *layers* of information superimposed on each other. A study might be conducted on just one layer, analysing the patterns formed by the objects in that layer and looking for clues as to the processes that created the pattern. In other cases the patterns of one layer might be analysed in relation to other layers. For example, the pattern of cases of cancer in an area might be studied to see if it bears any relationship to the pattern of electrical transmission lines in the area, or to the chemistry of drinking water.

Sometimes many different attributes are collected and published for the same sets of places. Census agencies do this when they compile hundreds of attributes for census tracts, counties, or states. In such cases comparing layers is a simple matter of comparing the different attributes assigned to the same places. But more generally the objects in two layers may not be the same, creating difficulties for analysis. For example, suppose a study is being conducted of people's perception of noise in a neighbourhood near an airport. Monitoring devices have been positioned at points around the neighbourhood, providing a layer of noise measurements. Residents have been interviewed, providing another layer of perceptual data. The analysis now moves to a comparison of the contents of the two layers, but it is hampered by the fact that the objects in the two layers are in different locations. A possible solution to this problem is discussed in Section 2.3.8.

2.2.6 Fields

The idea that the Earth's surface is a space littered with various kinds of objects certainly matches the ways humans tend to think about the world around them. Scenes perceived by the eye are immediately parsed by the brain into "tree", "car", "house", and other discrete objects. But in other cases the world may appear more continuous than discrete. Ground elevation (or ocean depth), for example, is in broad terms a continuously varying property that can be measured anywhere on the Earth's surface, as is atmospheric temperature, or level of noise. In essence there are two ways of thinking about phenomena on the Earth's surface. In the first, the *discrete-object* view, reality is like an empty table-top littered with discrete, countable objects that can be assigned to different classes. In the second, the *continuous-field* view, reality is a collection of continuous surfaces, each representing the variation of one property over the Earth's surface. When it is necessary to differentiate by height, the field becomes three-dimensional rather than two-dimensional, and time may add a fourth dimension. Mathematically, a field is a continuous function mapping every location to the value of some property of interest.

The distinction between continuous fields and discrete objects is merely conceptual, and there is no test that can be applied to the contents of a database to determine which view is operative. But it is a very important distinction that underlies the choices that analysts make in selecting appropriate techniques. Points, polylines, and polygons arise in both cases, and yet their meanings are very different. For example, points might represent the locations where a continuous field of temperature is measured, or they might represent isolated instances of a disease — in the latter case, which falls within the discrete-object view, there is no implication that anything happens in between the cases, whereas in the former case one would expect temperature to vary smoothly between measurement sites. Polylines might represent the connected pieces of a stream network, or they might represent the contours of an elevation surface — in the latter case, the polylines would collectively represent a field, and would consequently not be allowed to cross each other.

2.2.7 Networks

Mention has already been made of the lines that often constitute the objects of spatial data. They include streets, roads, railroads, and highways; rivers, canals, and streams; and the more abstract interpersonal networks studied by sociologists. Networks constitute one-dimensional structures embedded in two or three dimensions. Discrete point objects may be distributed on the network, representing such phenomena as landmarks or bridges on road networks, or observation points on rivers. Discrete line objects may also be found, in the form of tunnels, stretches of highway with a constant number of lanes, or river reaches. Continuous fields may also be distributed over the one-dimensional network, defining such variables as travel speed, railroad gradient, traffic density, or stream velocity.

Mathematically, a network forms a *graph*, and many techniques developed for graphs have applications to networks. These include various ways of measuring a network's connectivity, or of finding shortest paths between pairs of points on a network. These methods are explored in detail in Chapter 7.

2.2.8 Density estimation

One of the more useful concepts in spatial analysis is *density* — the density of humans in a crowded city, or the density of tracks across a

stretch of desert, or the density of retail stores in a shopping centre. Density provides an effective link between the discrete-object and continuous-field conceptualisations, since density expresses the number of discrete objects per unit of area, and is itself a continuous field. Mathematically, the density of some kind of object is calculated by counting the number of such objects in an area, and dividing by the size of the area. This is easily done when calculating population density from census data, for example, by dividing the population of each census tract or county by its physical area. But this leads to multiple values of density, depending on the object used to make the calculation. Techniques of *density estimation* try to avoid this problem and are discussed in detail in Section 4.3.4.1. The book by Silverman (1998) provides additional background reading on this issue.

2.2.9 Detail, resolution, and scale

The surface of the planet is almost infinitely complex, revealing more and more detail the closer one looks. An apparently simple concept such as a coastline turns out to be problematic in practice — should its representation as a polyline include every indentation, every protruding rock, even every grain of sand? It is clearly impossible to build a representation that includes every detail, and in reality decisions must be made about the amount of detail to include. *Spatial resolution* is the term given to a threshold distance, below which the analyst has decided that detail is unnecessary or irrelevant. For example, one might decide to represent a coastline at a spatial resolution of 100m, meaning that all crenulations, indentations, and protuberances that are no wider than 100m are omitted from the representation. In other words, the representation of the phenomenon appears simpler and smoother than it really is.

Cartographers use a somewhat different approach to the omission of detail from maps. A map is said to have a *scale* or *representative fraction*, being the ratio of distances on the map to their corresponding distances in the real world. For example, if a map shows Los Angeles and Santa Barbara as 10cm apart, and in reality they are 100km apart as the crow flies, then the map's scale is 1:1,000,000 (note that in principle, because of projection from a curved to a flat surface, no map's scale can be exactly uniform although this clearly does not apply to globes).

This approach to defining level of detail is problematic, however, if the dataset in question is not on paper. Data that are *born digital* do not have a representative fraction, and the representative fraction for data displayed on a computer screen depends on the dimensions of the screen. Project onto a larger screen and the representative fraction will change, irrespective of the value shown on the screen. So it is better to use spatial resolution to describe the level of detail of digital data. As a general rule of thumb the spatial resolution of a paper map is roughly 0.5mm at the scale of the map (500m on a 1:1,000,000 map, or 5m on a 1:10,000 map), a figure that takes into account the widths of lines on maps and the cartographic practice of smoothing what is perceived as excessive detail.

2.2.10 Topology

In mathematics, a property is said to be *topological* if it survives stretching and distorting of space. Draw a circle on a rubber sheet, and no amount of stretching or distorting of the sheet will make it into a line or a point. Draw two touching circles, and no amount of stretching or distorting will pull them apart or make them overlap. It turns out that many properties of importance to spatial analysis are topological, including:

- The distinction between point, line, area, and volume, which are said to have topological dimensions of 0, 1, 2, and 3 respectively
- Adjacency, including the touching of land parcels, counties, and nation-states
- Connectivity, including junctions between streets, roads, railroads, and rivers

- Containment, when a point lies inside rather than outside an area

Many phenomena are subject to topological constraints: for example, two counties cannot overlap, two contours cannot cross, and the boundary of an area cannot cross itself. Topology is an important concept therefore in constructing and editing spatial databases.

Figure 2-2 illustrates the concept of topology, using the example of two areas that share a common boundary. While stretching of the space can change the shapes of the areas, they will remain in contact however much stretching is applied.

Figure 2-2 Topological relationships

Topological properties are those that cannot be destroyed by stretching or distorting the space. In this case the adjacency of the two areas is preserved in all the illustrations, however much distortion is applied

2.3 Spatial Relationships

Despite the emphasis placed on it in the Section 2.2, and in spatial analysis generally, location is not in itself very interesting. While latitude does generally tend to predict average annual temperature, there are obvious exceptions (Iceland is much warmer in winter than Irkutsk, yet it is further north), and longitude is essentially arbitrary. The power of location comes not from location itself, but from the linkages or relationships that it establishes — from relative positions rather than absolute ones. This section looks at examples of those relationships, as fundamental concepts in spatial analysis. Some have already been mentioned as examples of topological properties, including adjacency, connectivity, and containment. Others are introduced in the subsections that follow, with examples of their significance.

Another way to think about this important point is to examine how the results of spatial analysis are affected by imagined changes in location. The patterns of crime in Los Angeles would be as interesting whether the city was analysed in its correct location, or if it were arbitrarily relocated to another place on the Earth's surface — patterns are *invariant under displacement*. Similarly the city could be rotated, or analysed in its mirror image, without substantially affecting the insights that spatial analysis would reveal. Under all of these operations — displacement, rotation, and mirror imaging — the *relative* locations of objects remain the same, and it is these relative locations that provide insight.

2.3.1 Co-location

Many useful insights can come from comparing the attributes of places. For example, a cluster of cases of lung cancer in a port city might be explained by knowing that asbestos was widely used in the construction of ships in that city in the Second World War. In Section 2.2.5 a distinction was made between cases where multiple attributes are compared for the same set of objects, and cases where the attributes to be compared are attached to distinct, overlapping objects. In the latter case analysis will require the use of special techniques, such as those discussed in Section 2.3.8. Analysts often use the term *overlay* to refer to the superimposition and analysis of layers of geographic data about the same place. Overlay may require the superimposition of area on area, or line on area, or point on area, or line on line, or point on line, depending on the nature of the objects in the database.

2.3.2 Distance and direction

Knowledge of location also allows the analyst to determine the distance and direction between objects. Distance between points is easily calculated using formulas for straight-line distance on the plane or on the curved surface of the Earth, and with a little more effort it is possible to determine the actual distance that would be travelled through the road or street network, and even to predict the time that it would take to make the journey. Distance and direction between lines or areas are often calculated by using representative points, but this can be misleading (as noted in Section 2.2.3). They can also be calculated using the nearest pair of points in the two objects, or as the average distance and direction between all pairs of points (note the difficulties of averaging direction, however, as mentioned in Section 2.2.2).

Many types of spatial analysis require the calculation of a table or matrix expressing the relative proximity of pairs of places, often denoted by **W** (a *spatial weights matrix*). Proximity can be a powerful explanatory factor in accounting for variation in a host of phenomena, including flows of migrants, intensity of social interaction, or the speed of diffusion of an epidemic. Software packages will commonly provide several alternative ways of determining the elements of **W**, including:

- 1 if the places share a common boundary, else 0

- the length of any common boundary between the places, else 0
- a decreasing function of the distance between the places, or between their representative points

In such analyses it will be **W** that captures the spatial aspects of the problem, and the actual coordinates become irrelevant once the matrix is calculated. Note that **W** will be invariant with respect to displacement, rotation, and mirror imaging.

2.3.3   Multidimensional scaling

Multidimensional scaling (MDS) is the general term given to the problem of reconstructing locations from knowledge of proximities. Tobler and Wineberg (1971) provided an excellent example in which the unknown locations of 33 ancient pre-Hittite settlements in Cappadocia were inferred from knowledge of their interactions with other settlements whose locations were known, based on the assumption that interaction declined systematically with distance. Scaling techniques have been used to create a wide range of specialised maps, including maps on which distances reflect travel times rather than actual distances, or the similarity of the species found in each place, or the perceptions of relative proximities in the minds of local residents (Golledge and Rushton, 1972).

2.3.4   Spatial context

Much useful insight can often be gained by comparing the attributes of objects with those of other objects in close proximity. The behaviour of a person on a crowded street might be explained in terms of the proximity of other people; the price of a house might be due in part to the existence of expensive homes in the immediate vicinity; and an area might find its homes losing value because of proximity to a polluting industrial plant. Location establishes context, by allowing distances between objects to be determined, and by providing information on their relevant attributes.

2.3.5   Neighbourhood

People think of themselves as living in neighbourhoods, or places that are sufficiently close to be experienced on a day-to-day basis. Very often neighbourhood is the basis of spatial context, characterising the nature of a person's surroundings. Neighbourhoods are often conceived as partitioning an urban space, such that every point lies in exactly one neighbourhood, but this may conflict with individual perceptions of neighbourhood, and by the expectation that neighbourhood extends in all directions around every individual's location.

Figure 2-3 shows three examples of possible neighbourhood definitions. In Figure 2-3A the neighbourhood is defined as a circle centred on the house, extending equally in all directions. In Figure 2-3B neighbourhood is equated with an existing zone, such as a census tract or precinct, reflecting the common strategy of using existing aggregated data to characterise a household's surroundings. In Figure 2-3C weights are applied to surroundings based on distance, allowing neighbourhood to be defined as a convolution with weight decreasing as a simple function of distance.

Figure 2-3 Three alternative ways of defining neighbourhood, using simple GIS functions

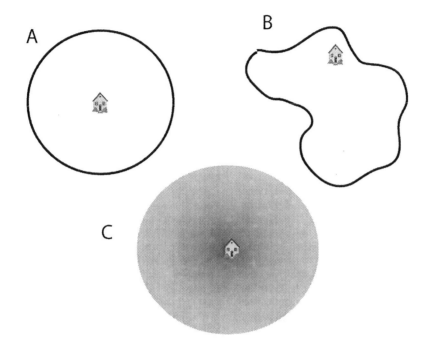

2.3.6 Spatial heterogeneity

The Earth's surface displays almost incredible variety, from the landscapes of the Tibetan plateau to the deserts of Australia and the urban complexity of London or Tokyo. Nowhere can be reasonably described as an *average place*, and it is difficult to imagine any subset of the Earth's surface being a representative sample of the whole. The results of any analysis over a limited area can be expected to change as that limited area is relocated, and to be different from the results that would be obtained for the surface of the Earth as a whole. These concepts are collectively described as *spatial heterogeneity*, and they tend to affect almost any kind of spatial analysis conducted on geographic data. Many techniques such as Geographically Weighted Regression (Fotheringham, Brunsdon, and Charlton, 2002, discussed in Section 5.6.3 of this Guide) take

spatial heterogeneity as given — as a universally observed property of the Earth's surface — and focus on providing results that are specific to each area, and can be used as evidence in support of local policies. Such techniques are often termed *place-based* or *local*.

2.3.7 Spatial dependence

Notwithstanding the comments in Section 2.3.6 about spatial heterogeneity, anyone examining the Earth's surface in detail would be struck by how conditions tend to persist locally, and how it is possible to divide the surface into regions that exhibit substantial internal similarity. For example, the desert regions are characterised by lack of rainfall, the tropical rainforests by abundant rainfall and dense vegetation, and the Polar Regions by extreme cold. Conditions at nearby points are not sampled independently and randomly from global distributions, but instead show remarkable levels of interdependence. Of course there are exceptions, where very rapid change in conditions over short distances, for example between the plains of India and the high Himalayas, or between coastal plains and adjacent ocean.

The general term for this is *spatial dependence*. Its pervasiveness was aptly captured by Tobler in what has become known as his First Law of Geography: "All things are related, but nearby things are more related than distant things". The magnitude of the effect can be measured using a number of statistics of *spatial autocorrelation* (Cliff and Ord, 1981). It also underlies the discipline known as *geostatistics*, which describes spatial variation in terms of a function (known as a *correlogram*) that shows how spatial autocorrelation decreases with increasing distance (for a general overview of geostatistics see Section 6.7, and the book by Isaaks and Srivastava, 1989). This correlogram reaches zero, or independence, at a distance known as the *range*.

The implications of Tobler's observation are profound. If it were not true, the full range of conditions anywhere on the Earth's surface could in principle be found packed within any small area. There would be no regions of approximately homogeneous conditions to be described by giving attributes to area objects. Topographic surfaces would vary chaotically, with slopes that were everywhere infinite, and the contours of such surfaces would be infinitely dense and contorted. Spatial analysis, and indeed life itself, would be impossible.

2.3.8   Spatial sampling

One of the implications of Tobler's First Law and spatial dependence is that it is possible to capture a reasonably accurate description of the Earth's surface with a few well-placed samples. Meteorologists capture weather patterns by sampling at measuring stations, knowing that conditions between these sample locations will vary systematically and smoothly. Geographers describe the contents of entire regions with a few well-chosen statements, provided that the regions have been defined to enclose approximately uniform conditions. Census data are published in spatially aggregated form to protect the confidentiality of individual returns, and Tobler's First Law ensures that the variance within each aggregate unit is not so large as to render the aggregated data meaningless.

Numerous approaches to spatial sampling exist, depending on the nature of the questions being asked. Perhaps the simplest occurs when sample observations are made in order to obtain a best estimate of average conditions within an area. For example, to show that average temperatures across the globe are increasing due to global warming, one would want to ensure that every location on the globe had an equal chance of being sampled (in reality weather stations are distributed far from randomly, often co-located with airports, and more common on land and in heavily populated areas). In other cases it might be more important to characterise various regions, by placing an equal number of sample points in each region, or to characterise each of a number of vegetation or soil types. If the objective is to characterise the pattern of spatial dependence in an area, then it may be more important to place samples so as to obtain equally reliable estimates of similarity in every range of distances. A discussion of spatial sampling can be found in the book by Longley *et al.* (2005) and in Section 5.1.2 of this Guide.

2.3.9   Spatial interpolation

If spatial sampling is an efficient way of capturing knowledge of spatial variation, then there must be reliable ways of filling in the unknown variation between sample points. Spatial interpolation attempts to do this, by providing a method of estimating the value of a field anywhere from a limited number of sample points. Spatial interpolation is no more and no less than intelligent guesswork, supported by Tobler's First Law.

Many methods of spatial interpolation exist, all of them based to some extent on the principle that conditions vary smoothly over the Earth's surface. Methods have even been devised to cope with exceptions to Tobler's First Law, by recognising the discontinuities that exist in some geographic phenomena, such as faults in geologic structures, or the barriers to human interaction created by international boundaries. Arguably the most rigorous is Kriging, a family of techniques based on the theory of geostatistics (Isaaks and Srivastava, 1989). Kriging and a wide range of other interpolation methods are described in Chapter 6 of this Guide.

2.3.10 Smoothing and sharpening

The Earth's surface is very dynamic, particularly in those aspects that fall in the social domain. People move, houses are constructed, areas are devastated by war, and severe storms modify coastlines. Patterns of phenomena on the Earth's surface reflect these processes, and each process tends to leave a distinctive footprint. Interpreting those footprints in order to gain insight into process is one of the most important motivations for spatial analysis.

The pervasiveness of Tobler's First Law suggests that processes that leave a smooth pattern on the Earth's surface are generally more prevalent than those that cause sharp discontinuities, but spatial analysts have developed many ideas in both categories. Several are reviewed in this subsection, which introduces some important concepts related to process.

Many processes involve some form of *convolution*, when the outcome at a point is determined by the conditions in some immediate neighbourhood. For example, many

social outcomes are responses to neighbourhood conditions – individual obesity may result in part from a physical design of the neighbourhood that discourages walking, alcoholism may be due in part to the availability of alcohol in the neighbourhood, and asthma incidence rates may be raised due to pollution caused by high local density of traffic. Mathematically, a convolution is a weighted average of a point's neighbourhood, the weights decreasing with distance from the point, and bears a strong technical relationship to density estimation. As long as the weights are positive, the resulting pattern will be smoother than the inputs. The blurring of an out-of-focus image is a form of convolution.

In the analysis of point patterns, a *first-order process* is one that produces a variation in point density in response to some causal variable. For example, the density of cases of malaria echoes the density of particular species of mosquito, which in turn reflects the abundance of warm stagnant water for breeding. Because both mosquitoes and people move around, the pattern of malaria density can be expected to be smoother than the patterns of either mosquito density or stagnant-water abundance.

A *second-order process* results from interactions, when the presence of one point makes others more likely in the immediate vicinity. Patterns of contagious disease reflect second-order processes, when the disease is passed from an initial carrier to family members, co-workers, sexual partners, and others who come into contact with infectious carriers. The result will be a pattern that clusters around carriers (Figure 2-4B and D). Both first- and second-order processes produce clustering and smooth variations in density, but the mechanisms involved and the relationships to other variables are very different.

Figure 2-4 Four distinct patterns of twelve points in a study area
(A) random, (B) clustered, (C) dispersed, (D) clustered

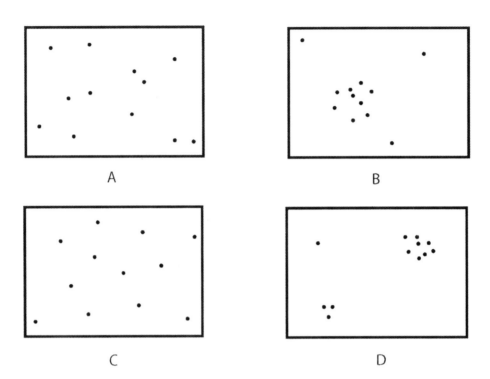

Competition for space provides a familiar example of a form of second-order process that results in an exception to Tobler's First Law. The presence of a shopping centre in an area generally discourages other shopping centres from locating nearby, particularly when the focus is on convenience goods and services. It is unusual, for example, to find two supermarkets next to each other, or two outlets of the same fast-food chain. These are examples of *dispersed* rather than clustered point patterns (Figure 2-4C), in which the presence of one point makes others less rather than more likely in the immediate vicinity.

2.4 Spatial Statistics

2.4.1 Spatial probability

Humans will never have a complete understanding of everything that happens on the Earth's surface, and so it is often convenient to resort to thinking in terms of probabilities. In principle one could completely characterise the physics of a human hand and a coin, but in practice it is much more productive to assign probabilities to the outcomes of a coin toss. In similar fashion spatial analysts may avoid the virtual impossibility of predicting exactly where landslides will occur by assigning them probabilities based on patterns of known causes, such as clay soils, rainfall, and earthquakes. A map of probabilities assigns each location a value between 0 and 1, forming a probability field.

Such a map considers only the probability of a single, isolated event, however. The probability that two points a short distance apart will both be subject to landslide is not simply the product of the two probabilities, as it would be if the two outcomes were independent, a conclusion that can be seen as another manifestation of Tobler's First Law . For example, if the probability of a landslide at Point A is ½ and at Point B a short distance away is also ½, the probability that both will be affected is more than ¼, and possibly even as much as ½. Technically, the *marginal* probabilities of isolated events may not be as useful as the *joint* probabilities of related events — and joint probabilities are properties of pairs of points and thus impossible to display in map form unless the number of points is very small.

2.4.2 Probability density

One of the most useful applications of probability to the Earth's surface concerns uncertainty about location. Suppose the location of a point has been measured using GPS, but inevitably the measurements are subject to uncertainty, in this case amounting to an average error of 5m in both the east-west and north-south directions. Standard methods exist for analysing measurement error, based on the assumption, well justified by theory, that errors in a measurement form a bell curve or *Gaussian* distribution. Spatially, one can think of the east-west and north-south bell curves as combining to form a bell. But the surface formed by the bell is not a surface of probability in the sense of Section 2.4.1 — it does not vary between 0 and 1, and it does not give the marginal probability of the presence of the point. Instead, the bell is a surface of *probability density*, and the probability that the point lies within any defined area is equal to the volume of the bell's surface over that area. The volume of the entire bell is exactly 1, reflecting the fact that the point is certain to lie somewhere.

It is easy to confuse probability density with spatial probability, since both are fields. But they have very different purposes and contexts. Probability density is most often encountered in analyses of positional uncertainty, including uncertainty over the locations of points and lines.

2.4.3 Uncertainty

Any geographic dataset is only a representation of reality, and it inevitably leaves its user with uncertainty about the nature of the real world that is being represented. This uncertainty may concern positions, as discussed in the Section 2.4.2, but it may also concern attributes, and even topological relationships. Uncertainty in data will propagate into uncertainty about conclusions derived from data. For example, uncertainty in positions will cause uncertainty in distances computed from those positions, in the elements of a **W** matrix, and in the results of analyses based on that matrix.

Uncertainty can be due to the inaccuracy of measuring instruments, since an instrument that measures a property to limited accuracy leaves its user uncertain about the true value of the property. It can be due to vagueness in definitions, when land is assigned to classes that are not rigorously defined, so that different observers may classify the same land differently. Uncertainty can also be due to missing or inadequate documentation, when the user is left to guess as to the meaning or definitions of certain aspects of the data. Clearly it is important to spatial analysts to know about the uncertainties present in data, and to investigate how those uncertainties impact the results of analysis. A range of techniques have been developed, and there is a rich literature on uncertainty in spatial data and its impacts (Zhang and Goodchild, 2002).

2.4.4 Statistical inference

One of the most important tools of science is statistical inference, the practice of reasoning from the analysis of samples to conclusions about the larger populations from which the samples were drawn. Entire courses are devoted to the topic, and to its detailed techniques — the t, F, and Chi-Squared tests, linear modelling, and many more. Today it is generally accepted that any result obtained from an experiment, through the analysis of a sample of measurements or responses to a survey, will be subjected to a *significance test* to determine what conclusions can reasonably be drawn about the larger world that is represented by the measurements or responses.

The earliest developments in statistical inference were made in the context of controlled experiments. For example, much was learned about agriculture by sowing test plots with new kinds of seeds, or submitting test plots to specific treatments. In such experiments it is reasonable to assume that every sample was created through a random process, and that the samples collectively represent what might have happened had the sample been much larger — in other words, what might be expected to happen under similar conditions in the universe of all similar experiments, whether conducted by the experimenter or by a farmer. Because there is variation in the experiment, it is important to know whether the variation observed in the sample is sufficiently large to reach

conclusions about variation in the universe as a whole. Figure 2-5 illustrates this process of statistical inference. A sample is drawn from the population by a random process. Data are then collected about the sample, and analysed. Finally, inferences are made about the characteristics of the population, within the bounds of uncertainty inherent in the sampling process.

Figure 2-5 The process of statistical inference

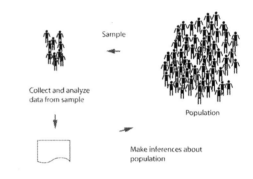

These techniques have become part of the standard apparatus of science, and it is unusual for scientists to question the assumptions that underlie them. But the techniques of spatial analysis are applied in very different circumstances from the controlled experiments of the agricultural scientist or psychologist. Rather than creating a sample by laying out experimental plots or recruiting participants in a survey, the spatial analyst typically has to rely on so-called *natural* experiments, in which the variation among samples is the result of circumstances beyond the analyst's control.

In this context the two fundamental principles of statistical inference raise important questions: (i) were the members of the sample selected randomly and independently from a larger population, or did Tobler's First Law virtually ensure lack of independence, and/or did the basic heterogeneity of the Earth's surface virtually ensure that samples drawn in another location would be different? (ii) what universe is represented by the samples? and (iii) is it possible to reason from the results of the analysis to conclusions about the universe?

All too often the answers to these questions are negative. Spatial analysis is often conducted on all of the available data, so there is no concept of a universe from which the data were drawn, and about which inferences are to be made. It is rarely possible to argue that sample observations were drawn independently, unless they are spaced far apart. Specialised methods have been devised that circumvent these problems to some degree, and they will be discussed at various points in the book. More often, however, the analyst must be content with results that apply only to the sample under analysis, and cannot be generalised to some larger universe.

2.5 Spatial Data Infrastructure

This final section introduces some of the basic concepts associated with what has become known as *spatial data infrastructure*, the set of techniques, institutions, standards, and other arrangements that facilitate the exchange of spatial data and tools, and in this way support spatial analysis. Until the 1980s it was generally accepted that responsibility for the acquisition and dissemination of spatial data lay with national mapping agencies, who produced topographic maps and other information at public expense. This system began to fall apart around 1990, however, for a number of reasons. Governments were less and less willing to foot the steadily increasing bill; developments in computing technology and GPS had made it possible for virtually anyone to become a creator of spatial data at low cost; and the commercial sector was pushing for new markets for its products, arguing that subsidised government production of spatial data constituted unfair competition. Today, many countries now have extensive spatial data infrastructures, and responsibility for the acquisition and dissemination of spatial data is now shared between numerous agencies, companies, and individuals.

2.5.1 Geoportals

A *geoportal* is a Web site providing a single point of access to spatial data, allowing its users to browse through extensive collections of data and related materials, evaluating their potential fitness for use. Geoportals may include data that are in the public domain or available free of charge, and may also support licensing and purchasing of proprietary data. They include the ability to search through catalogues, examining the contents and quality of data. Geoportals include standard layers of data on topography, roads and streets, rivers and streams, named places, imagery, and political boundaries; and may also provide access to more specialised data. A number of geoportals are listed in the Web links Annex to this Guide, "Spatial data, test data and spatial information sources" section.

2.5.2 Metadata

Fundamental to the operation of a geoportal are *metadata*, the catalogue entries that allow users to search for datasets and to assess their contents. Metadata may include information on contents, quality, the data production process, details of formats and coordinate systems, and many other topics. For more details see the UK Geospatial Information Gateway: http://www.gigateway.org.uk/ or the US Nationa Spatial Data Infrastructure site: http://www.fgdc.gov/nsdi/nsdi.html

2.5.3 Interoperability

Over the past four decades many hundreds of formats have been devised and implemented for spatial data. Some are open, while others are proprietary. The development of an effective spatial data infrastructure has required extensive efforts to achieve *interoperability*, in other words the ability of systems to exchange and use data without major effort. The standards that make this possible have been devised and promulgated by several organisations, including national mapping agencies, international standards bodies, and the Open Geospatial Consortium (OGC): www.opengeospatial.org

2.5.4 Conclusion

This chapter has provided a brief overview of some of the more important and fundamental concepts that underpin spatial analysis. Many more will be encountered at various points in this book, and many more can be found in the background readings cited in Section 1.2.3. These concepts and others together frame the theory of (geo)spatial analysis, just as the field of statistics is framed by the concepts of statistical analysis. While much of spatial analysis will appear intuitive, it is important to recognise that it is underpinned by concepts and terms whose meanings are often far from intuitive, and require explicit recognition.

3 Historical and Methodological Context

3.1 Historical context

The term 'GIS' is widely attributed to Roger Tomlinson and colleagues, who used it in 1963 to describe their activities in building a digital natural resource inventory system for Canada — Tomlinson (1967, 1970). The history of the field has been charted in an edited volume by Foresman (1998) containing contributions by many of its early protagonists. A timeline of many of the formative influences upon the field is available via the CASA website: www.casa.ucl.ac.uk/gistimeline/; a printed summary is provided by Longley *et al.* (2005, pp19-21), and useful background information may be found at the GIS History Project web site based at the NCGIA (Buffalo): http://www.ncgia.buffalo.edu/gishist/. Each of these sources makes the unassailable point that the success of GIS as an area of activity has fundamentally been driven by the success of its applications in solving real world problems.

Figure 3-1 is a simplified illustration of the way in which today's GIS is configured, in terms of physical and human resources. Innovations in *hardware* diffuse rapidly, and the physical configuration of a system at any particular point in time will be a function of recent and past investment decisions. The *software tools* that characterise a GIS are similarly likely to be a function of budget considerations and the availability of suitable toolsets. *Networking*, both local- and wide-area, is essential to the functioning of a GIS. The near universal availability of the Internet in developed countries, the use of websites to showcase products and services, and the popularity of downloads as a means of acquisition make the wide-area network key to the acquisition of *data* and *software tools* often identified using search engines. However, ubiquity of Internet

connections and search engines is not alone sufficient to successful GIS application. In the case of data, it is necessary to establish the provenance of the different data options that are available, by reference to the metadata that should accompany each of them.

Figure 3-1 The components of a GIS

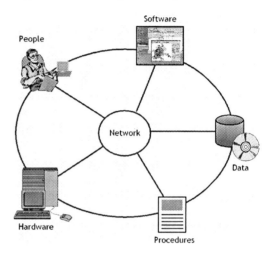

In the case of software tools, the sometimes vast choices that are presented as the result of Internet searches rarely establish fitness for particular purposes, or identify the most cost-effective option for a particular application. This highlights an important difference between data and software: while the veracity of metadata can often be checked (e.g. with respect to other datasets, or sometimes through direct field measurement in the real world), there are no such safeguards available for most software tools. Commercial software is very rarely 'open source', for a range of reasons including protecting copyright and maintaining user support. Even 'public domain' software is often not open source. Thus, although the Internet makes it possible to assemble information about software, Internet searches do not necessarily enable objective comparison of software functions. One of the important contributions of the present Guide is to raise awareness of the range of software options that are available and the quality of the results that may be produced. In this way we hope to assist the *people* that use GIS, and to help users to develop improved *procedures* for undertaking spatial analysis.

Our primary concern is with developing adequate understanding of the ways in which computer *software* can be used to solve geographical problems. Some of this software has its roots in particular applications sectors and the technological setting in which it was developed, but these distinctions are blurring with time. The field of spatial analysis has a rich and multidisciplinary history, extending back well before the advent of cheap, high performance computing. Today, however, it is the suites of software commonly described as 'Geographic Information Systems' (GIS) that provide the computer environment for much spatial analysis research and practice.

Updates and developments in hardware and networking are widely discussed in much of the GIS trade press (see, for example, the "Trade sites" links at the end of this Guide). The *people* requirements of GIS and the way that the GIS industry meets them are considered in Longley *et al.* (2005, pp24-31). The importance of people and *procedures* are usually specific to the context in which an application is developed and define the organisational setting to GIS. This has received quite wide attention in the literature – e.g. Bernhardsen (2002); Masser *et al.* (1996). The issues of GIS data hold an intermediate position, in that classes of applications share some very common data problems, yet these can be alleviated or accentuated by the particular context in which GIS analysis is performed. At the end of the day, any GIS application can only be as good as the data that are used to create it and its relevance to the organisation that commissioned it. Thus the relationship between data issues and the organisational setting in which people and procedures operate will be core to much of our discussion of a range of techniques and representative applications areas.

There are some caveats to this heavy emphasis upon the importance of context in GIS analysis. The wide availability of low cost Global Positioning System (GPS) receivers means that primary data relating to physical features can be collected wherever uninterrupted signals can be received to standards that are generally understood. Yet many GIS applications build upon the legacy of spatial data that have previously been collected, and there is considerable variability in the amount, relevance and quality of secondary data that can be assembled for different parts of the world.

Chapter 2 described the established scientific elements that govern 'typical' GIS applications. In practice, applications are likely to be governed by the organisational practices and *procedures* that prevail with respect to particular *places* (Section 2.2.1). Not only are there wide differences in the volume and remit of data that the public sector collects about population characteristics in different parts of the world, but there are differences in the ways in which data are collected, assembled and disseminated (e.g. general purpose censuses versus statistical modelling of social surveys, property registers and tax payments). There are also differences in the ways in which different data holdings can legally be merged and the purposes for which data may be used – particularly with regard to health and law enforcement data. Finally, there are geographical differences in the cost of geographically referenced data. Some organisations, such as the US Geological Survey, are bound by statute to limit charges for data to sundry costs such as media used for delivering data while others, such as most national mapping organisations in Europe, are required to exact much heavier charges in order to recoup much or all of the cost of data creation. Analysts may already be aware of these contextual considerations through local knowledge, and other considerations may become apparent through browsing metadata catalogues. GIS applications must by definition be sensitive to context, since they represent unique locations on the Earth's surface.

A goal of this Guide is to focus in greater detail upon the characteristics of data and how they support particular applications in particular organisational settings. We illustrate this here by using a wide range of examples from datasets that come from all parts of the world. However, in addition we will be providing a separate series of Case Studies accessible via the associated web site, www.spatial-literacy.org initially using London as our 'applications laboratory'. We are not seeking to present a 'London centric' view of the world — rather, through selected case studies, to illustrate how different applications fields have developed.

3.2 Methodological context

3.2.1 Spatial analysis as a process

In many instances the process of spatial analysis follows a number of well-defined (often iterative) stages: problem formulation; planning; data gathering; exploratory analysis; hypothesis formulation; modelling and testing; consultation and review; and ultimately final reporting and/or implementation of the findings. GIS and related software tools that perform analytical functions only address the middle sections of this process, and even then may be relatively "blunt" instruments.

Having first identified and formulated the problem to be addressed (often a significant task in itself), and developed an outline plan, the first task typically involves obtaining the data which are to be the subject of analysis. This immediately raises many questions that have an important bearing on subsequent stages: what assumptions have been invoked in order to represent the "real-world" and what are the implications of this for subsequent analysis? How complete are the data — spatially and temporarily? How accurate are the data (spatially, temporarily and in terms of measured attributes)? Are all of the datasets compatible and consistent with one another — what sources are they drawn from and how do they compare in terms of scale, projection, accuracy, modelling, orientation, date of capture and attribute definition? Are the data adequate to address the problem at hand?

The second stage, once the data have been obtained and accepted as fit for purpose (and/or as the best available), is often exploratory. This may involve: simple mapping of the data, points, lines, regions, grids, surfaces, computation of rates, indices, densities, slopes, directional trends, level sets, classifications etc.; or more complex and dynamic exploration of the data, such as brushing and linking. One or more analytical techniques and tools may be utilised at this and subsequent stages.

The third stage will depend upon the objective of the analysis. In many instances presentation of the results of exploratory analysis in the form of commentary, maps, descriptive statistics and associated documents will complete the process. In others it will involve the development and testing of hypotheses about the patterns observed, and/or modelling of the data in order to undertake some predictive or optimisation exercise. Frequently the result of this process is a series of possible outcomes (scenarios) which then need to be summarised and presented for final analysis and decision-making by stakeholders, interest groups, policy makers or entrepreneurs. The process may then iterate until an agreed or stable and robust flow is achieved, from problem specification to data selection, and thence to analysis and outcomes.

This kind of process can be formalised and may be implemented as a standard procedure in operational systems or as part of a planning process. Such procedures may involve relatively lengthy decision cycles (e.g. identifying the next location for a new warehouse as customer demand grows) or highly dynamic environments, for example controlling road traffic lights and routing to reflect the type and density of traffic in real time.

To a substantial degree spatial analysis can be seen as part of a decision support infrastructure, whether such decision-making reflects purely academic interest or commercial, governmental or community interests. Increasingly, to reflect this trend, GIS vendors have developed tools to facilitate such processes. This has been principally through the provision of Internet-enabled input, output and processing facilities, but also through improved usability, speed, interactivity, decision-support and visualisation techniques. With these improvements the range and experience of users has grown and will continue to develop. For this reason it is essential that the underlying principles are well-understood and not misused, accidentally or deliberately, leading to ill-informed or invalid decision-making. All providers and users of such systems have a responsibility to ensure that information generated is communicated in an unambiguous and unbiased manner, and to ensure that limitations and assumptions made are transparent and well-understood. These concerns are of particular importance and relevance to those in the print and broadcast media, especially television, where visualisation is so important and timeslots for presentation so brief.

3.2.2 Analytical methodologies

As with all scientific disciplines, spatial analysis has its fair share of terminology, abbreviations, methods and tools. In this subsection we start by examining spatial analysis in the broader context of analytical methodologies, and attempt to highlight the similarities with such methods and to identify some of the distinctive aspects of spatial and GIS analysis.

Earlier we suggested that the process of spatial analysis often proceeds in a simple sequence from problem specification to outcome. In reality this is an over-simplification — the analytical process is far more complex and iterative than this summary of steps suggests. So, for example, Mitchell (2005) suggests the sequence for spatial *data* analysis shown in Figure 3-2.

Again, it appears that there is a natural flow from start to finish, but this is rarely the case. Not only is the process iterative, but at each stage one often looks back to the previous step and re-evaluates the validity of the decisions made. Explicit recognition of this need is incorporated in the approach of Haining (2003, p.359) to "Data-driven statistical modelling", which is otherwise very similar to Mitchell's model (Figure 3-2).

The need for a more iterative approach is partly a reflection of adopting scientific methods, and a recognition that most analytical tasks take place within a much broader context.

Figure 3-2 Analytical process — after Mitchell (2005)

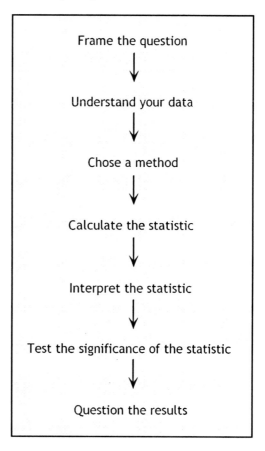

For example, pragmatic decisions often have to be made based on a series of common questions:

- how much time, money and resource can I afford to apply to this problem?

- what research has previously been carried out on problems of this type, and what strengths and weaknesses have these shown?

- if I commence by examining the requirements of the outcome closely, what implications does this have on my selection of techniques and data, and what caveats should I apply to the scope and validity of my analyses — e.g. in relation to sample/region size, timespan, attribute set, validation etc.?

- who will be the recipient of the results, and what are their expectations and requirements?

- how will I deal with data inadequacies — e.g. missing data, unsuitable data, delays in receiving or obtaining access to key datasets?

- how will I deal with limitations and errors in the software I have chosen to use?

- what are the implications of producing wrong or misleading results?

- are there independent and verifiable means for validating results obtained?

This list is by no means exhaustive but serves to illustrate the kind of questioning that should form part of any analytical exercise, in GIS and many other fields.

In supplier-client relationships it is often useful to start with the customer expectations and requirements, including the form of presentation of results (visualisations, data analyses, reports, public participation exercises etc.), and work *back* through the process stages identifying what needs to be included and might be excluded in order to have the best chance of meeting those expectations and requirements within the timescale and resourcing available.

A good example of the modern analytical process is a recent study of the incidence of childhood cancer in relation to distance from high voltage power lines by Draper *et al.* (2005). The client in this instance was not a single person or organisation, but health policy makers and practitioners in the UK and elsewhere, together with academic and medical specialists in the field. With such scrutiny, political sensitivity and implications

for the health of many children, great care in conducting the analysis and reporting the findings was essential. In this instance some of their initial findings were reported by the press and radio prior to the intended publication date, so the authors took the decision to publish ahead of schedule despite the issues this might raise. The researchers summarised their approach in the form shown in Figure 3-3:

Figure 3-3 Analytical process — after Draper et al. (2005)

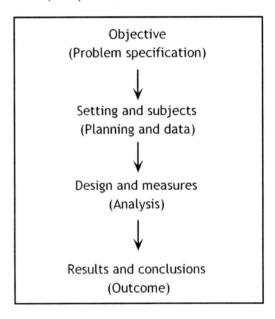

Each step of this process is described clearly, qualifying the information and referring to earlier work where appropriate. In addition the paper clarifies the roles of the various participants in the research, addresses issues relating to competing interests (one of the researchers involved is an employee of the main power grid company), and deals with issues of funding and ethical approval. All of these elements provide transparency and enable "clients" to evaluate the quality and importance of the research. The form of publication (via the British Medical Journal) also facilitated extensive online discussion, with comments on the paper by acknowledged experts from around the world being published electronically, as well as the author's responses to these comments.

An interesting aspect of this research, and a significant number of similar studies, is that it deals with rare events. In general it is far more difficult to be confident about drawing definitive conclusions about such events, but this does not detract from the importance of attempting to carry out research of this kind. What it does highlight is the need to identify and report on the scope of the study, and attempt to identify and highlight any obvious weaknesses in the data or methodology. In the cancer study just cited, for example, the datasets used were details of the location of power grid lines (by type, including any changes that had occurred during the 30+ year study period) together with around 30,000 records stored in the national cancer registry. No active participation from patients or their families was involved, and homes were not visited to measure actual levels of Electro-Magnetic (EM) radiation. These observations raise important but *unstated* questions: how accurate are the grid line datasets and the cancer registry records? is home address *at birth* an appropriate measure (surrogate for exposure to EM radiation)? is vertical as well as horizontal proximity to high voltage lines of importance? were the controls similarly located relative to the power lines as the cases — if not, then this could account for some aspects of the results reported? given that the research findings identified a pattern of increased risk for certain cancers that was *not* a monotonic function of distance, can this result be explained, are the observations purely chance occurrences, or are there some other factors at work?

The latter issue is often described as *confounding*, i.e. identifying a result which is thought to be explained by a factor that has been identified, but is in reality related to one or more other factors that did not form part of the analysis. For example, in this instance one might also consider that: (i) socio-economic status and lifestyle factors may be associated with proximity to overhead power lines;

(ii) population densities may vary with distance from power lines; (iii) pre-natal exposure may be important; and (iv) the location of nurseries and playgroups may be relevant. Any of these factors may be important to incorporate in the research. In this instance the authors did analyse some of these points but did not include them in their reporting, whilst other points remain for them to consider further and comment on in a future publication. The "distance from power lines" study is an example of modern scientific research in action — indeed, it has provided an important input to the debate on health and safety, leading to national recommendations affecting future residential building programmes.

Mackay and Oldford (2002) have produced an excellent discussion of the context and role of (statistical) analysis within the broader framework of scientific research methods: "Scientific method, statistical method and the speed of light". Their paper studies the work carried out in 1879 by A A Michelson, a 24-year old Naval ensign, and his (largely successful) efforts to obtain an accurate estimate of the speed of light. The paper is both an exercise in understanding the role of statistical analysis within scientific research, and a fascinating perspective on a cornerstone of spatial analysis, that of distance determination. They conclude that statistics is defined in large measure by its methodology, in particular its focus on seeking an understanding of a population from sample data. In a similar way spatial analysis and GIS analysis are defined by the methods they adopt, focusing on the investigation of spatial patterns and relationships as a guide to a broader understanding of spatial patterns and processes. As such spatial analysis is defined by both its material (spatial datasets) and its methods. Within this set of methods are the full range of univariate and multivariate statistical methods, coupled with a wide range of specifically spatial tools, and a broad mix of modelling and optimisation techniques.

Mackay and Oldford describe the *statistical method* in terms of a sequence of steps they label *PPDAC*: Problem; Plan; Data; Analysis; and Conclusions. This methodology is very similar to those we have already described, although a little broader in scope. They then expand upon each step and in the paper relate each step to the process adopted by Michelson. This approach emphasises the place of formal analysis as very much a part of a process, rather than a distinct discipline that can be considered in isolation. Daskin (1995, Chapter 9) makes a very similar case within the context of planning facility locations, and sees this as an iterative process culminating in implementation and ongoing monitoring. Spatial analysis as a whole may be considered to follow very similar lines — however the steps need clarification and some amendment to Mackay and Oldford's formulation. A

summary of a revised PPDAC approach is shown in Figure 3-4. As can be seen from the diagram, although the clockwise sequence (1→5) applies as the principal flow, each stage may and often will feed back to the previous stage. In addition, it may well be beneficial to examine the process in the reverse direction, starting with Problem definition and then examining expectations as to the format and structure of the Conclusions (without pre-judging the outcomes!). This procedure then continues, step-by-step, in an anti-clockwise manner (e→a) determining the implications of these expectations for each stage of the process. A more detailed commentary on the various stages, particularly as they relate to problems in spatial analysis, is provided in Section 3.2.3.

Figure 3-4 PPDAC as an iterative process

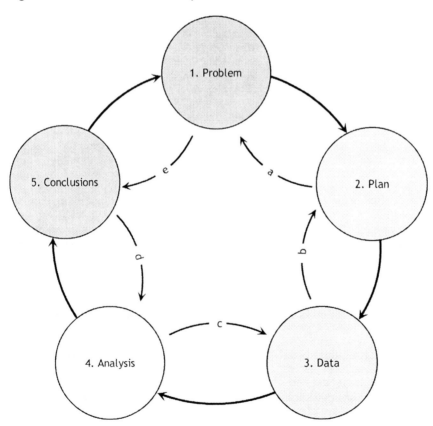

3.2.3 Spatial analysis and the PPDAC model

Mackay and Oldford's perspective is perhaps a little narrow and somewhat less exploratory and iterative than applies for many real-world problems. In respect of problems that the GIS user is likely to meet there are a number of reasons for these differences, most notably:

- Spatial analysis is particularly concerned with problems that have an explicit spatial context, and frequently data at one location is not independent of data at other locations (see Sections 2.3 and 2.4). Indeed such associations (spatial correlations) are the norm, especially for measurements taken at locations that are near to one another. Identifying and analysing such patterns is often the goal of analysis, at least at the early stages of investigation

- Many problems must be considered in a spatio-temporal context rather than simply a spatial context. Time of day/week/month/year may have great relevance to obtaining an understanding of particular spatial problems, especially those of an environmental nature and problems relating to infrastructure usage and planning

- The theoretical foundations of statistics rely on a set of assumptions and sampling procedures that are often more applicable to experimental datasets than purely observational data. Very few problems addressed by spatial analysis fall into the category of truly experimental research

- Often the purpose of spatial analysis is not merely to identify pattern, but to construct models, if possible by gaining an understanding of process. But spatial patterns are rarely if ever uniquely determined by a single process, hence spatial analysis is often the start of further investigations into process and model

building, and rarely an end in itself. Where explicitly spatial factors, such as distance or contiguity, are identified as important or significant, the question why? must follow: is the problem under consideration directly or indirectly affected by purely spatial factors; or is the spatial element a surrogate for one or more explanatory variables that have not be adequately modelled or are unobtainable?

- Spatial datasets are often provided by third parties, such as national mapping agencies, census units and third party data vendors. Metadata provided with such material may or may not provide adequate information on the quality, accuracy, consistency, completeness and provenance of the information. In many areas of spatial research these elements are pre-determined, although they are often augmented by corporate datasets (e.g. customer databases, crime incident records, medical case details) or field research (e.g. georeferenced collection of soil samples or plant locations, market research exercises, bathymetric surveys etc.)

Each of these factors serves to distinguish spatial analysis from analysis in other disciplines, whilst at the same time recognising the considerable similarities and overlap of methodologies and techniques. In subsections 3.2.3.1 to 3.2.3.5 we examine each step of the PPDAC model in the context of spatial analysis. In this "revised" version of the PPDAC approach the Plan stage is much broader than in Mackay and Oldford's model. In their approach the Plan stage focuses largely on the data collection procedures to be adopted. In our case it covers all aspects of preparation for data acquisition and analysis, including considerations of feasibility within given project constraints.

For many problems that arise in spatial analysis there will be multiple instances of the process described, especially for the data and analysis stages, where several different but related datasets are to be studied in order to address the problem at hand. Finally, in instances where the geospatial analyst is simply presented with the data and asked to carry out appropriate analyses, it is essential that the context is first understood (i.e. Problem, Plan and Data) even if this cannot be influenced or revisited.

3.2.3.1 Problem: Framing the question

Understanding and defining the problem to be studied is often a substantial part of the overall analytical process — clarity at the start is obviously a key factor in determining whether a programme of analysis is a success or a failure. Success here is defined in terms of outcomes rather than methods. And outcomes are typically judged and evaluated by third parties — customers, supervisors, employers — so their active involvement in problem specification and sometimes throughout the entire process is essential. Breaking problems down into key components, and simplifying problems to focus on their essential and most important and relevant components, are often very effective first steps. This not only helps identify many of the issues to be addressed, likely data requirements, tools and procedures, but also can be used within the iterative process of clarifying the customer's requirements and expectations. Problems that involve a large number of key components tend to be more complex and take more time than problems which involve a more limited set. This is fairly obvious — but perhaps less obvious is the need to examine the interactions and dependencies between these key components. The greater the number of such interactions and dependencies the more complex the problem will be to address, and as the numbers increase complexity tends to grow exponentially.

Analysis of existing information, traditionally described as "desk research", is an essential part of this process and far more straightforward now with the advantage of online/Internet-based resources. Obtaining relevant information from the client/sponsor

(if any), interested third parties, information gatekeepers and any regulatory authorities, forms a further and fundamental aspect to problem formulation and specification.

Spatial analysts also need to be particularly aware of a number of well-known problems associated with grouped data in general, and spatial data in particular. This applies to both problems in which the analyst contributes to the entire PPDAC process as well as the common situation in which the analyst is presented with step 1, Problem formulation and maybe the (broad) approach to be taken by another party (for example an academic or business supervisor, a technical committee, or even a client). Issues requiring particular attention in the case of spatial problems include:

- spatial scale factors: over what study region is the work to be carried out, and what are the implications of altering this for some or all datasets? Do the same scale factors apply for all the data of interest?

- statistical scale factors: at what levels of grouping are data to be analysed and reported?

- spatial arrangement factors: does the specification of the spatial arrangement or re-arrangement of subsections of the study area have an impact on the analysis?

- does the problem formulation require data of types, sizes or quality standards that are available, within the time, budget and resources available? if not, compromises will be necessary

- are conclusions regarding spatially grouped data being sought that imply the grouping (e.g. at the county level, at the farm level) is truly representative of all the components in the group (e.g. individuals living within the county, fields within the farm)? If so, the grouped regions must be entirely or largely homogeneous in order to avoid the so-called *ecological fallacy* — ascribing characteristics to members of a group when only the overall group characteristics are known

- are conclusions regarding spatial grouped data being sought based on the measured characteristics of sampled individuals? If so, the sample must be entirely or highly representative of the grouping in order to avoid the so-called *atomistic fallacy* — ascribing characteristics to members of a group based on a potentially unrepresentative sample of members

Framing the question is not always a one-time exercise. Once an initial problem specification has been drafted, it may be altered in the light of preliminary investigations, technical or commercial considerations, or unforeseen events (Figure 3-4, feedback loops). As far as possible, however, only essential changes should be made (and documented as such) once problem formulation has been defined and documented, and all interested parties have agreed on problem content. GIS has a particular role to play here, in providing tools (principally mapping related) for storing and visualising existing data and facilitating discussion of aspects of the problem prior to subsequent stages in the process. This role may continue throughout the various stages of a project, assisting in the interpretation, analysis and presentation of results. Careful consideration should be given to change control management, documentation and reporting, especially in commercial and governmental environments.

3.2.3.2 Plan: Formulating the approach

Having agreed on the problem definition the next stage is to formulate an approach that has the best possible chance of addressing the problem and achieving answers (outcomes) that meet expectations. The approach adopted involves consideration of issues such as:

- the nature of the problem and project — is it purely investigative, or a formal research exercise; is it essentially descriptive, including identification of structures and relationships, or more concerned with processes, in which clearer understanding of causes and effects may be required, especially if predictive models are to be developed and/or prescriptive measures are anticipated as an output?

- does it require commercial costings and/or cost-benefit analysis?

- are particular decision-support tools and procedures needed?

- what level of public involvement and public awareness is involved, if any?

- what particular operational needs and conditions are associated with the exercise?

- what time is available to conduct the research and are there any critical deadlines?

- what funds and other resources are available?

- is the project considered technically feasible, what assessable risk is there of failure and how is this affected by problem complexity?

- what are the client (commercial, governmental, academic) expectations?

- are there specifications, standards, quality parameters and/or procedures that must be used (for example to comply with national guidelines)?

- how does the research relate to other studies on the same or similar problems?

- what data components are needed and how will they be obtained (existing sources, collected datasets)?

- are the data to be studied (units) to be selected from the target population, or will the sample be distinct in some way and applied to the population subsequently (in which case one must consider not just sampling error but so-called study error also)?

When deciding upon the design approach and analytical methods it is essential to identify available datasets, examine their quality, strengths and weaknesses, and carry out exploratory work on subsets or samples in order to clarify the kind of approach that will be both practical and effective. There will always be unknowns at this stage, but the aim should be to minimise these at the earliest opportunity, if necessary by working through the entire process, up to and including drafting the presentation of results based on sample, hypothetical or simulated data.

The application of a single analytical technique is to be avoided unless one is extremely confident of the outcome, or it is the analytical technique or approach itself that is the subject of investigation. If a series of approaches, visualisations, techniques and tests all suggest a similar outcome then confidence in the findings tends to be greatly increased. If such techniques suggest different outcomes the analyst is encouraged to explain the differences, by re-examining the design, the data and/or the analytical techniques applied. Ultimately the original problem definition may have to be reviewed.

The impact on research of exceptions — rare events, spatial outliers, extreme values, unusual clusters — is extremely important in geospatial analysis. Exploratory methods, such as mapping and examining cases and producing box-plots (see further, Section 5.2.2.2), help to determine whether these observations are valid and important, or require removal from the study set.

Some analytical techniques are described as being more *robust* than others. By this is meant that they are less susceptible to data extremes or unusual datasets — for example

the median or middle value of a dataset is generally regarded as more robust than the mean or average value, because it is unaffected by the specific values of the set. However, the *spatial* mean and median exhibit different properties from those applied to individual tabulated attributes, and other measures of centrality (e.g. the central feature of a set) may be more appropriate in some instances. Likewise, statistical tests that make no assumptions about the underlying distribution of the dataset tend to be more robust than those that assume particular distributional characteristics — for example non-parametric versus parametric tests. However, increasing robustness may result in loss of *power* in the sense that some methods are described as being more powerful than others, i.e. they are less likely to accept hypotheses that are incorrect or reject hypotheses that are correct.

3.2.3.3 Data: Data acquisition

Spatial analysis is somewhat unusual, in that key datasets are normally provided by third parties rather than being produced as part of the research. Analysis is often of pre-existing spatial datasets, so understanding their quality and provenance is extremely important. It also means that in many instances this phase of the PPDAC process involves selection of one or more existing datasets from those available. In practice not all such datasets will have the same quality, cost, licensing arrangements, availability, completeness, format, timeliness and detail. Compromises have to be made in most instances, with the over-riding guideline being fitness for purpose. If the only datasets available are unsuitable for addressing the problem in a satisfactory manner, even if these are the only data that one has to work with, then the problem should either not be tackled or must be re-specified in such a way as to ensure it is possible to provide an acceptable process of analysis leading to worthwhile outcomes. For many problems important components of the project involve intangible data, such as perceptions or concerns relating to pollution, risk or noise levels. Quantifying such components in a manner that is accepted by the key participants, and in a form suitable for analysis, is a difficult process. This is an area in which a number of formal methodologies have been established and widely tested (including spatial decision support tools and cost-benefit analysis tools). Such tools are almost always applied separately and then utilised to generate or qualify input to GIS software (e.g. weightings).

Almost by definition no dataset is perfect. All may contain errors, missing values, have a finite resolution, include distortions as a result modelling the real world with discrete mathematical forms, incorporate measurement errors and uncertainties, and may exhibit deliberate or designed adjustment of positional and/or attribute data (e.g. for privacy reasons, as part of aggregation procedures). Spatial analysis tools may or may not incorporate facilities for explicitly handling some of the more obvious of these factors. For example, special GIS tools exist for handling issues such as:

- boundary definition and density estimation

- coding schemes that provide for missing data and for masking out invalid regions and/or data items

- modelling procedures that automatically adjust faulty topologies and poorly matched datasets, or datasets of varying resolutions and/or projections

- a wide range of procedures exist to handle difficulties in classification

- data transformation, weighting and normalisation facilities exist to facilitate comparison and combination of datasets of differing data types and extents

- lack of continuity in field data can be explicitly handled via breaklines and similar methods

- a range of techniques exist for modelling data problems and generating error bounds, confidence envelopes and alternative realisations

In other words, many facilities now exist within GIS packages that support analysis even when the datasets available are less than ideal. It is worth emphasising that here we are referring to the analytical stage, not the data cleansing stage of operations. For many spatial data this cleansing has been conducted by the data supplier, and thus is outside of the analyst's direct control. Of course for data collected or obtained within a project, or supplied in raw form (e.g. original, untouched hyperspectral imagery) data cleansing becomes one element in the overall analytical process.

An additional issue arises when data are generated as part of a research exercise. This may be as a result of applying a particular procedure to one or more pre-supplied datasets, or as a result of a simulation or randomisation exercise. In these latter cases the generated datasets should be subjected to the same critical analysis and inspection as source datasets, and careful consideration should be given to generated and magnified distortions and errors produced by the processing carried out. A brief discussion of such issues is provided in Burrough and McDonnell (1998, Ch.10) whilst Zhang and Goodchild (2002) address such issues in considerable detail.

3.2.3.4 Analysis: Analytical methods and tools

One of the principal purposes of this Guide is to assist readers in selecting appropriate analytical methods and tools from those readily available in the GIS marketplace. Since analysis sits well down the chain in what we have described as the analytical methodology, initial selection of methods and tools should have been made well before the analytical stage has been reached. Simplicity and parsimony — using the simplest and clearest tools, models and forms of visualisation — and fit to the problem and objectives are the key criteria that should be applied. Other factors include: the availability of tools; time and cost constraints; the need to provide validity and robustness checks, which could be via internal and/or external checks on consistency, sensitivity and quality; the use of multiple techniques; and the use of independent and/or additional datasets or sampling.

There are a very large number of software tools available for performing spatial analysis. All products describing themselves as geographic information systems have a basic or core level of analytical facilities, including many of those we describe in this Guide. We cannot hope to draw examples from even a representative sample of products, so we have made our own selection, which we hope will provide sufficient coverage to illustrate the range of analytical techniques implemented in readily available toolsets. In addition to mainstream GIS products we have included a number of specialist tools that have specific spatial analysis functionality, and which provide either some GIS functionality and/or GIS file format input and/or output.

A recurrent theme in spatial analysis is the notion of *pattern*. Frequently the objective of analysis is described as being the identification and description of spatial patterns, leading on to attempts to understand and model the processes that have given rise to the observed patterns. Unfortunately the word 'pattern' has a very wide range of meanings and interpretations. One way of defining whether a particular set of observations constitute a spatial pattern is by attempting to define the opposite, i.e. what arrangements of objects are not considered to constitute a pattern. The generally agreed notion of *not a pattern* is a set of objects or an arrangement that provides no information to the observer. From a practical point of view no information is often the situation when the arrangement is truly random, or is indistinguishable from a random arrangement. An alternative definition might be an even arrangement of objects, with deviations from this uniformity considered as patterns. Thus spatial pattern is a relative concept, where a model of *not a pattern* (e.g. Complete Spatial Randomness or CSR) is a pre-condition (see for example, Figure 2-4).

Observed spatial arrangements are frequently of indirect or *mapped data* rather than direct observations — the process of data capture and storage (e.g. as points and lines or remote sensed images) has already imposed a model upon the source dataset and to an extent pre-determined aspects of the observable arrangements. The method(s) of visualisation and the full and/or sampled extent of the dataset may also impose pre-conditions on the interpretation and investigation of spatial patterns.

Identification of spatial pattern is thus closely associated with a number of assumptions or pre-conditions: (i) the definition of what constitutes *not a pattern* for the purposes of the investigation; (ii) the definition of the dataset being studied (events/observations) and the spatial (and temporal) extent or scale of the observations; and (iii) the way in which the observations are made, modelled and recorded. Observed patterns may *suggest* a causal relationship with one or more principal processes, but they do not provide even a remotely secure means of inference about process.

For example, consider the case of the distribution of insect larvae. Imagine that an insect lays its eggs at a location in a large region at least 200 metres from the egg sites of other insects of the same species, and then flies away or dies. These other insects of the same species do likewise, at approximately the same time, each laying a random number of eggs, most of which hatch to form larvae that slowly crawl away in random directions from the original site for up to 100 metres. At some point in time shortly thereafter an observer samples a number of sites and records the pattern of all larvae within a given radius, say 10 metres. This pattern for each sampled site is then individually mapped and examined. The observer might find that the mapped patterns appear entirely random or might have a gradient of larvae density across the sampled regions. Zooming out to a 100m radius (i.e. using a larger region for sampling) a different pattern might have been observed, with a distinct centre and decreasing numbers of larvae scattered in the sampled region away from this point. However, if observations had been made over 1km squares in a 10kmx10km region it might only have been practical to identify the centres or egg sites. At this scale the pattern may appear regular since we implied that each egg site is not randomly distributed, but is influenced by the location of other sites. However, it could equally well be the case that egg sites are actually random, but only eggs on those sites that are laid on suitable regularly distributed vegetation survive and go on to produce live larvae. Zooming out again to 100kmx100km the observer might find that all egg sites are located in particular sub-regions of the study area, thus appearing strongly clustered, a pattern perhaps related to an attraction factor between insects or uneven large-scale vegetation cover or some other environmental variable. The mapped patterns at any given scale may not be sufficient to enable the analyst to determine whether a given set of observations are random, evenly spread, clustered or exhibit some specific characteristic such as radial spread, nor to infer with any reliability that some particular process is at work (as discussed in Section 2.2.9: Detail, resolution, and scale).

On the other hand it is fairly straightforward to generate particular spatial patterns, in the manner described above, using simple (stochastic) process models. These patterns are specific realisations of the models used, but there is no guarantee that the same pattern could not have been generated by an entirely different process. A specific mapped dataset *may* be regarded as a specific outcome of a set of (known and unknown) processes, and as such is but one of many possible outcomes or *process realisations*. To this extent, and perhaps to this extent only, such datasets can be thought of as samples (see further, Sections 2.3.8 and 2.4.4)

3.2.3.5 Conclusions: Delivering the results

The last stage of the PPDAC process is that of reaching conclusions based upon the analyses conducted, and communicating these to others. Note that implementation of findings (e.g. actually proceeding with building a bypass, designating an area as unfit for habitation, or implementing a vaccination programme) does not form part of this model process, but lies beyond its confines.

"The purpose of the Conclusion stage is to report the results of the study in the language of the Problem. Concise numerical summaries and presentation graphics [tabulations, maps, visualisations] should be used to clarify the discussion. Statistical jargon should be avoided. As well, the Conclusion provides an opportunity to discuss the strengths and weaknesses of the Plan, Data and Analysis especially in regards to possible errors that may have arisen" Mackay and Oldford (2002)

For many problems treated in spatial analysis this summary is quite sufficient. For others the conclusions stage will be the start of additional work: re-visiting the problem and iterating the entire process or parts of the process; a new project; implementing proposals; wider consultation; and/or the development of models that help to understand or explain the observations, guide future data gathering, and predict data values at locations/times not included within the analysis.

3.2.4   The changing context of PPDAC

The preceding discussion of PPDAC has illustrated how the data used for decision-making in a wide range of applications areas often have a spatial component. The general context to the supply and provision of spatially referenced data has changed profoundly over the last 20 years (see Section 2.5) and this has had far-reaching implications for the practice of applications-led science. Specifically:

• Most developed countries now have advanced spatial data infrastructures, comprising 'framework' data. Such data typically record the variable nature of the Earth's surface, landscape characteristics (e.g. land cover), and a range of artificial structures ranging from roads to mail delivery points. This provides valuable context for decision-making (e.g. see the range of Great Britain's Ordnance Survey's case studies on their web site). Even in parts of the world where this is not the case, high resolution remote sensing imagery can today provide an acceptable framework from which environmental, population and infrastructure characteristics can be ascertained

- Advances in data dissemination technologies, such as geoportals, have encouraged governments to disseminate public sector data on social, health, economic, demographic, labour market and business activity to the widest possible audiences (e.g. www.statistics.gov.uk; www.geodata.gov). Many geoportals include metadata ('data about data') on their holdings, which can be used to establish the provenance of a dataset for a given application. The advent of geoportals is having a catalytic role in stimulating public participation in GIS (PPGIS), for example amongst amateur genealogists when historic census data are put online. Of more central importance to professional users, geoportals empower very many more users to assemble datasets across domains and create added value information products for specific uses

- Advances in data capture and collection are leading to conventional data sources being supplemented, and in some cases partially replaced, with additional measures of physical, social and environmental information. This is creating challenges for those national mapping agencies that are required to recover most of their costs through user charges and copyright enforcement. One scenario is that the monolithic structure of parts of the data industry may become obsolete. In the socio-economic realm, Longley and

Harris (1999) have written about the ways in which routine customer transactions in retailing, for example, are captured and may be used to supplement census-based measures of store catchment characteristics

- There is now a greater focus upon metadata. As suggested above, the proliferation of data sources from formal and informal sources is leading to greater concern with the quality and provenance of data, and the situations in which such data are fit for purpose. This is particularly important post the innovation of the Internet, since data collectors, custodians and users are often now much more dispersed in space and time, and Internet GIS enables much greater apparent ease in conflating and concatenating remotely located data

- Allied to the previous two points, there is a wider realisation that the potential of data collected by public sector agencies at the local level or at fine levels of granularity, remains under-exploited. There exist very good prospects for public sector organisations to 'pool' data pertaining to their local operations, to the good of the populations that they serve. Innovations such as Google Earth are making it easy for organisations with no previous experience of GIS to add spatial context to their own data

- The practice of science itself is changing. Goodchild and Longley (1999) have written of the increasingly interdisciplinary setting to the creation of GIS applications, of the creation and maintenance of updates of GIS databases in real time, and of the assembly and disbanding of task-centred GIS teams — most obviously in emergency scenarios, but increasingly across the entire domain of GIS activity

Each of these developments potentially contributes to the development of GIS applications that are much more data rich than hitherto. Yet such applications may be misguided if the analyst is unaware of the scientific basis to data collection, the assumptions implicitly invoked in conflation and concatenation of different datasets, or general issues of data quality. Additionally, the creation of GIS applications and the dissemination of results may be subject to a range of ethical considerations, and issues of accountability and data ownership. The remaining sections of this Guide give the flavour of these issues, and illustrate the ways in which GIS based representations are sensitive to the properties of software and the data that are used.

In an ideal world, all data and software would be freely available, but in practice such a world would be unlikely to prioritise investments for future applications development. In the case of software, public domain sources are by no means all open source, but it is possible to form a view as to the quality of a software product by examining the nature of the results that it yields compared against those obtained using similar software offerings and known results with test datasets. Gauging the provenance of data sources is often an altogether trickier proposition, however. In seeking to validate one dataset with reference to a second, as in the comparison of two digitised street network products or the classification of two remotely sensed data resources, analysis of mismatches is more likely to suggest uncertainties than pinpoint precisely measurable inaccuracies. Discrepancies are only ultimately reconcilable through time-consuming field measurement, although models of error propagation and understanding of data lineage or classification method may optimise use of scarce validation resources.

While it is possible to validate some of the data used in representations of built or natural environments, validation of social measurements is an altogether more problematic task (Kempf-Leonard, 2005). Repeated measurement designs are usually prohibitively expensive, and some of the

attitudes and behaviours that are under investigation may in any case be transient or ephemeral in nature. In general terms it is helpful to distinguish between *direct* versus *indirect* indicators of human populations. A direct indicator bears a clear and identifiable correspondence with a focus of interest: net (earned plus unearned) income, along with measures of wealth accumulated in assets provides a direct indicator of affluence. Direct indicators may be time-consuming or difficult to measure in practice (how affluent are you?), as well as being invasive of personal privacy. Thus many data sources rely upon indirect indicators, which are easier to assemble in practice, but which bear a less direct correspondence with the phenomenon of interest — housing tenure, car ownership and possession of selected consumer durables might be used in a composite indicator of affluence, for example.

Many of these issues are crystallised in the current focus upon 'spatial literacy', which has been defined by the National Research Council (NRC), part of the US National Academy of Sciences, as comprising the following integrated components:

- understanding spatial relationships

- developing knowledge about how geographic space is represented; and

- the ability to reason and make key decisions about spatial concepts

In this Guide we seek to demonstrate how the *practice* of GIS is fundamentally about finding out about the real (geographic) world and how, to quote the NRC (2006): "spatial thinking is a cognitive skill that can be used in everyday life, the workplace, and science to structure problems, find answers, and express solutions using the properties of space". Additionally, in the spirit of modification to the PPDAC model, a common theme is the sourcing of data that are fit for purpose, and their geographical integration in ways that are efficient, effective and safe to use. With an eye to the needs of those that are exploring the usefulness of GIS for the first time, here and on the Spatial Literacy web site, we also discuss the level of resolution and availability of appropriate spatial units and whether public domain data are available and readily downloadable.

4 Building Blocks of Spatial Analysis

4.1 Spatial data models and methods

Spatial datasets make it possible to build operational models of the real world based upon the field and object conceptions discussed in Section 2.2.6 and the use of coordinate geometry to represent the object classes described in Section 2.2.3. These include: discrete sets of point locations; ordered sets of points (open sets forming composite lines or *polylines*, and closed, non-crossing sets, forming *simple polygons*); and a variety of representations of continuously varying phenomena, sometimes described as surfaces or fields. The latter are frequently represented as a continuous grid of square cells, each containing a value indicating the (estimated) average height or strength of the field in that cell. In most of the literature and within software packages the points/ lines/ areas model is described as vector data, whilst the grid model is described as raster (or image) data.

Longley *et al.* (2005) provide a summary of spatial data models used in GIS and example applications (Table 4.1). The distinctions are not as clear-cut as they may appear, however. For example, vector data may be converted (or transformed) into a raster representation, and vice versa. Transformation in most cases will result in a loss of information (e.g. resolution, topological structure) and thus such transformation may not be reversible. For example, suppose we have a raster map containing a number of distinct zones (groups of adjacent cells) representing soil type. To convert this map to vector form you will need to specify the target vector form you wish to end up with (polygon in this example) and then apply a conversion operation that will locate the boundaries of the zones and replace these with a complex jagged set of polygons

following the outline of the grid form. These polygons may then be automatically or selectively smoothed to provide a simplified and more acceptable vector representation of the data. Reversing this process, by taking the smoothed vector map and generating a raster output, will generally result in a slightly different output file from the one we started with, for various reasons including: the degree of boundary detection and simplification undertaken during vectorisation; the precise nature of the boundary detection and conversion algorithms applied both when vectorising and rasterising; and the way in which special cases are handled, e.g. edges of the map, "open zones", isolated cells or cells with missing values.

Table 4.1 Geographic data models

Data model	Example application
Computer-aided design (CAD)	Automated engineering design and drafting
Graphical (non-topological)	Simple mapping
Image	Image processing and simple grid analysis
Raster/grid	Spatial analysis and modelling, especially in environmental and natural resource applications
Vector/Geo-relational topological	Many operations on vector geometric features in cartography, socio-economic and resource analysis, and modelling
Network	Network analysis in transportation, hydrology and utilities
Triangulated irregular network (TIN)	Surface/terrain visualisation
Object	Many operations on all types of entities (raster/vector/TIN etc.) in all types of application

Similar issues arise when vector or raster datasets are manipulated and/or combined in various ways (e.g. filtering, resampling). In the following sections we describe a large variety of such operations that are provided in many of the leading software packages. We concentrate on those operations which directly or indirectly form part of analysis and/or modelling procedures, rather than those relating to data collection, referencing and management. These processes include the various "methods" that form part of the OGC simple feature specifications (Table 4.2) and test protocols, including the procedures for determining convex hulls, buffering, distances, set-like operators (e.g. spatial intersection, union etc.) and similar spatial operations. In each case it is important to be aware that data manipulation will almost always alter the data in both expected and unexpected ways, in many instances resulting in some loss of information. For this reason it is usual for new map layers and associated tables, and/or output files, to be created or augmented rather than source datasets modified.

Table 4.2 OGC OpenGIS Simple Features Specification — Methods

Method	Description
Note: a and b are two geometries (one or more geometric objects or features — points, line objects, polygons, surfaces including their boundaries); $I(x)$ is the interior of x; $dim(x)$ is the dimension of x, or maximum dimension if x is the result of a relational operation	
Spatial relations	
Equals	spatially equal to: $a=b$
Disjoint	spatial disjoint: equivalent to $a \cap b = \varnothing$
Intersects	spatially intersects: $[a \cap b]$ is equivalent to [not a disjoint(b)]
Touches	spatially touches: equivalent to $[a \cap b \neq \varnothing$ and $I(a) \cap I(b) = \varnothing]$; does not apply if a and b are points
Crosses	spatially crosses: equivalent to $[dim(I(a) \cap I(b)) < max\{dim(I(a)), dim(I(b))\}$ and $a \cap b \neq a$ and $a \cap b \neq b]$
Within	spatially within: within(b) is equivalent to $[a \cap b = a$ and $I(a) \cap I(b) \neq \varnothing]$
Contains	spatially contains: $[a$ contains(b)] is equivalent to $[b$ within(a)]
Overlaps	spatially overlaps: equivalent to $[dim(I(a) \cap I(b)) = dim(I(a)) = dim(I(b))$ and $a \cap b \neq a$ and $a \cap b \neq b]$
Relate	spatially relates, tested by checking for intersections between the interior, boundary and exterior of the two components
Spatial analysis	
Distance	the shortest distance between any two points in the two geometries as calculated in the spatial reference system of this geometry
Buffer	all points whose distance from this geometry is less than or equal to a specified distance value
Convex Hull	the convex hull of this geometry (see further, Section 4.2.13.1)
Intersection	the point set intersection of the current geometry with another selected geometry
Union	the point set union of the current geometry with another selected geometry
Difference	the point set difference of the current geometry with another selected geometry
Symmetric difference	the point set symmetric difference of the current geometry with another selected geometry (logical XOR)

4.2 Geometric and Related Operations

Individual vector features (e.g. polygons) and groups of grid cells exhibit a range of distinctive spatial attributes — for example the *length* of a polyline or polygon boundary and the *area* of a polygon or grid patch. In this section we examine a number of basic geometric attributes that apply to such features. These basic attributes may be included in the attribute table record associated with each feature, or may be treated as an *intrinsic* attribute that may be computed on demand and optionally added as a field in the associated table. If it is necessary to perform operations on these attributes, for example sorting areas by size, or calculating the ratio of areas to perimeters, it is helpful and often more efficient to add the data as an explicit field if it is not already available.

4.2.1 Length and area for vector datasets

Since the majority of GIS mapping involves the use of datasets that are provided in projected, plane coordinates, distances and associated operations such as area calculations are carried out using Euclidean geometry. Polygon areas (and the areas of sections of polygons) are calculated using integration by the trapezoidal or Simpson's rule, taking advantage of the fact that progressing clockwise around all the polygon vertices generates the desired result. For example, in Figure 4-1 the area between the x-axis and the the two vertical lines at x_1 and x_2 is simply the width between the two lines, i.e. (x_2-x_1) times the average height of the two verticals, i.e. $(y_2+y_1)/2$, giving a rectangle like the first one shown to the right of the polygon. The area between the x-axis and the two verticals at x_2 and x_3 is shown by the second narrow rectangle, and if we then subtract the shorter and wider rectangle, which represents the area outside of the polygon between the two vertical lines of interest, we obtain the area of

the part of the overall polygon we are interested in (A,B,C,D).

More generally, if we take the $(n-1)$ distinct coordinates of the polygon vertices to be the set $\{x_i,y_i\}$ with $(x_n,y_n)=(x_1,y_1)$ then the total area of the polygon is given by:

$$A = \frac{1}{2}\sum_{i=1}^{n-1}(x_{i+1} - x_i)(y_i + y_{i+1})$$

Traversing a polygon in a clockwise direction generates a sequence of positive and negative rectangular sections (negative when x_i is less than x_{i+1}) whose total sums to a positive value, the area of the polygon. Multiplying out the brackets the formula can also be written as

$$A = \frac{1}{2}\sum_{i=1}^{n-1}(x_{i+1}y_i - x_iy_{i+1})$$

This is an expression we use again in Section 4.2.5.1, in connection with the determination of polygon centroids.

Figure 4-1 Area calculation using Simpson's rule

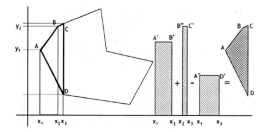

If the order is reversed the total is the same but has a negative sign. This feature can be used to remove unwanted polygons contained within the main polygon (e.g. areas of water or built-up zones within a district when calculating available land area for farming or the area to be included and excluded in density or similar areal computations). The procedure fails if the polygon is self-crossing (which should not occur in well-structure GIS files) or if some of the y-coordinates are negative — in this latter case either positive-only coordinate systems should be used or a fixed value added (temporarily) to all y-coordinate values to ensure they all remain positive during the calculation process.

Lengths and areas are normally computed using Cartesian (x,y) coordinates, which is satisfactory for most purposes where small regions (under 100 kms x 100 kms) are involved. For larger areas, such as entire countries or continents, and where data are stored and manipulated in latitude/longitude form, lengths (distances) and areas should be computed using spherical or similar measures (e.g. ellipsoidal, geoidal). If one uses Cartesian calculations over large regions the results will be incorrect — for example, using the ArcGIS measure tool to show the distance between Cairo and Capetown on a map of Africa will give a value substantially below the "true", great circle distance (depending on the projection applied). Computation of polygon areas using spherical or ellipsoidal models is complex, requiring careful numerical integration. Functions such as AREAINT in the Mapping Toolbox of MATLab provide such facilities, and the source code for the module which can be inspected and amended if required. It should also be noted that in most cases only two coordinates are utilised, whereas on surfaces with variable heights three dimensional distances, areas and volumetric analysis may be more appropriate (see further, Section 4.2.3).

In some cases GIS software products will select the appropriate calculation method based on the dataset, whilst in others (e.g. MapInfo) the option to specify Cartesian or Spherical computation may be provided where relevant. If we identify two mapped locations as having Cartesian coordinates (x_i,y_i) and (x_j,y_j) then the distance between them, d_{ij}, is simply the familiar:

$$d_{ij} = \sqrt{\left(x_i - x_j\right)^2 + \left(y_i - y_j\right)^2}$$

The length of a polyline or perimeter length of a polygon is then computed as the sum of the individual segments comprising the feature. Distances calculated using this formula, or the 3D equivalent, we shall denote d_E, or Euclidean distance.

A straight line, l, in the x-y plane can be expressed as an equation: $ax+by+c=0$, where a, b and c are constants. Using this formulation the Euclidean distance of a point $p(x_p,y_p)$ from this line is given by:

$$d_{(p,l)} = \frac{\left|ax_p + by_p + c\right|}{\sqrt{\left(a^2 + b^2\right)}}$$

Within GIS the distance from a point to a line segment or set of line segments (e.g. a polyline) is frequently needed. This requires computation of one of three possible distances for each (point, line segment) pair: (i) if a line can be extended from the sample point that is orthogonal to the line segment, then the distance can be computed using the general formula $d_{p,l}$ provided above; (ii) and (iii) if case (i) does not apply, the sample point must lie in a position that means the closest point of the line segment is one of its two end points. The minimum of these two is taken in this instance. The process is then repeated for each part of the polyline and the minimum of all the lengths computed selected.

In spherical coordinates the (great circle) distance equation often quoted is the so-called Cosine formula:

$$d_{ij} = R\cos^{-1}\left[\sin\phi_1 \sin\phi_1 + \cos\phi_1 \cos\phi_2 \cos(\lambda_1 - \lambda_2)\right]$$

where: R is the radius of the Earth (e.g. the average radius of the WGS84 ellipsoid); (ϕ,λ) pairs are the latitude and longitude values for the points under consideration; and $-180\le\lambda\le180$, $-90\le\phi\le90$ (in degrees, although such calculations normally are computed using radians. nb: 180 degrees=Pi radians and one radian=57.30 degrees). This formula is sensitive to computational errors in some instances (very small angular differences). A safer equation to use is the Haversine formula:

$$d_{ij} = 2R\sin^{-1}\left(\sqrt{\sin^2(A) + \sin^2(B)\cos\phi_i\cos\phi_j}\right)$$

$$where : A = \frac{\phi_i - \phi_j}{2}, B = \frac{\lambda_i - \lambda_j}{2}$$

Most GIS packages do not specify how they carry out such operations. Distances calculated using this formula we shall denote d_S, or Spherical distance.

4.2.2 Length and area for raster datasets

Length and area computations may also be meaningful for grid datasets, and in this case the base calculations are very simple, reflecting the grid cell size or resolution (summing the number of cells and multiplying by the edge length or edge length squared, as appropriate). The difficulty in this case is in the definition of what comprises a line or "polygon". A set of contiguous cells (a clump or patch) with a common attribute value or classification (e.g. deciduous woodland) has an area determined by the number of cells in the clump and the grid resolution. The perimeter length of the clump is a little less obvious. Taking the perimeter as the length of the external boundary of each cell in the clump appears to be the simplest definition, although this will be much longer than a vectorised (smoothed) equivalent measure (typically around 50% longer). Furthermore, depending on the software used, differences in definition of what a clump is (for example, whether a clump can include holes, open edges that are not part of the perimeter, cells whose attributes do not match those of the remaining cells) will result in variations in both area and perimeter calculation.

Many environmental science datasets (e.g. soil type, land use, vegetation cover) show a near continuous spread of attribute values across cells, with few if any distinct sharp breaks. In this case boundaries are less definitive, determined more by common consent or other procedures (see further, Section 4.2.9) than by clear demarcation of cell values. The determination of areas and perimeters of

zones in such cases is less clear, although some progress has been made using the concept of *membership functions* (MFs).

A further issue with gridded data is that the gridding model (or imaging process) results in distortions or constraints on the way in which data are processed. The distortions are primarily of three kinds — Orientation, Metrics and Resolution:

Orientation — with a single rectangular grid the allocation of source data values to cells and the calculations of lengths and areas will alter if the grid is first rotated; with two or more grids common orientation and resolution are essential if data are to be combined in any way. This may mean that source data (e.g. remote sensed images) will have been manipulated prior to provision of the gridded data to ensure matching is possible, with each source item receiving separate treatment to achieve this end, and/or that data will require resampling during processing

Metrics — in order to compute distances for gridded data it is common to add up the number of cells comprising a line or boundary and multiplying this number by the cell edge length, E. This is essentially a "rook's move" calculation, zig-zagging across the grid, although it may be adjusted by allowing diagonal or "bishop's moves" also (see diagram below). Diagonal lengths are then taken as E√2.

These methods only provide correct values for the distance between cells in 4 or 8 directions, with distances to all other directions (for example, those highlighted in grey below) being in error by up to 41% for the 4-point rook's move model and 7.6% for the 8-point queen's move model. Each of the squares shown in grey below are 2.24 units from the central cell but would be calculated as either 3 units away (rook's move) or 2.41 units (queen's move). The position of these grey squares is sometimes described as Knight's move, again by reference to the movement of chess pieces.

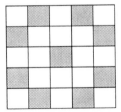

Similar issues arise in computations of cost distance, path alignment and the calculation of surface gradient, aspect and related measures (e.g. material flows). This is not always a question of what is correct or incorrect in some absolute sense, since this will be determined by the application and adequacy or appropriateness of the approach adopted, but it is a significant issue to be aware of

Resolution — the cell size (and shape) of a grid or image file and the range of attribute values assigned to cells has an important role to play in spatial measures and analysis. Amongst the more obvious is the fact that finer resolution images/grids provide more detailed representation of points, lines and areas, as well as a finer breakdown of associated attribute data. But there are many additional factors to be considered, including the increase in data and processing overhead with finer resolutions. Another factor is the implication that an attribute value assigned to a cell applies throughout the cell, so larger cells imply a greater degree of averaging, including the possibility of missing important variations within the cell. At one extreme, assuming cells are assigned a single attribute value, a grid consisting of a single huge cell will show no variation in attribute values at all, whilst variation between cells will tend to increase as cell sizes are reduced and the number of cells increases. At the other extreme, with cell sizes that are extremely small (and thus very large in number) every cell may be different and contain a unique value or a single observation (presence/

absence). Likewise, with attributes showing binary presence/ absence all cells will show either presence or absence, whilst with continuous variables (e.g. terrain height, soil moisture levels) entries may be continuous (e.g. any real positive number) or categorised (e.g. coded as 0-255 in an 8-bit attribute coding). Cell entries may also be counts of events, in which cases the earlier comments on cell size have a direct bearing — with one large cell all N observations will fall in the single grid cell, whilst with smaller cell sizes there will be a range of values $0 \leq n \leq N$ until with very small cell sizes the only values for n will be 0 and 1 unless events are precisely co-located

4.2.3 Surface area

4.2.3.1 Projected surfaces

Most GIS packages report planar area, not surface area. Some packages, such as ArcGIS and Surfer provide surface area, although they may not specify in detail how these are calculated. Computation of surface areas will result in a value that is greater than or equal to the equivalent planar projection of the surface. The ratio between the two values provides a crude index of surface roughness.

If a surface is represented in TIN form, the surface area is simply the sum of the areas of the triangular elements that make up the TIN. Let $T_j=\{x_{ij},y_{ij},z_{ij}\}$ i=1,2,3 be the coordinates of the corner points of the jth TIN element. Then the surface area of the planar triangular (or *planimetric* area) determined when z=constant, can be calculated from a series of simple cross-products:

$$A_j = \frac{1}{2}\begin{pmatrix} + x_{2j}y_{1j} + x_{3j}y_{2j} + x_{1j}y_{3j} \\ - x_{1j}y_{2j} - x_{2j}y_{3j} - x_{3j}y_{1j} \end{pmatrix}$$

Note that this expression is the same as the polygon area formula we provided earlier, but for a $(n-1)$=3-vertex polygon:

$$A = \frac{1}{2} \sum_{i=1}^{3} (x_{i+1} y_i - x_i y_{i+1})$$

where $i+1=4$ is defined as the first point, $i=1$.

The term inside the brackets is the entire area of a parallelogram, and half this gives us our triangle area. This expression can also be written in a convenient matrix determinant form (ignoring the j's for now) as:

$$A = -\frac{1}{2} \begin{vmatrix} x_1 & y_1 & 1 \\ x_2 & y_2 & 1 \\ x_3 & y_3 & 1 \end{vmatrix}$$

If the slope of T_j, θ_j, is already known or easily computed, the area of T_j can be directly estimated as $A_j / \cos(\theta_j)$. Alternatively the three dimensional equivalent of the matrix expression above can be written in a similar form, as:

$$A = -\frac{1}{2} \sqrt{\begin{vmatrix} x_1 & y_1 & 1 \\ x_2 & y_2 & 1 \\ x_3 & y_3 & 1 \end{vmatrix}^2 + \begin{vmatrix} y_1 & z_1 & 1 \\ y_2 & z_2 & 1 \\ y_3 & z_3 & 1 \end{vmatrix}^2 + \begin{vmatrix} z_1 & x_1 & 1 \\ z_2 & x_2 & 1 \\ z_3 & x_3 & 1 \end{vmatrix}^2}$$

If $z=0$ for all i, the second and third terms in this expression disappear and we are left with the result for the plane. Suppose that we have a plane right angled triangle with sides 1,1 and $\sqrt{2}$, defined by the 3D coordinates: (0,0,0), (1,0,0) and (1,1,0). This plane triangle has area of one half of a square of side length 1, so its area is 0.5 units, as can be confirmed using the first formula above. If we now set one corner, say the point (1,1,0) to have a z-value (height) of 1, its coordinate becomes (1,1,1) and the area increases to 0.7071 units, i.e. by around 40%.

If the surface representation is not in TIN form, but represented as a grid or DEM, surface area can again be computed by TIN-like computations. In this case the 8-position immediate neighbourhood of each cell can be remapped as a set of 8 triangles connecting the centre-points of each cell to the target cell (Figure 4-2).

Figure 4-2 Triangular approximation of surface area

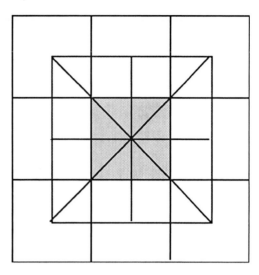

In each case the triangle area can be computed, as above, and 25% of each area assigned to the central cell (shaded grey in Figure 4-2). Research has suggested that the resulting total surface area for the cell is close to that produced by TIN models for grids of 250+ cells. Grid surface areas can also be computed using the slope adjustment method described above assuming that a slope grid has been first been computed. Note that grid resolution is generally coarser than the vertical or z-dimension resolution. For example, a DEM might be defined in 25mx25m squares, with a vertical resolution of 1m. If generating your own slope values, areas etc. from such data the different scales must be included within the calculation. If the units for the vertical resolution do not match the horizontal scale then it may also be necessary to apply a scaling factor to the z-dimension.

Surface areas are computed relative to a reference plane, usually z=0. In some instances it may be necessary to specify the value to be associated with the reference plane, and whether the absolute values, positive values and/or negative values are to be computed. Surfer, for example, supports all three options which it describes as positive planar and surface area, negative planar and surface area and total planar area. For example, consider the surface illustrated in Figure 4-3. This shows a surface model of a GB Ordnance Survey DEM for tile TQ81NE, which we shall use as a test surface in various sections of this

Guide. This is a 5000mx5000m area provided as a 10mx10m grid of elevations, with an elevation range from around 10m-70m. Using a reference plane of z=30m the surface area above this reference plane is roughly 1.3sq kms and below 30m is roughly 2.1sq kms (3.4sq kms in total compared to 2.5sq kms for the planar area). For details of volume computations, which also use reference surfaces, see Section 6.2.6.

Figure 4-3 Surface model of DEM for OS TQ81NE tile

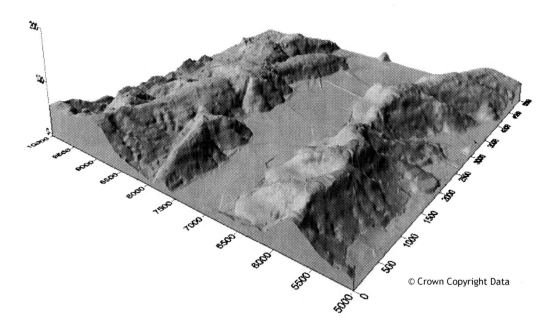

© Crown Copyright Data

4.2.3.2 Terrestrial (unprojected) surface area

For large regions (e.g. "rectangular" regions with sides greater than several hundred kilometres) surface areas will be noticeably affected by the curvature of the Earth. Such computations apply at large state and continental levels, or where the areas of ocean, ice or cloud cover are required over large regions. Computation of areas using spherical trigonometry or numerical integration is preferable in such cases.

The spherical area of a rectangular region defined by fixed latitude and longitude *coordinates* (e.g. determined by 100°W30°N and 90°W40°N) is greater than the area bounded by *lines* of constant latitude — the latter are not great circles, although the lines of longitude are. This effect increases rapidly nearer the poles and is weakest close to the equator. The MATLab Mapping Toolbox (MMT) function AREAQUAD provides a convenient way of computing the area of a latitude-longitude "quadrangle", although it can be computed without too much difficulty using spherical trigonometry. The area north of a line of latitude, ϕ, is the key factor in the calculation (the surface area of an entire sphere is $4\pi R^2$). The former area is:

$$A = 2\pi(1 - \sin\phi)R^2$$

where R is the radius of the spherical Earth, e.g. 6378.137 kms. The area of the quadrangle is then simply the difference between the areas, A_1 and A_2, north of each of the two lines of latitude, ϕ_1 and ϕ_2, adjusted by the proportion of the Earth included in the difference in longitude values, λ_1 and λ_2:

$$A = \frac{\pi}{180}\left|A_1 - A_2\right|\left|\lambda_1 - \lambda_2\right|R^2$$

A (simple) polygon with n vertices on the surface of the Earth has sides that are great circles and area:

$$A = \left(\sum_i \theta_i - (n-2)\pi\right)R^2$$

where again R is the radius of the Earth and the θ_i are the internal angles of the polygon in radians measured at each polygon vertex. The simplest spherical polygon is a triangle (n=3 so n-2=1) and on a sphere the sum of its internal angles must be greater than 180° (π), so the formula equates to 0 when the triangle is vanishingly small. A spherical triangle with internal angles that are all 180° is the largest possible, and is itself a great circle. In this case the formula yields A=$2\pi R^2$, as expected.

All polygon arcs must be represented by great circles and internal angles are to be measured with respect to these curved arcs. The term in brackets is sometimes known as the *spherical excess* and denoted with the letter E, thus A=ER^2.

Note that every simple polygon on a sphere divides it into two sections, one smaller than the other unless all vertices lie on a great circle. If only the latitude and longitude values

are known, the internal angles must be computed by calculating the true great circle bearing (or *azimuth*) from each vertex to the next — this can be achieved using spherical trigonometry or, for example, the MMT function AZIMUTH, which supports explicit geoid values. Azimuth values are reported from due north, so computations of internal angles need to adjust for this factor. Alternatively the MMT function AREAINT may be used, ideally with high point density (i.e. introducing additional vertices along the polygon arcs). This latter function computes the enclosed area numerically. The OpenSource GIS, GRASS, provides a number of similar functions, including the source code modules area_poly1, which computes polygon areas on an ellipsoid where connecting lines are grid lines rather than great circles, and area_poly2 which computes planimetric polygon area.

4.2.4 Line Smoothing and point-weeding

Polylines and polygons that have a large number of segments may be considered over-complex or poor representations of the original features, or unsuitable for display and printing at all map scales. A range of procedures exist to achieve simplification of lines in such cases, including those illustrated in Figure 4-4.

A. Point weeding (Figure 4-4a) is a procedure largely due to Douglas and Peuker (1973) — for a fuller discussion, see Whyatt and Wade (1988) and Visvalingam and Whyatt (1991). In this example the original black line connecting all the vertices 1 to 10 is progressively simplified to form the green/pale grey line shown. The procedure commences by connecting point 1 (the anchor) to point 10 (the floating point) with a temporary line (the initial segment). This enables point 8 to be identified as the vertex that deviates from this initial segment by the greatest distance (and more than a user-defined tolerance level, shown here by the bar symbol on the left of the upper diagram). Euclidean distance squared is sufficient for this algorithm and is significantly faster to compute. Point 10 is then directly connected to point 8 and point 9 is dropped. A new temporary line is drawn between point 8 (the new floating point) and 1 and the process is repeated. This time points 5, 6 and 7 are dropped. Finally point 3 is dropped and the final line contains only 4 segments and 5 nodes (1, 2, 4, 8 and 10). This process can be iterated by moving the initial anchor point to the previous floating point and repeating the process for further simplification. Some implementations of this algorithm, including those of the original authors, fail to handle exceptions where the furthest point from the initial segment is one end of a line that itself lies close to the initial segment. Ebisch (2002) addresses this error and provides corrected code for the procedure. The procedure is very sensitive to the tolerance value selected, with larger values resulting in elimination of more intermediate points.

A number of GIS products provide a range of simplification methods. The default in ArcGIS is point-weeding (POINT_REMOVE) but an alternative BEND_SIMPLIFY which removes extraneous bends is also available.

B. Simple smoothing utilises a family of procedures in which the set of points is replaced by a smooth line consisting of a set of curves and straight line segments, closely following but not necessarily passing through the nodes of the original polyline. This may be achieved in many ways with Figure 4-4b showing one example.

C. Spline smoothing typically involves replacing the original polyline with a curve that passes through the nodes of the polyline but does not necessarily align with it anywhere else, or approximates the point set rather like a regression fit (Figure 4-4c). A range of spline functions and modes of operation are available (including in some cases Bézier curves and their extension, B-splines), many of which involve using a series of smoothly connected cubic curves. In some cases additional temporary nodes or "knots" are introduced (e.g. half way between each existing vertex)

to force the spline function to remain close to the original polyline form.

Figure 4-4 Smoothing techniques

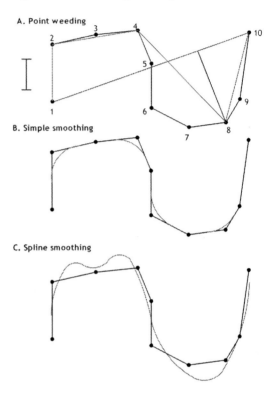

Each method and implementation (including the selection of parameters and constraints for each variant) will have different results within and across different GIS packages and the process will not generally be reversible. Furthermore, different GIS packages tend to offer only one or two smoothing procedures, so obtaining consistent results across implementations is almost impossible. Note that once a polyline or polygon has been subject to smoothing its length will be altered, becoming shorter in many cases, and enclosed areas will also be altered though generally by a lesser amount in percentage terms. The orientation of the line segments also changes dramatically, but not the relative positions of the two end nodes, so the end-node to end-node orientation is retained. Determination of the length and area of amended features may require numerical integration utilising fine division of the smoothed objects rather than the simple procedures we have described in Section 4.2. Further discussion of smoothing, including additional approaches, is provided in Chapter 6.

These methods have been described by McMaster and Shea (1992) as forms of *feature generalisation*. Other forms of generalisation include: *amalgamation* (replacing a large number of separate area features or cells by a smaller number, e.g. a single feature); *merging* (replacing several line features with a single line feature); and *collapse* (replacing an area object by a point, cells or a combination of points and lines, or replacing dual lines by a single centreline).

4.2.5 Centroids and centres

The terms centre and centroid have a variety of different meanings and formulas, depending on the writer and/or software package in which they are implemented. There are also specific measures used for polygons, point sets and lines so we treat each separately below. Centres or centroids are provided in many GIS packages for multiple polygons and/or lines, and in such cases the combined location is typically computed in the same manner as for point sets, using the centres or centroids of the individual objects as the input points. In almost all cases the metric utilised is d_E or d_S, although other metrics may be more appropriate for some problems. Centroids are often defined as the nominal centre of gravity for an object or collection of objects, but many other measures of centrality are commonly employed within GIS and related packages. In these latter instances the term centre (qualified where necessary by how this centre has been determined) is more appropriate than centroid, the latter being a term that should be reserved for the centre of gravity.

4.2.5.1 Polygon centroids and centres

Polygon centres have many important functions in GIS: they are sometimes used as "handle points", by which an object may be selected, moved or rotated; they may be the default position for labelling; and for analytical purposes they are often used to "represent" the polygon, for example in distance calculations (zone to zone) and when assigning zone attribute values to a point location (e.g. income per capita, disease incidence, soil composition). The effect of such assignment is to assume that the variable of interest is well approximated by assigning values to a single point, which essentially involves loss of information. This is satisfactory for some problems, especially where the polygons are small, homogeneous and relatively compact (e.g. unit ZIP-code or postcode areas, residential land parcels), but in other cases warrants closer examination.

With a number of raster-based GIS packages the process of vector file importing may utilise polygon centres as an alternative to the polygons themselves, applying interpolation to the values at these centroids in order to create a continuously varying grid rather than the common procedure of creating a gridded version of discretely classified polygonal areas (see further Chapter 6). Similar procedures may be applied in vector-based systems as part of their vector-to-raster conversion algorithms.

If (x_i,y_i) are the coordinate pairs of a point set or defining a single polygon, we have for the Mean Centre, M1:

$$M1 = \left(\sum_i \frac{x_i}{n}, \sum_i \frac{y_i}{n} \right) = (\overline{x}, \overline{y})$$

The Mean Centre is not the same as the centre of gravity for a polygon (although it is for a point set, see subsection 4.2.5.2). A weighted version of this expression is sometimes used:

Mean Centre (weighted), M1*:

$$M1^* = \left(\frac{\sum_i w_i x_i}{\sum_i w_i}, \frac{\sum_i w_i y_i}{\sum_i w_i} \right)$$

The RMS variation of the point set $\{x_i, y_i\}$ about the mean centre is known as the *standard distance*. It is computed using the expressions:

$$SDis = \sqrt{\sum_i \frac{(x_i - \bar{x})^2}{n} + \sum_i \frac{(y_i - \bar{y})^2}{n}},$$

or weighted:

$$SDis^* = \sqrt{\frac{\sum_i w_i (x_i - \bar{x})^2}{\sum_i w_i} + \frac{\sum_i w_i (y_i - \bar{y})^2}{\sum_i w_i}}$$

As noted above, the term *centroid* for a polygon is widely used to refer to its assumed centre of gravity. By this is meant the point about which the polygon would balance if it was made of a uniform thin sheet of material with a constant density, such as a sheet of steel or cardboard. This point, M2, can be computed directly from the coordinates of the polygon vertices in a similar manner to that of calculating the polygon area, A, provided in Section 4.2.1. Indeed, it requires computation of A as an input. Using the same notation as before, the formulas required for the x- and y-components are ($x \geq 0$, $y \geq 0$):

$$M2_x = \sum_{i=1}^{n-1} (x_{i+1}y_i - x_i y_{i+1})(x_i + x_{i+1}) / 6A$$

$$M2_y = \sum_{i=1}^{n-1} (x_{i+1}y_i - x_i y_{i+1})(y_i + y_{i+1}) / 6A$$

Figure 4-5 shows a sample polygon with 6 nodes or vertices, A-F, together with the computed locations of the mean centre (M1), centroid (M2) as defined by the centre of gravity, and the centre (M3), of the Minimum Bounding Rectangle (or MBR) which is shown in grey.

Figure 4-5 Polygon centroid (M2) and alternative polygon centres

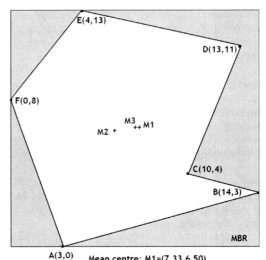

Mean centre: M1=(7.33,6.50)
Centre of gravity: M2=(5.82,6.33)
MBR centre: M3=(7.00,6.50)

Each of these three points fall within the polygon boundary in this example, and are fairly closely spaced. The MBR centre is clearly the fastest to compute, but the most subject to outliers — e.g. a single vertex location that is very different from the majority of the elements being considered. For example, if point B had (valid or invalid) coordinates of B(34,3) then we would find M1=(10.67,6.5), M2=(8.67,5.36) and M3=(17,6.5). M3 is now well outside of the polygon, M1 is close to the polygon boundary and M2 remains firmly inside. None of these points minimises the sum of distances to polygon vertices.

Despite the apparent robustness of M2, with a polygon of complex shape the centroid may well lie outside of its boundary. In the example illustrated in Figure 4-6 we have generated centroids for a set of polygons using the X-Tools add-in for ArcGIS, and these produce slightly different results from those created by ArcGIS itself.

Figure 4-6 Centre and Centroid positioning

Census tracts and centroids

Combined centroid of the two selected census tracts

This centroid lies in a lake - no census data apply to this polygon so it must be excluded in subsequent computations

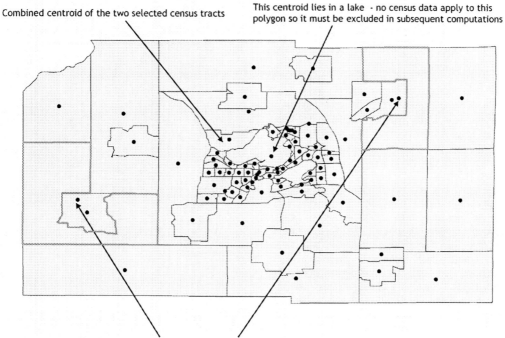

These centroids relate to the census tracts that are highlighted, in both cases being outside of their own tracts and inside another tract

Data source: US Census Bureau

ArcGIS includes a function, Features to Points, which will create polygon centres that are guaranteed to lie inside the polygon if the option INSIDE is included in the command. Manifold includes a Centroids|Inner command which performs a similar function.

If the polygon is a triangle, which is a widely used form in spatial analysis, the centre of gravity lies at the intersection of straight lines drawn from the vertices to the mid-points of the opposite sides (Figure 4-7).

Figure 4-7 Triangle centroid

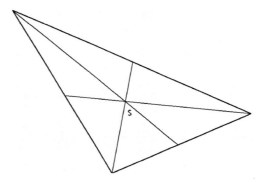

An alternative to using the MBR is to find the smallest circle that completely encloses the polygon, taking the centre of the circle as the

polygon centre (Figure 4-8, M4). This is the default in the Manifold GIS (viewable simply by selecting the menu option to View|Structure|Centroids) and/or using the Transform function "Centroids". This procedure suffers similar problems to that of using the MBR. Yet another alternative is to find the largest inscribed circle, and take its centre as the polygon centre (Figure 4-8, M5). This approach has the advantage that the centre will definitely be located inside the polygon, although possibly in an odd location, depending on the polygon shape. If the polygon is a triangle then M5 and M2 will coincide. Manifold supports location of centres of types M2, M3 and M4.

As we can also see in Figure 4-6 multiple polygons may be selected and assigned combined centres. In this example the combined centre appears to have been computed from the MBR of the two selected census tracts, but different packages and toolsets may provide alternative procedures and thus positions. In such cases the centre may be nowhere near any of the selected features, and if it is important to avoid such a circumstance the GIS may facilitate selection of the most central feature (e.g. the most central point) if one exists. In some instances the combined centre calculation may be weighted by the value of an attribute associated with each polygon. This procedure will result in a weighted centre whose location will be pulled towards regions with larger weights. However, in such cases the distribution of polygons used in the calculation may produce unrepresentative results.

This is apparent again in Figure 4-6 where a polygon centre calculation for all census tracts in the western half of the region, weighted by farm revenues, would be pulled strongly to the east of the sub-region where many small urban tracts are found, even though these may have relatively low weights associated with them.

Figure 4-8 Polygon centre selection

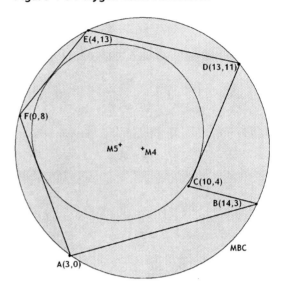

4.2.5.2 Point sets

If (x_i, y_i) are the coordinate pairs of a point set then the Mean Centre, $M1(x_0, y_0)$, is simply the average of the coordinate values, as we saw earlier:

$$M1(x_0, y_0) = \left(\sum_i w_i x_i / \sum_i w_i, \sum_i w_i y_i / \sum_i w_i \right)$$

In this version of the expression for M1 there is an additional (optional) component, w_i, representing weights associated with the points included in the calculation. For points sets the weights might be the number of crimes recorded at a particular location, or the number of beds in a hospital; for multiple polygons, where these are represented by their individual centres or centroids, the weights would typically be an attribute value associated with each polygon. Clearly if all weights=1 the formula simplifies to a calculation that is purely based on the coordinate values, with the number of points, $n = \sum w_i$.

M1 in this case is the centre of gravity, or *centroid*, of the point set assuming the locations used in the calculation are treated as point masses. M1 is the location that minimises the sum of (weighted) *squared distance* to all the points in the set, is simple and fast to compute, and is widely used. However, it is not the point that minimises the sum of the (weighted) *distance* to each of the points. The latter is known sometimes as the (bivariate) "median centre" (although Crimestat, for example, uses the term median centre to be simply the middle values of the *x*- and *y*-coordinates). It is best described as the MAT point (the centre of Minimum Aggregate Travel, M6). This point may be determined using the following iterative procedure, with M1 as the initial pair (x_0, y_0) and $k=0,1...$:

$$x_{k+1} = \sum_{i=1}^{n} w_i x_i / d_{i,k} \Big/ \sum_{i=1}^{n} w_i / d_{i,k}$$

$$y_{k+1} = \sum_{i=1}^{n} w_i y_i / d_{i,k} \Big/ \sum_{i=1}^{n} w_i / d_{i,k}$$

In this expression $d_{i,k}$ is the distance from the ith point to the kth estimated optimal location. It is usual to adjust distances in this, and similar formulas involving division by interpoint distances, by a small increment and/or to apply code checks to avoid divide-by-zero or close to zero situations. Iteration is continued until the change in the objective function (the cumulative distance) or both coordinates is less than some pre-specified small value. These formulas for M1 and M6 can be derived by taking the standard equation for distance, d_E, or distance squared, d_E^2, and partially differentiating first with respect to *x* and then with respect to *y*, and finally equating the results to 0 to determine the minimum value. An extension to this type of weighted mean is provided by the Geographically Weighted Regression (GWR) software package, and is described further in Section 5.6.3. With the GWR software it is possible to compute a series of locally defined (geographically weighted) means, and associated variances, which may then be analysed and/or mapped.

The positions of M1, M3 and M6 are illustrated in Figure 4-9, using the same set of coordinates as for the vertices A-F of the polygon described in subsection 4.2.5.1. Note that M1=M2 in this case, and for both M1 and M3 (the MBR centre) their position is unchanged from the polygonal case. Also note that if we move point B to B′ the position of M6 is unchanged (although the cumulative distance will be greater). This observation is true for each of the points A-F — they can be moved away from the MAT point to an arbitrary distance along a line connecting them to M6, and the position of M6 will be unaffected. This would not generally be the case with M1 or M3.

Figure 4-9 Point set centres

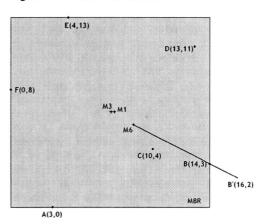

Mean centre/Centre of gravity: M1=(7.33,6.50)
MAT: M6=(8.58,5.61)
MBR centre: M3=(7.00,6.50)

The locations described for different types of centre, both here and in subsection 4.2.5.1, often assume that all points (or polygons) are weighted equally. If the set of weights, w_i, are not all equal, the locations of M1 and M6 will be altered, but M3 will be unchanged. The affect of unequal weights is to "pull" the location of M1 and M6 towards the locations with higher weights. For example, if point B in our previous example had a weight of 3, M1 would be moved to (9.00,5.63) just to the right of M6 in Figure 4-9, M3 would be unaltered, and M6 would be altered to (10.41,4.29) which is very close to point C. Any weights in the dataset that are zero or missing will generally result in the point being removed from the calculation. Such occurrences require checking to ensure valid data is not being discarded due to incomplete information or errors.

There is no difference between weighted calculations where the weights are integer values and the standard calculations for unit weights if some points in a set are co-located. For example, if point B is recorded in a dataset 3 times, its effective weight would be 3, and instead of 6 points there would be 8 to consider. Co-located point recording is very common, especially in crime and medical datasets where each incident is associated with a nominal rather than precise location. Examples of co-located data might be the closest street intersection to an incident, the nominal coordinates of a shopping mall, the location of the doctor's surgery where a patient is registered, or a location that recurs because incidents or cases have been rounded to the nearest 50 metres for data protection reasons. It is often a good idea to execute queries looking for duplicate and unweighted locations before conducting analyses of this type, since mapped datasets may not reveal the true underlying patterns. Likewise it is important to check that co-located data are meaningful for the analysis to be undertaken — surgery location is not generally a substitute for a patient's home address.

The preceding calculations have all been carried out using the standard Euclidean metric, d_E i.e. L_2. As previously stated, this is the standard for GIS packages, but other metrics may be more appropriate depending on the problem at hand (see further, Section 4.4.1). Specialist packages like LOLA and Crimestat support a range of other metrics. Using the city block metric (L_1) and the minimax metric (L_∞) the MAT point, M6, is no longer guaranteed to be unique — all points within a defined region may be equally close to the input point set (Crimestat only provides a single location, so LOLA or similar facilities are preferable for such computations). For example, the point set in Figure 4-9 with L_1

metric has an MAT solution point set (a rectangle) bounded by (4,4) and (10,8). Clearly, if the point set lay on a network the location of M6 would again been different.

To add to the confusion of the above Crimestat provides three further measures of centrality for point sets. Each involves variations on the way the mean is calculated (see further, Table 1.4): the geometric mean (the x- and y-coordinates are calculated using the sum of the logarithms of the coordinates, averaging and then taking antilogs); the harmonic mean (the x- and y-coordinates are calculated using the reciprocal of the coordinates, averaging and then taking the reciprocal of the result); and the triangulated mean (a Crimestat "special"). The first two alternative means are less sensitive to outliers (extreme values) than the conventional mean centre, whilst the latter measure is claimed to represent the directionality of the data better. The Crimestat manual, Chapter 4, provides more details of each measure, with examples. Note that the harmonic mean is vulnerable to coordinate values which are 0 or close to 0.

4.2.5.3 Lines

In the case of a line (single segment, polyline or curve) the common notion of the centre is simply the point on the line that is equidistant from the two endpoints. This provides both the (intrinsic) centre of gravity and the mean centre, although when a polyline is viewed as embedded in a plane its "centre" should be considered to lie in the plane and not on the line.

For collections of lines there is no generally applied formula and a common central point might be selected from the centres or centroids of the individual elements, or from the MBR, or utilising a central feature if one exists. Combinations of points, lines and polygons are treated in a similar manner. Because lines and line segments have a well-defined orientation, labelling tools that utilise line or segment centroids may also use the orientation of that element to align associated labels.

4.2.6 Point (object) in polygon (PIP)

One of the most basic of spatial operations is that of determining whether a given point lies inside a polygon. More generally the problem extends to the case of multiple points and polygons, with the problem being to assign every point to the correct polygon. Related problems include line in polygon and polygon in polygon tests. A fast initial screening approach is to check whether a point (or line, polygon) lies in the polygon's MBR. However, the standard algorithm for determining PIP in the vector model (known as the semi-line algorithm) is to extend a line vertically upwards (or horizontally, e.g. to the right) and then count the number of times this line crosses the polygon boundary. If the line crosses the boundary an odd number of times it lies inside the polygon. This is true even when the polygon is concave or contains holes. A number of special cases need to be considered, as illustrated in Figure 4-10. These include points that lie on the boundary or at a vertex, and points that lie directly below a vertical segment of the polygon boundary. A useful discussion of PIP algorithms and related procedures is provided in Worboys and Duckham (2004, pp 197-202).

If the given point lies on the boundary or at a vertex of the polygon there is a further difficulty, since a unique assignment rule could prevent the point from being allocated to any polygon. Solutions to such cases include: assigning the point to the first polygon tested and not to any subsequent polygons; randomly assigning the point to a unique polygon from the set whose boundary it lies on; reporting an exception (non-assignment); assigning a weight value to the point (e.g. 0.5 on a boundary or $1/n$ at a vertex, where n is the number of polygons meeting at that vertex).

Figure 4-10 Point in polygon — tests and special cases

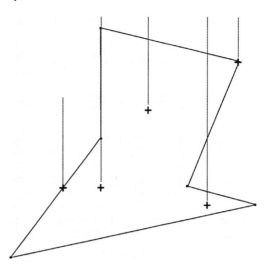

Where the PIP process is one of overlaying, the procedure within a GIS will typically append attribute data from the polygon to the point, although this may be in accordance with user-selectable transfer rules (e.g. sum or average a particular field). ArcGIS includes two PIP-related topology rules: (i) point must be properly inside polygon; and (ii) point must be covered by boundary of polygon (i.e. must be exactly on the boundary). These are two of 16 main vector topology rules supported within ArcGIS for combinations of point, line and polygon features.

A second PIP algorithm of similar computational complexity, but requiring trigonometric operations in its standard form (so more processor intensive) is known as the *winding number* method. In this procedure a line is extended to each vertex of the polygon from the sample point in turn (traversing anticlockwise). If the sum of the angles from the point to the vertices, v_i, is 0 the point lies outside the polygon, otherwise it lies inside or on the polygon. To clarify this procedure, let the polygon have n vertices $v_i = v_1, v_2, v_3, \ldots v_n = v_0$ and consider a unit vector from a sample point, p, to each vertex in turn. Define the (signed) angle at p between the vectors v_i and v_{i+1}, as θ_i. Then compute the winding number:

$$wn = \sum_{i=0}^{n-1} \theta_i \text{ , or}$$

$$wn = \sum_{i=0}^{n-1} \cos^{-1}\left(\frac{\mathbf{pv}_i \cdot \mathbf{pv}_{i+1}}{\|\mathbf{pv}_i\|\|\mathbf{pv}_{i+1}\|} \right)$$

If $wn=0$ the point lies outside the polygon, otherwise it lies inside or on the polygon boundary. With some adjustment the algorithm can avoid calculating the angle and result in a computation that is as fast, or faster than crossing methods (tests suggest 20-30% faster for many non-convex polygons). It is also more general, in that it applies to cases where polygons are not simple, i.e. self-crossing of the polygon boundary is permitted.

The above addresses the issue for the case of a single polygon and for multiple non-overlapping contiguous polygons. With overlapping polygons, which is supported for example within the Manifold GIS, then unique allocation is not assured. Furthermore, if the sample region is not completely covered by polygons sample points may fall into no polygon. Similar issues arise with raster representation of points and polygons (as zones of continuous fields) but the problem of locating points inside zones is far simpler, whilst the allocation issues remains.

4.2.7 Polygon decomposition

Polygons may be divided into smaller sections in many different ways, for a wide variety of reasons. Most vector GIS packages support some decomposition procedures, through overlay of a template or *cookie cutter* polygons (e.g. the Clip operation in ArcGIS) and/or through the provision of explicit decomposition procedures. For example, complex polygons with convoluted boundaries may be divided up into a series of non-overlapping convex polygons. These polygons will have centres (e.g. centres of gravity) that are guaranteed to lie inside the original polygon. The multiple centres thus generated may then be used to assist labelling or in assigning data in spatial overlay operations. Polygons can also be divided into non-overlapping triangles, in a variety of different ways. Manifold, for example, provides commands to perform both such decompositions: Transform|Decompose to convex parts, and Transform|Decompose to triangles.

A (simple) polygon with n vertices or nodes can be triangulated into n-2 triangles using a procedure known as diagonal triangulation. This form of triangulation utilises existing polygon vertices, but other triangulations are possible if additional internal vertices are permitted. With m such internal vertices (or Steiner points) a total of $n+2m$-2 triangles will be generated.

A polygon can also be *skeletonised*, providing an additional view of its form and key internal points. In Figure 4-11 bands of increasing distance in constant steps from the external boundary are shown. These terminate in a single central point and generate one intermediate point. Connecting these internal points to the polygon nodes creates a skeleton of lines known as the *medial axis* of the polygon. The medial axis is not necessarily comprised of straight line segments. In practice skeletonisation is carried out using algorithms known generically as *medial axis transforms*, and may be computed in linear time. The single central point in the example shown in Figure 4-11 is the centre of the largest inscribed circle.

Applications of skeletonisation include label placement, especially useful for complex polygons which are concave and/or contain holes and polygon generalisation/smoothing, which has been applied to closed polygons and to open regions such as contour maps – see for example Gold and Thibault (2001).

Figure 4-11 Skeletonised convex polygon

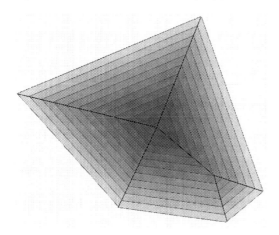

The reverse process, "given a skeleton (e.g. stream pattern) reconstruct a landscape", has also been the subject of much research. Furthermore, the dual of skeletonisation is one form of polygon decomposition, although not generally the version implemented within currently available GIS software. Neither does such software tend to provide skeletonisation facilities, but asymmetric buffering (see Section 4.4.4), which is a commonly available feature, enables first-stage skeletonisation to be carried out.

Raster-based spatial analysis makes limited use of region centres, where they occur mainly in connection with patch analysis. Patches are typically collections of contiguous cells that have the same or similar attribute values, although in some instances a patch may include holes. The commonest notion of centre in such cases is the unweighted or weighted mean value of the cell coordinates of the patch under consideration. In cases where cell values (weights) are counts of points falling in each cell the weighted mean can be regarded as equivalent to the simple mean value (minimum of squared Euclidean distances) for point patterns.

4.2.8 Shape

A number of the sections in this Guide have touched on the issue of polygon shape: how this affects where the centre lies; how compactness of polygon form may be related to standard shapes such as a circle or rectangle; and how polygons may be decomposed into simpler shapes, such as triangles. In practice defining some form of shape measure that provides an adequate description is difficult and many indices and procedures have been proposed. Shape measures may be applied to polygon forms or to grid patches (sets of contiguous grid cells that have common or similar attributes or are regarded as being a patch for analytical purposes). Measures relating to grid structures are described in Section 5.3.4, Landscape Metrics. The comments in this latter section apply equally well to shape measures for polygonal (vector) regions:

"The most common measures of shape complexity are based on the relative amount of perimeter per unit area, usually indexed in terms of a perimeter-to-area ratio, or as a fractal dimension, and often standardized to a simple Euclidean shape (e.g., circle or square). The interpretation varies among the various shape metrics, but in general, higher values mean greater shape complexity or greater departure from simple Euclidean geometry."

Vector-based GIS software packages do not normally provide shape index measures directly – it is up to the user to create index values. In many instances such values are functions of the area and perimeter of the polygon, and may be generated as a new calculated field from these (intrinsic or explicit) attributes. If we define A_i as the Area of polygon i, and L_i as its perimeter length, and B_i as the area of a circle with perimeter L_i, then example measures include:

Perimeter/Area ratio (P1A):

$$P1A_i = \frac{L_i}{A_i}$$

Note that this ratio is unsatisfactory for most applications as its value changes with the size of the figure (i.e. irrespective of resolution issues).

Perimeter²/Area ratio (P2A):

$$P2A_i = \frac{L_i^{2}}{A_i}$$

This second ratio was used in the UK 2001 census Output Area generation exercise (see further, Section 4.2.11). If preferred, this index, or its square root may be adjusted so that its value equals 1 when applied to a circle of the same area as A_i.

Shape Index or Compactness ratio (C):

$$C_i = \sqrt{\frac{A_i}{B_i}}$$

Note that this ratio is dimensionless (i.e. not affected by the size of the polygon) and has a value of 1 for a circular region and a range of [0,1] for all plane shapes. Note that sometimes this index is computed as (1-C_i]. Grid-based packages (e.g. Idrisi, CRATIO command) and specialised pattern-analysis software (e.g. Fragstats) may compute ratios of this type as inbuilt functions.

Related bounding figure (RBF):

$$RBF_i = 1 - \frac{A_i}{F_i}$$

where F_i is the area of a bounding figure. Typically the bounding figure will be the minimum bounding circle, but could equally well be the MBR or convex hull. In each case the ratio lies in the range [0,1], with 0 being the value if the polygon matches the bounding

figure. When index values of this type are computed for multiple polygons (e.g. as a new column in an attribute table) a *shape distribution* is created, that may be examined with conventional univariate statistics (e.g. mean, median, standard deviation, range etc.) and/or graphically examined as a histogram.

More complex shape measures, such as those based on multiple characteristics of polygon form (e.g. the length of the major and minor internal axes, or the set of possible straight lines computable within the polygon) are more difficult to compute. At present all such measures typically require separate programmed computation, either within a GIS environment or externally.

Shape measures may also be computed for other vector objects, such as sets of points or lines. In the case of point sets the common procedure is to compute an RBF-type measure, i.e. essentially generating a polygon such as the convex hull of a cluster of points and analysing its form. The standard deviational ellipse (see further, Section 4.5.3 and Section 5.4) also provides an option for point-set shape analysis, utilising the ellipse area and major and minor axes.

For linear objects or collections of linear objects, many options exist, a number of which are directly supported within standard GIS packages, particularly those with hydrological analysis components (e.g. providing ratios such as stream length/basin area). Many such measures are indirect measures of linear component shapes. Perhaps the simplest measure for individual polylines is to compute the ratio of polyline length to the Euclidean (or spherical) distance between start and end points. This kind of measure is highly dependent on scale, line representation and point pair selection.

A *circuity index* (devised by Kansky) for transport networks based on this kind of idea involves comparing the inter-vertex network distance, *N*, with the Euclidean distance, *E*:

$$I = \frac{1}{V} \sum_{i=1}^{n} (N_i - E_i)^2$$

where V is the number of vertices in the network and n is the number of distinct pairwise connections, i.e. $n=V(V-1)/2$. This overall index can be disaggregated into separate indices for each vertex, providing separate measures of vertex based circuity. If information is available on user preferences for travel between vertex pairs, then so-called "desire lines" can be utilised in index computations of this kind (e.g. as weightings). As with many of the other measures described in this subsection the computation of index components and totals must utilise generic calculation facilities within the GIS or be computed externally.

4.2.9 Overlay and combination operations

Overlay operations involve the placement of one map layer (set of features) A, on top of a second map layer, B, to create a map layer, C, that is some combination of A and B. C is normally a new layer, but may be a modification of B. Layer A in a vector GIS will consist of points, lines and/or polygons, whilst layer B will normally consist of polygons. All objects are generally assumed to have planar enforcement, and the resulting object set or layer must also have planar enforcement. The general term for such operations is topological overlay, although a variety of terminology is used by different GIS suppliers, as we shall see below. In raster GIS layers A and B are both grids, which should have a common origin and orientation — if not, resampling is required.

The process of overlaying map layers has some similarity with point set theory, but a large number of variations have been devised and implemented in different GIS packages. The principal operations have previously been outlined as the spatial analysis component of the OGC simple features specification (Table 4.2). The OpenSource package, GRASS, is a typical example of a GIS that provides an implementation of polygon overlay which is very similar to conventional point set theory (Figure 4-12), with functions provided including:

- *Intersection*, where the result includes all those polygon parts that occur in both A and B

- *Union*, where the result includes all those polygon parts that occur in either A or B, so is the sum of all the parts of both A and B

- *Not*, where the result includes only those polygon parts that occur in A but not in B (sometimes described as a Difference operation), and

- *Exclusive or* (XOR), which includes polygons that occur in A or B but not both, so is the same as (A Union B) minus (A Intersection B)

Note: (a) these operations may split up many of the original polygons (and/or lines) to create a new set of features, each with their own (inherited and derived) attributes (lengths, areas, perimeters, centroids, and associated attributes from A and/or B). New composite regions (with associated combined attribute datasets) are often the desired end result, in which case borders between adjacent polygons may be dissolved and datasets merged in order to achieve this end result; (b) operations are not generally symmetric, i.e. A<operation>B is not necessarily the same as B<operation>A, both from a point of view of the geometry and associated attribute data held in tables.

TNTMips provides similar functionality and uses much the same terminology as GRASS (AND, OR, XOR, SUBTRACT) under the heading of vector combinations rather than overlay operations, and permits lines as well as polygons as the "operator" layer (Figure 4-12).

Figure 4-12 GRASS overlay operations, v.overlay

Source: *http://grass.itc.it/grass60/screenshots/vector.php*

In ArcGIS operations of this type are known by the general term "Overlay", whilst in Manifold V6 the ArcGIS Overlay operations are described as *Topological Overlays* to distinguish these rather more composite operations from the more component based operations that Manifold describes as *Spatial Overlays*. Topological overlays either modify the target set or create a new set. In the former case the source set retains all of its existing fields (columns) and acquires all columns from the target, whilst in the latter case all fields of both source and target are included in the new set. Other vector GIS packages use descriptions such as cookie cutters (or EXTRACT operations) and *spatial joins*, and without doubt the terminology is confusing — the exact consequences of such operations applied using any particular GIS package will be different and should be examined very carefully. The ArcGIS implementation provides four main Overlay functions: Intersection and Union, which are essentially similar to those illustrated in Figure 4-12, and two additional operations: Identity, which splits up the

elements in *B* based on the position of elements in *A*; and Update, which amends those parts of *B* that intersect with A such that in these regions *B*'s geometry and associated data reflects those of *A*.

The Manifold GIS generalises the notion of overlay to include spatial proximity, for example lines that touch areas, and areas that are adjacent to (immediate neighbours of) other areas. The full set of permitted methods is shown in Table 4.3. Manifold also permits polygons within a layer to be overlapping, which can result in more complex output patterns. In all cases the Manifold Spatial Overlay process transfers data (fields) from a source object (set of features of the same type) to a target object (also a set of features of the same type) according to transfer rules. These rules specify how the data fields are to be transferred (e.g. copied, summed) conditional upon the spatial overlay method. As such they are rather like an SQL "Update table" or "Create table" command, subject to "Where" conditions.

Table 4.3 Spatial overlay methods, Manifold GIS

	Contained	Containing	Intersecting	Boundary	Touching	Neighbouring
Areas, A	A,L,P	A	A,L	P	A,L	A,L
Lines, L	L,P	A,L	A,L	P	A,L	A,L
Points, P		A,L			P	A,L

Polygon-on-polygon overlay, which is a relatively common procedure, frequently results in the creation of very small thin polygons known as *slivers*. These may be the genuine result of an overlay operation (i.e. valid objects and data) or they may be artefacts, created as a result of differences in the original data capture, manipulation or storage process associated with *A* and *B*, even though they should exactly match in places. Such slivers may be removed automatically during the overlay operation, usually by setting tolerance levels, or by post-processing, e.g. removing all polygons with a width less than a tolerance value and replacing these with an arc along the sliver centreline or assigning them to the largest adjacent polygon (e.g. in

ArcGIS use the ELIMINATE function in ArcToolbox; in Manifold use the Normalize Topology function).

Overlays involve operations where two distinct sets of spatial object are combined in particular manner, visualised as overlaying one upon the other. Many other forms of combination are possible. For example, with a single layer of polygons, boundaries between polygons could be removed (*dissolved*) if the polygons are adjacent/share a common boundary and meet certain criteria. The attributes of the newly formed polygon will be the sum or some other measure of the component parts. If a new map layer is created by this process rather than an existing layer modified the process is sometimes described as *merging* rather than dissolving.

Manifold permits the use of the Dissolve operation on lines and points as well as polygons. In the case of lines, multiple line segments are replaced with a single line segment, whilst with points multiple points are replaced by a single centre point. The attribute used to guide the dissolve process may be transferred to the new object(s) as: Blank (i.e. do not transfer values for this field); Average; Count; Maximum; Minimum; Sample (randomly chosen value from the original object); or Sum. Other fields are transferred according to transfer rules set for each field.

ArcGIS supports two additional composite operations: Erase and Symmetric difference. In the Erase operation objects in A overlaid on B result in the removal of those parts of B that intersect with A (essentially the NOT operation of GRASS). Symmetric difference is similar to Erase, but those parts of A that do not intersect with B are included in the output set also (essentially the XOR operation we described earlier for GRASS). In addition ArcGIS supports operations such as Split, Append (a merge operation) and Integrate, which combine or separate features according to well-defined rules. Collectively within ArcGIS these various facilities are described as Geoprocessing Tools, and in ArcGIS V8 were available via a "Geoprocessing Wizard" but in ArcGIS V9 are provided in ArcToolbox as separately identified functions.

4.2.10 Areal interpolation

The polygon-on-polygon overlay process highlights an important problem: "how should attributes be assigned from input polygons to newly created polygons whose size and shape differ from the original set?" This is a specific example of a closely related problem, that of re-districting. Suppose that we have a set of non-overlapping polygons that represent census tracts, and a second set of polygons that represent hospital service districts, generated (for example) as simple Voronoi

polygons around each hospital (see further, Section 4.2.14.3). We wish to estimate the population numbers and age/sex mix for the service districts. Based on this information we might assign budgets to the hospitals or iteratively adjust the service districts to provide a more even set of service areas (see further, Section 4.2.11).

In a simple overlay procedure we calculate the proportion of each census district that intersects with each separate service zone (by area) and then assign this proportion of the relevant attribute to the hospital in question. This assumes that the distribution of each attribute is constant, or uniform, within each census tract while also ensuring that the total counts for each attribute remain consistent with the census figures (so-called volume-preserving or *pycnophylactic* assignment). The standard intersection operator in many GIS packages will not carry out such proportional assignment, but simply carry over the source attributes to the target polygons. Assuming the source attribute table includes an explicit area value, and the intersected target provides an explicit or intrinsic measure of the intersection areas, then the proportional allocation procedure may be carried out by adding a calculated field containing the necessary adjustments. Note that this discussion assumes spatially extensive attributes (see Section 2.2.2) whereas with spatially intensive attributes alternative procedures must be applied (for example, initially using kernel density estimation — see Section 4.3.4 for more details).

Figure 4-13 and Figure 4-14 illustrate this process for test UK census Output Areas in Manchester. The population totals for each source area are shown in Figure 4-13, together with the area of overlap with a sample 1 km square region and a selected source polygon intersection (highlighted in red/darker grey).

Figure 4-13 Areal interpolation from census areas to a single grid cell

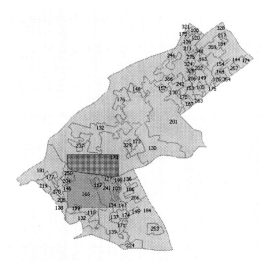

Figure 4-14 Proportionally assigned population values

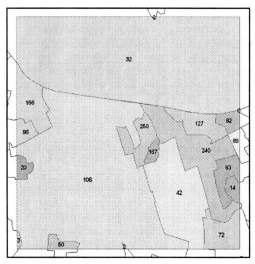

Figure 4-14 shows the result of applying area-based proportional assignment to the square region in question. The region highlighted in red/darker grey in the Figure 4-13 has its population estimate of 173 reduced to 32 in this case, with the sum of these proportional assignments (1632) providing the estimated total population for the entire square.

Variants of this procedure attempt to correct for the often unrealistic assumption of uniformity in the spatial distribution within polygons. If data are available for smaller sample regions (e.g. unit postcode areas in the above example) then this may be utilised in a similar manner to that already described. Another alternative is to model the distribution of the variable of interest (e.g. population) as a continuous surface. Essentially this procedure assigns the attribute value of interest to a suitable polygon centre and then calculates the attribute value over a fine grid within the polygon by taking into account the values of the attribute at adjacent or nearby polygon centres.

For example, if the source attribute values were much higher east of the target polygon than west, the values assigned within the source polygon might be assumed to show a slope from east to west rather than be distributed uniformly. Having estimated values for the *source* polygons on the fine grid, these values are then summed for each target polygon and adjusted to make sure that the totals match those for the original polygon. The individual grid cells may then be re-assigned to the *target* polygons (the hospital service areas in our example) giving a hopefully more accurate picture of the attribute values for these service areas. Current mainstream GIS packages do not tend to support the latter procedure, although simple sequences of operations, scripts or programs can be written to make such field assignments. The SURPOP online software utility is an example of such a program. Details of the algorithm applied and software may be obtained from the UK Census Dissemination Unit (CDU). In this instance the population, P_i, of each grid cell is approximated as:

$$\hat{P}_i = \sum_{j=1}^{N} w_{ij} P_j$$

where P_j is the population of "centroid" j, and w_{ij} is the weight applied to centroid j for grid cell i. The number N is determined by the windowsize used, i.e. the size of a moving NxN grid cell region. The weights distribute a proportion of each centroid's population value to the cells within the window and are determined as a form of distance decay function (see also, Section 4.4.6) with a decay parameter, α:

$$w_{ij} = \left(\frac{d^2 - s_{ij}^2}{d^2 + s_{ij}^2} \right)^\alpha$$

where d is the average inter-centroid distance within the sampled window, and s_{ij} is the distance between cell i and centroid j. As a result of tests on UK enumeration districts, typically α=1 in this model. The values chosen for the grid resolution, windowsize and decay parameter, together with the selection of centroid locations, all affect the resulting surface model produced.

The above, adjusted volume-preserving assignment approaches, have significant drawbacks where the geography of the study region is highly variable and/or where attributes do not vary in a similar manner (do not exhibit strong positive covariance). For example, population is typically concentrated in urban areas and not spread evenly across arbitrary zones. In urban-rural borders and rural areas with distinct settlements a far better approach would be to utilise ancillary information, such as land use, road network density (by type) or remote sensed imagery, to adjust attribute allocation. This ancillary data, which is often selected on the basis that it is readily available in a convenient form for use within GIS packages, can then be used as a form of weighting. Recent tests (see for example, Hawley and Moellering, 2005) have shown that the quality of areal interpolation can be substantially improved by such methods.

4.2.11 Districting and re-districting

A common analysis problem involves the combination of many small zones (typically stored as polygons) into a smaller number of merged larger zones or districts. This merging process is usually subject to a set of spatial and attribute-related constraints. Spatial constraints might be:

1. districts must be comprised of adjacent (coterminous) regions

2. districts must be sensible shapes e.g. reasonably compact

Attribute constraints might be:

1. no district may have less than 100 people

2. all districts must have a similar number of people (e.g. within a target range)

Several GIS packages now facilitate the automatic or semi-automatic creation of districts, or the re-organisation of existing districts. These operations have many applications, from the designation of service areas to the definition (and re-definition) of census and electoral districts. For example, within Manifold (with the Business Tools extension) there are facilities to enable an existing set of districts to be re-assigned automatically to one of N new districts, such that each new district has a similar (balanced) value by area (an intrinsic attribute) or by a selected attribute (optionally weighted). This automated process may yield unsatisfactory results (for example, very convoluted districts) and an alternative, visual interface facility, is provided to enable the user to make manual alterations and see the results in terms of the attribute or attributes of interest. In the case of ArcGIS Districting is provided as a free downloadable add-in, and is essentially a manual operation supported by map and statistical windows.

Districting and re-districting are generally processes of agglomeration or construction.

The initial set of regions is reduced to a smaller set, according to selected rules. Automating this process involves a series of initial allocations, comparison of the results with the constraints and targets, and then re-allocation of selected smaller regions until the targets are met as closely as possible. As with many such problems, it may not be possible to prove that a given solution is the best possible (i.e. is optimal), but only that it is the best that has been found using the procedures adopted. An ArcGIS-compatible automated zone design system, ZDES, has been developed at the University of Leeds and may be downloaded from the associated web site (see details of web site addresses at the end of this Guide).

Creating new districts can be a confusing process. In addition to the kinds of issue we have already discussed there are two important affects to be aware of: scale (grouping or statistical) effects; and zoning arrangement effects. For example, consider the employment statistics shown in Table 4.4.

Areas *A* and *B* both contain a total of 100,000 people who are classified as either employed or not. In area *A* 10% of both Europeans and Asians are unemployed (i.e. equal proportions), and likewise in Area *B* we have equal proportions (this time 20% unemployed). So we expect that combining areas *A* and *B* will give us 200,000 people, with an equal proportion of Europeans and Asians unemployed (we would guess this to be 15%), but it is not the case — 13.6% of Europeans and 18.3% of Asians are seen to be unemployed! The reason for this unexpected result is that in Area *A* there are many more Europeans than Asians, so we are working from different total populations.

The second issue is due to the way in which voting and census areas are defined — their shape, and the way in which they are aggregated, affects the results and can even change which party is elected. This is not to say that a particular arrangement *will* have such effects, but that it is possible to deliberately produce a districting plan that

meets specific criteria, at least for a single attribute.

Table 4.4 Regional employment data — grouping affects

	Employed (000s)	Unemployed (000s)	Total (000s) (Unemployed %)
Area *A*			
European	81	9	90 (10%)
Asian	9	1	10 (10%)
Total	90	10	100 (10%)
Area *B*			
European	40	10	50 (20%)
Asian	40	10	50 (20%)
Total	80	20	100 (20%)
A* and *B			
European	121	19	140 (13.6%)
Asian	49	11	60 (18.3%)
Total	170	30	200 (15%)

Figure 4-15 illustrates this issue (which is sometimes referred to by the awkward name Modifiable Areal Unit Problem, or MAUP) for an idealised region consisting of 9 small voting districts. The individual zone, row, column and overall total number of voters are shown in diagram A, with a total of 1420 voters of whom roughly 56% will vote for the first party listed/the red party (R) and 44% for the second party listed/ the blue party (B). With 9 voting districts we expect roughly 5 to be won by the reds and 4 by the blues on a "first past the post" voting system (majority in a voting district wins the district), as is indeed the case in this example. However, if these zones are actually not the voting districts themselves, but combinations of adjacent zones are used to define the voting areas, then the results may be quite different. As diagrams B to F show, with a first past the post voting system then we could have a result in which every district was won by the reds (case C), to one in which 75% of the districts were won by the blues (case F). Note that the solutions shown are not unique, and several arrangements of adjacent zones will give the same voting results.

Figure 4-15 Grouping data — Zone arrangement effects on voting results

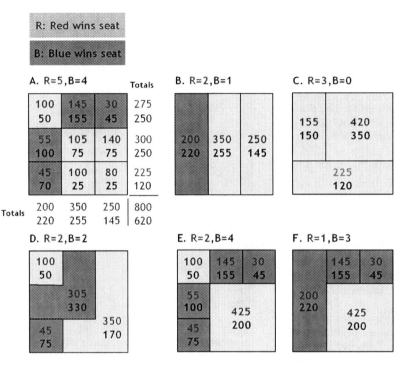

So it is not just the *process* of grouping that generates confusing results, but also the *pattern* of grouping, which is of great interest to those responsible for defining and revising electoral district boundaries. And this is not just a problem confined to voting patterns. For example, if the information being gathered relates to the proportions of trace metals (for example lead and zinc) in the soil, similar issues arise. Samples based on different field boundaries would show that in some arrangements the proportion of lead exceeded that of zinc, whilst other arrangements would show the opposite results.

In practice, for example in the commercial sphere, the MAUP may not be as significant an issue as it might appear. As Professor Richard Webber, the noted expert on geodemographics recently remarked (private communication): "I have yet to come across any real world example of a conclusion being invalidly reached as a result of this hypothetical possibility". A wise strategy is to examine the data at various levels of aggregation, including the lowest (finest level) possible, and check that scale and zoning effects are understood and are unlikely to distort interpretations.

Automated zoning was used to create the current UK 2001 Census of Population areas — see further Martin (2000) and Openshaw (1977). The procedure (known as AZP) was based on a 7-step approach, as follows:

- **Step 1:** Start by generating a random zoning system of N small zones into M regions, M<N

- **Step 2:** Make a list of the M regions.

- **Step 3:** Select and remove any region *K* at random from this list

- **Step 4:** Identify a set of zones bordering on members of region *K* that could be moved into region *K* without destroying the internal contiguity of the donor region(s)

- **Step 5:** Randomly select zones from this list until either there is a local improvement in the current value of the objective function or a move that is equivalently as good as the current best. Then make the move, update the list of candidate zones, and return to step 4 or else repeat step 5 until the list is exhausted

- **Step 6:** When the list for region *K* is exhausted return to step 3, select another region, and repeat steps 4-6

- **Step 7:** Repeat steps 2-6 until no further improving moves are made

In the UK case the initial small zones were taken as unit postcodes, which identify roughly 10-14 addresses. Unfortunately these individual addresses are stored as a list of georeferenced point data (nominal centres) rather than postcode areas, so the first stage of the analysis involved generating Voronoi polygons from these points, for the whole of the UK. These initial polygons were then merged to form unit postcode areas (Figure 4-16), based on the generated polygons and road alignments.

Once these basic building blocks had been constructed the automated zoning process could begin. This is illustrated as a series of assignments in Figure 4-17, staring with an arbitrary assignment of contiguous unit postcode areas (Figure 4-17a), and then progressively selecting such areas for possible assignment to an adjacent zone (e.g. the pink/darkest grey regions in Figure 4-17b and Figure 4-17e). Assignment rules included target population size and a measure of social homogeneity, coupled with a measure of shape compactness and enforced contiguity.

This process was then applied to a number of test areas across the UK, as is illustrated in the two scenarios, A and B, for part of Manchester, shown in Figure 4-18. These scenarios were as follows:

- A: Threshold population=100; Target population=150; Shape constraint (P2A)=(perimeter squared)/area; no social homogeneity constraint applied

- B: Threshold population=100; Target population=250; Shape constraint (P2A)=(perimeter squared)/area; social homogeneity constraint = tenure and dwelling type intra-area correlations

If scenario B is modified to exclude the shape constraint the assignment of areas is broadly similar, with detailed variations resulting in less compact final areas. Based on this methodology the Great Britain Ordnance Survey developed a full set of Output Areas (OAs) with a target population of 125 households with a minimum of 40. These OAs were used for the 2001 census enabling OA data and postcodes to be linked, and providing a set of OAs that can be grouped into higher level units, such as wards and counties, as required. A total of 175,434 OAs were created in this manner, with 37.5% containing 120-129 households, 79.6% containing 110-139 households, and just 5% containing 40-99 households.

Figure 4-16 Creating postcode polygons

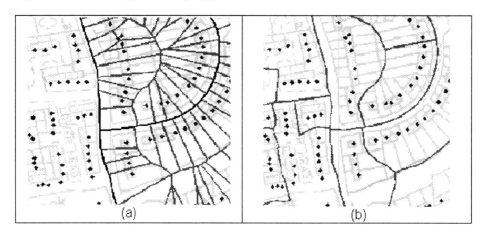

Figure 4-17 Automated Zone Procedure (AZP)

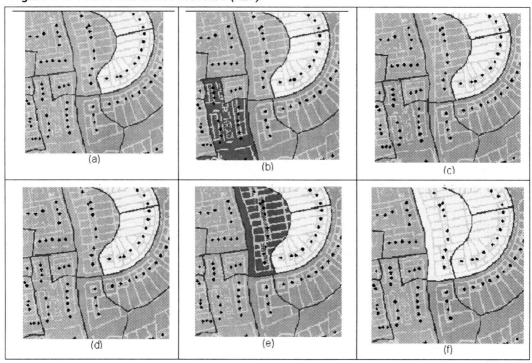

Figure 4-18 AZP applied to part of Manchester, UK

A. Target population 150	B. Target population 250

4.2.12 Classification and clustering

"Classification is, perhaps, *the* basic procedure by which we impose some sort of order and coherence upon the vast inflow of information from the real world." Harvey (1969)

Classification needs to be seen in terms of purpose as well as method. However, GIS software places almost no constraints upon a user's selection of classification method, so it is the responsibility of the analyst to choose appropriate procedures and settings that are fit for purpose. Classification is by no means a peculiarly spatial problem, and most methods apply to almost any kind of data, i.e. there is rarely a specific spatial aspect, such as a contiguity condition, applied. Such facilities are more likely to be found in the spatial autocorrelation and pattern analysis features of GIS toolsets.

4.2.12.1 Univariate classification schemes

Within GIS software classification facilities are found as: basic tools to aid in the production of choropleth or thematic maps; in the

exploration of untransformed or transformed datasets; in the analysis of image data (classification and re-classification); and in the display of continuous field data. These procedures perform classification based purely on the input dataset, without reference to separate external evaluation criteria. In almost all instances the objects to be classified are regarded as discrete, distinct items that can only reside in one class at a time. Separate schemes exist for classifying objects that have unclear boundaries (as discussed briefly in Section 4.2.13.4), or which require classification on the basis of multiple attributes (see Section 4.2.12.2). Typically the attributes used in classification have numerical values that are Real or Integer type. In most instances these numeric values represent interval or ratio-scaled variables. Purely nominal data are effectively already classified (see Section 2.2.2 for a brief discussion of commonly used scales).

Harvey (1969, Chapter 18) provides a very useful background to the field of classification, especially in a spatial context, but for the basics of classification within a GIS context Mitchell (1999, pp.46-55) and Longley *et al.* (2005, Ch.12) are probably a more useful starting point. These latter works address the question of how to classify and map attribute data associated with polygonal regions or provided as continuous fields. The manuals and online help facilities of the main GIS packages are also a good source of material for deciding which scheme might be the most appropriate for the dataset you have. Most of these schemes provide a range of standard alternatives for classifying univariate attributes, for example mapping population density by census tract. The attributes selected may be transformed in advance, for example: by producing spatially intensive variables such as densities rather than population totals; and/or by mathematically transforming data, e.g. by classifying log transformed data rather than the original dataset; and/or by normalising datasets (as discussed in Section 4.3.3). Alternatively the classification procedure selected may impose some form of normalisation on the untransformed data (e.g. standard deviation based classification).

Table 4.5 provides details of a number of such schemes, with comments on their use. Most of the main GIS packages provide classification options of the types listed, although some (such as the box and percentile methods) are only available in a limited number of software tools (e.g. GeoDa). A useful variant of the box method, known as hybrid equal interval in which the inter-quartile range is itself divided into equal intervals, does not appear to be implemented in mainstream GIS packages. Schemes that take into account spatial contiguity, such as the so-called minimum boundary error method described by Cromley (1996), also do not appear to be readily available as a standard means of classification.

Brewer and Pickle (2002) examine many of these methods, with particular attention being paid to ease of interpretation and comparison of map series, e.g. mapping time series data, for example of disease incidence by health service area over a number of years. They concluded that simple quantile methods were amongst the most effective, and had the great advantage of consistent legends for each map in the series. The latter was found to be very important in assisting choropleth map comparisons, whatever choice of method was made.

Table 4.5 Selected univariate classification schemes

Classification scheme	Description/application
Unique values	Each value is treated separately, for example mapped as a distinct colour
Manual classification	The analyst specifies the boundaries between classes required as a list, or specifies a lower bound and interval or lower and upper bound plus number of intervals required
Equal interval, Slice	The attribute values are divided into n classes with each interval having the same width=Range/n. For raster maps this operation is often called *slice*
Defined interval	A variant of manual and equal interval, in which the user defines each of the intervals required
Exponential interval	Intervals are selected so that the number of observations in each subsequent interval increases exponentially
Equal count or quantile	Intervals are selected so that the number of *observations* in each interval is the same. If each interval contains 25% of the observations the result is known as a quartile classification. Ideally the procedure should indicate the exact numbers assigned to each class, since they will rarely be exactly equal.
Percentile	Percentile plots are a variant of equal count or quantile plots. In the standard version equal percentages (percentiles) are included in each class. In GeoDa's implementation of percentile plots unequal numbers are assigned to provide

Classification scheme	Description/application
	classes that contain 6 intervals: ≤1%, 1% to <10%, 10% to <50%, 50% to <90%, 90% to <99% and ≥99%
Natural breaks	Widely used within GIS packages, these are forms of variance-minimisation classification. Breaks are typically uneven, and are selected to separate values where large changes in value occur. May be significantly affected by the number of classes selected and tends to have unusual class boundaries. Typically the method applied is due to Jenks, as described in Jenks and Caspall (1971), which in turn follows Fisher (1958). See also, Figure 4-19, for more details
Standard deviation	The mean and standard deviation of the attribute values are calculated, and values classified according to their deviation from the mean (z-transform). The transformed values are then mapped, usually at intervals of 1.0 or 0.5 standard deviations. Note that this results in no central class, only classes either side of the mean
Box	A variant of quartile classification designed to highlight outliers, due to Tukey (1977, Section 2C). Typically six classes are defined, these being the 4 quartiles, plus two further classifications based on outliers that may exist within the lower and upper quartiles. These outliers are defined as being data items (if any) that are more than 1.5 times the inter-quartile range (IQR) from the median. An even more restrictive set is defined by 3.0 the IQR. A slightly different formulation is sometimes used to determine these box ends or *hinge* values. Box plots (see Section 5.2.2.2) are implemented in GeoDa and STARS, but are not generally found in mainstream GIS software. They are commonly implemented in statistics packages, including the MATLab Statistics Toolbox

In addition to selection of the scheme to use, the number of breaks or intervals and the positioning of these breaks are fundamental decisions. The number of breaks is often selected as an odd value: 5, 7 or 9. With an even number of classes there is no central class, and with a number of classes less than 4 or 5 the level of detail obtained may be too limited. With more than 9 classes gradations may be too fine to distinguish key differences between zones, but this will depend a great deal on the data and the purpose of the classification being selected. In a number of cases the GIS package provides linked frequency diagrams with breaks identified and in some cases interactively adjustable. In other cases generation of frequency diagrams should be conducted in advance to help determine the ideal number of classes and type of classification to be used. For data with a very large range of values, smoothly varying across a range, a graduated set of classes and colouring may be entirely appropriate. Such classification schemes are common with field-like data and raster files.

Positioning of breaks may be pre-determined by the classification procedure (e.g. Jenks Natural breaks), but these are often manually adjustable for some of the schemes provided — this is particularly useful if specific values or intervals are preferred, such as whole numbers or some convenient values such as 1000s, or if comparisons are to be made across a series of maps. In some instances options are provided for dealing with zero-valued and/or missing data prior to classification, but if not provided the data should be inspected in advance and subsets selected prior to classification where necessary. Some authors recommend that maps with large numbers of zero or missing data values should have such regions identified in a class of their own.

The Jenks method, as implemented in a number of GIS packages such as ArcGIS, is not always well-documented so a brief description of the algorithm follows (Figure 4-19). This description also provides an initial model for some of the multivariate methods described in subsection 4.2.12.2.

Figure 4-19 Jenks Natural Breaks algorithm

Step 1: The user selects the attribute, x, to be classified and specifies the number of classes required, k

Step 2: A set of k-1 random or uniform values are generated in the range [min{x},max{x}]. These are used as initial class boundaries

Step 3: The mean values for each initial class are computed and the sum of squared deviations of class members from the mean values is computed. The total sum of squared deviations (TSSD) is recorded

Step 4: Individual values in each class are then systematically assigned to adjacent classes by adjusting the class boundaries to see if the TSSD can be reduced. This is an iterative process, which ends when improvement in TSSD falls below a threshold level, i.e. when the within class variance is as small as possible and between class variance is as large as possible. True optimisation is not assured. The entire process can be optionally repeated from Step 1 and TSSD values compared

4.2.12.2 Multivariate classification and clustering

There is an enormous variety of classification methods that utilise two or more attributes for class selection and assignment. Within GIS packages most such methods are based on some form of multivariate model in which objects are assigned to one of M classes in a manner that minimises some measure of within-class variation and/or maximises between-class variation. Typically they are non-hierarchical and are simply criteria-driven assignment algorithms rather than carefully constructed taxonomic classification systems. Essentially they provide grouping or clustering based on multiple attributes in order to identify and map regions that are similar (close) in n-dimensional attribute space.

Classification analysis of (spatial) datasets based purely on attribute information (i.e. ignoring spatial relationships) is best conducted using general purpose statistical and mathematical toolsets. Most such packages and code libraries provide a range of tools for this purpose. These include:

- facilities that can be used to reduce the complexity (dimensionality) of the source datasets, such as *factor analysis*, *principal components analysis* and *multidimensional scaling*

- facilities to identify clusters within the dataset, either on a single level for a fixed number of clusters (e.g. K-means clustering) or on a hierarchical basis (typically techniques that rely on some form of linkage function, usually based on a function of the simple n-dimensional Euclidean distance between points within separate clusters), and

- facilities for optimally assigning new observations to existing classes (e.g. *discriminant analysis*)

Some techniques, such as clustering based on multivariate analysis of variance, require that input datasets be Normally distributed, but many methods make no such assumptions. K-means analysis (and developments of this method) is one of the most widely used and commonly available in GIS packages, so we shall describe this method in more detail before reviewing the range of facilities available within GIS generally. Readers familiar with location-allocation methods (see further, Section 7.4) will also recognise the similarity with these techniques.

K-means clustering attempts to partition a multivariate dataset into K distinct (non-overlapping) clusters such that points within a cluster are as close as possible in multi-dimensional space, and as far away as possible from points in other clusters. The input dataset can be viewed as a set of objects with an associated set of n (typically real-valued) measures or attributes. Each object can then be viewed as a single point in n-dimensional space. The cluster procedure is then as follows:

- The method starts with a set of K initial cluster centres. These might be assigned: (i) as K random locations in n-space; (ii) as K uniformly sited locations; (iii) as a user-defined set of K locations; or (iv) as a set of K locations obtained by clustering a small subset of the data using random or uniform allocation. A minimum separation distance for starting points may be specified

- The distance between every point and each of the K means is computed, based on some prior defined metric (typically squared Euclidean, L_2^2, or City block, L_1, as these are both fast to compute – see further, Section 4.4.1). Points are assigned to the nearest centre. Any entirely empty clusters may be discarded and (optionally) a new start point added

- The location of the centre of each cluster is then re-computed based on the set of points allocated to that centre, and the previous step is then repeated. This sequence is iterated until no further reassignments occur, or a preset number of iterations have been completed, or no cluster centre is moved by more than a pre-specified small amount. The total distance (DSUM) of all points to the centres of the K clusters to which they have been assigned is calculated

- Each point is then re-examined in turn (Phase 2 of the process) and checked to see if DSUM is reduced if the point is assigned to another cluster. If DSUM is reduced the point is reassigned to the cluster that results in the maximum overall reduction

- The entire process is then repeated with a different choice of starting points. This may result in an overall reduction in the objective function, DSUM, since the algorithm is not guaranteed to provide an optimal partitioning

Within GIS software multivariate classification methods are almost exclusively applied to remote sensing datasets – typically multi-band image files. The objective is to identify distinct areas or features (collections of pixels or cells) such as forest, water, grassland and buildings and assign all occurrences of such features to distinct classes. This is either performed semi-automatically (supervised) or automatically (unsupervised). Most of the more sophisticated raster-based GIS packages provide such facilities, for example TNTMips (see the TNTMips Image Classification Guide) and Idrisi (see the Idrisi Guide to Image processing), as well as the more generic ArcGIS Spatial Analyst toolset and the specialised image processing suites, ERDAS from Leica and ENVI from RSI.

Harvey (1969, p.346) describes the key steps of such classification as follows:

1. quantitative analysis of the inter-relationships among the attributes or among the objects

2. transformation or reduction of the correlations to a geometric structure with known properties (usually Euclidean)

3. grouping or clustering of the objects or attributes on the basis of the distance measured in this transformed space, and

4. once the classes have been firmly identified, the development of rules for assigning phenomena to classes

Within GIS software the first of these steps involves multivariate statistical analysis of the input rasters and for supervised classification, any training datasets that the analyst has prepared. For example, manual analysis of selected images may be used to identify features of interest and assign polygonal regions to user-defined classes. The statistical characteristics of these regions (within and across rasters, sometimes described as *signatures*) can then be used to guide subsequent supervised or automated assignment. However, it may be necessary to normalise the datasets (to ensure the range of values in each band is similar), and/or to (dimensionally) transform the datasets in some way before proceeding to the formal clustering and assignment processes (for example using techniques such as principal components analysis, PCA, or factor analysis, FA).

As an example of the range of classification methods available in GIS we outline a number of methods provided with packages such as TNTMips, Idrisi, GRASS and ArcGIS Spatial Analyst (Table 4.6). Note that most GIS packages only support a subset of these methods, commonly including *K*-means, ISODATA and Maximum Likelihood (ML) procedures. With unsupervised classification it may be helpful to specify more clusters than you would ultimately like or expect and then combine some of those produced if they seem essentially the same in terms of your own objectives rather than purely computational factors. For example separate clusters may be produced for deciduous and coniferous vegetation cover, but if the objective is to classify areas as woodland without separating these by type, post classification combination may be effective. Details of many of the methods described may be found in Tou and Gonzales (1974, Chapter 3), Ball and Hall (1965) and Burrough and McDonnell (1998, Ch.11).

In addition to the methods described above, a number of fundamentally different models are becoming available; principally those based on so-called Artificial Neural Network concepts. These methods include Self Organisation, Adaptive Resonance and Back Propagation. Each of these techniques involves some form of training dataset and learning process, resulting in a set of class allocation weights or rules that seek to ensure the assignment of the cells in the entire dataset is as close as possible to that identified in the training dataset.

Table 4.6 Multivariate classification methods

Method (S=supervised, U=unsupervised)		Description
Simple one-class clustering	U	A technique that generates up to *M* clusters by assigning each input cell to the nearest cluster if its Euclidean distance is less than a given threshold. If not the cell becomes a new cluster centre. It principal merit is speed, but its quality of assignment may not be acceptable
K-means	U	Partition-based algorithm, as described above. Phase 2 of this process may be omitted
Fuzzy *c*-means (FCM)	U	Similar to the *K*-means procedure, originally developed by Dunn (1973) but uses weighted distances rather than unweighted distances. Weights are computed from prior analysis of sample data for a specified number of classes. These cluster centres then define the classes and all cells are assigned a membership weight for each cluster. The process then proceeds as for *K*-means but with distances weighted by the prior assigned membership coefficients (see also, Section 4.2.13.4)
Minimum distribution angle	U	An iterative procedure similar to *K*-means but instead of computing the distance from points to selected centres this method treats cell centres and data points as directed vectors from the origin. The angle between the data point and the cluster centre vector provides a measure of similarity of attribute mix (ignoring magnitude). This concept is similar to considering mixes of red and blue paint to produce purple. It is the proportions that matter rather than the amounts of paint used
ISODATA/ISOCluster (Iterative Self-Organising)	U	Again, similar to the *K*-means procedure but at each iteration the various clusters are examined to see if they would benefit from being combined or split, based on a number of criteria: (i) combination − if two cluster centres are closer than a pre-defined tolerance they are combined and a new mean of means calculated as the cluster centre; if the number of members in a cluster is below a given level the cluster is discarded and the members re-assigned to the closest cluster; and (ii) separation − if the number of members, or the standard deviation, or the average distance from the cluster centre exceed pre-defined values than the cluster may be split
Minimum distance to mean	S	Essentially the same as Simple one-pass clustering but cluster centres are pre-determined by analysis of a training dataset. Fast but subject to similar problems as the Simple method
Maximum likelihood	S	A method that uses statistical analysis (variance and covariance) of a training dataset, whose contents are assumed to be Normally distributed. It seeks to determine the probability (or likelihood) that a cell should be assigned to a particular cluster, with assignment being based on the Maximum Likelihood value computed.
Stepwise linear/Fisher	S	This is essentially a Discriminant Analysis method, which attempts to compute linear functions of the dataset variables that best explain or *discriminate* between values in a training dataset. New linear functions are added incrementally, orthogonal to each other, and then these functions are used to assign all data points to the classes. The criterion function minimised in such methods is usually Mahalanobis distance, or the D^2 function.

4.2.13 Boundaries and zone membership

As noted in the commentary on polygons, the definition and selection of boundaries is often a difficult, but nevertheless essential element in many areas of spatial analysis. Application areas include: modelling the real world; classification of regions; and pattern analysis.

One group of problems where this is especially important is in the analysis of point patterns — sets of points in the plane representing cases of a particular disease, or the location of ancient artefacts, or a particular species of plant. If the density of points (frequency of occurrence per unit area) is required then the total area under examination needs to be defined. Definition of this area, A, requires identification of a boundary, although there is now a variety of special procedures (known as kernel density methods) available for computation of densities without pre-defined fixed boundaries (see further, Section 4.3.4).

A further problem relates to the process of modelling the real world with abstracted linear forms (i.e. polylines and polygons). As noted elsewhere in this Guide, there are a variety of ways of modelling linear forms — mathematical, statistical, fractal, zonal — and each of these models will have implications for analysis. Data recorded in a GIS database, either as vector or grid elements, have already been modelled and to some extent the kinds of analyses that may be carried out are pre-determined. For example, most GIS packages do not provide facilities for analysing the complexity or "texture" of modelled lines and surfaces, although some, such as Idrisi and Fragstats do support such analyses by examining grid datasets at various scales/using variable sized windows. The ArcGIS add-in, Hawth's Tools, provides a *line metrics* tool for generating measures of estimated fractal dimension and sinuosity of stored polylines. For fractal dimension this software computes, for each polyline:

$$D_c = \frac{\log_e(n)}{\log_e(n) + \log_e(d/L)}$$

where n is the number of line segments that make up the polyline, d is the distance between the start and end points of the polyline, and L is the total length of the polyline, i.e. the cumulative length of all polyline segments. Sinuosity is computed as a simple ratio: d/L, for each polyline. In both cases the results are added as new fields to the feature attribute table.

4.2.13.1 Convex hulls

For point sets in the plane a simple and often useful boundary is the convex hull (Figure 4-20). This is taken as the convex polygon of smallest area that completely encloses the set (it may also be used for line and polygon sets).

Figure 4-20 Convex hull of sample point set

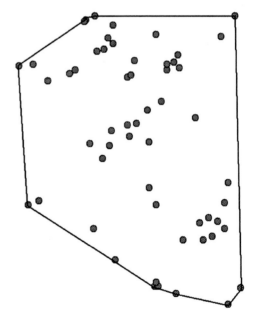

If the data you are working with consist solely of a single point set then the convex hull of this set encloses all the information you have available, even though you may know that the data extend beyond these limits.

The convex hull may also be useful in the analysis of optimum location and networking problems. For example the centre of gravity of a set of points in the plane will always lie inside or on the convex hull of the point set, and the sequence of points which comprises the convex hull form part of the solution to certain network routing problems (systematic minimal tours of a set of locations). However, it is important to note that the convex hull of a set of objects (points, lines etc.) is affected markedly by extreme values, so outliers and/or errors can have a substantial impact on the shape, area, perimeter length and centroid of a convex hull. For example, one or two extreme values can substantially increase the area enclosed by the convex hull.

By definition the convex hull includes a number of points as part of its boundary. For some problems this may be satisfactory whilst for others it may not be suitable. For example, when analysing the distribution of points in a study region it may be preferable to minimise the potential distortions to sample statistics by avoiding points that lie on or close to the convex hull. Creating an inner buffer of the convex hull and analysing the subset of points lying within the core of the region, with or without consideration of points lying in the buffer zone, may be preferable. Alternatively the convex hull may be uniformly increased (buffered) in size, for example by a given percentage of area or by a specified distance such as half the mean distance to the nearest-neighbours of enclosed points.

4.2.13.2 Non-convex hulls

The computation of convex hulls is a fast and useful facility, but does not necessarily match the requirements of the analyst attempting to define a region enclosing a set of objects. We have already noted, for example, that the method is strongly affected by the location of outliers, and as a result may enclose large regions that are of no interest or relevance to the analysis. Whilst the convex hull of a point set in the plane is unique, any number and variety of non-convex hulls may be generated. Such alternative hulls may meet certain pre-defined application-related criteria. Examples might include: the hull must be a polygon; the polygon must be as near convex as possible; the polygon must have the smallest possible area enclosing the set as possible; the polygon must reflect the density of points in the sample; the polygon must assign each point its own nominal region and include all points.

There are three principal methods for generating non-convex polygonal hulls (NCPHs) that meet specified criteria. These are: (a) expansion; (b) contraction; and (c) density contouring. Some of these methods may be implemented using most GIS packages, but the majority require some programming. TNTMips includes facilities for computing several of the variants described within its Polygon Fitting facility. These algorithms were originally developed to help identify animal "home ranges" but may be utilised in many areas — see for example, Harvey and Barbour (1965), White and Garrott (1990, pp145-82, "Home range estimation") and Mitchell (2006).

In case (a), expansion, the non-convex hull is created by assigning each point a surrounding region and then growing this region until all points are covered and a continuous external hull is obtained. This may be achieved in several ways. One method is to compute Voronoi polygons for all the points and use the external boundary of the (finite) set as the NCPH (see for example Figure 4-25B). This leaves the problem of how to treat the points that make up the convex hull. An alternative non-polygonal approach is to buffer all the points and increase the buffer width until a single merged zone is formed. This process may highlight the need to consider the benefits of using several sub-regions rather than a single region — for example, in cases where many of the observations lie near the region boundary with a large central region containing few observations. Yet another alternative is to overlay the study region with a rectangular grid and define the NCPH with reference to this (for example, requiring that the grid includes every point and a minimum number of cells surrounding every point). Where necessary interstitial gaps in the grid are grown to ensure a contiguous overall region is achieved without necessarily growing the external boundary.

Case (b), contraction, involves reduction of the convex hull according to selected rules. The simplest of these is a systematic area minimisation procedure. First the area of the convex polygon is computed. Next a point on the convex hull is removed and the new convex hull recomputed. The area of this new region is calculated and stored. The process is repeated for all points on the original convex hull and the point that results in the greatest decrease in area is permanently removed. The procedure then iterates until only a pre-defined percentage of the original points remain (e.g. 90%) or the area has been reduced by a given percentage. The result is a convex hull of an optimised subset of the original points. A second procedure, which results in a NCPH, involves a form of shrinkage of the convex hull around the point set, rather like a plastic bag full of small balls having the air removed. In this method the longest linear segment of the convex hull is selected and replaced with two segments connecting the original two end points via an intermediate point that is the closest to the original line — almost the reverse of point-weeding (Section 4.2.4). The process continues until the overall area enclosed has been reduced by a pre-determined amount, or a number of iterations have been reached — 10 is a typical number to use. Hybridisation of this method with the preceding method is possible.

Note that in the majority of contraction and expansion methods the result is either a grid or polygon form, with the latter having straight line edges. Procedures also exist in which the restriction to straight lines is relaxed, and some degree of edge curvature is permitted, e.g. forming so-called *alpha hulls* where concavities of curvature alpha are permitted. Alpha hulls are generalisations of the concept of a convex hull and are defined by the intersection of a set of discs of radius $1/\alpha$ placed over all the points in the set. Different hull shapes are formed depending on the sign and size of the parameter α. With $\alpha<0$ the hull will exhibit concavity, and as α tends to 0 from above or below the alpha hull will approach the convex hull. Conversely, as α increases the point set hull is characterised by increasing indentations, and ultimately the hull shrinks to nothing. Specialised tools exist for constructing such hulls in 2D and 3D and are of particular application in visualisation of environmental and similar datasets.

Case (c), density contouring, involves overlaying the study region with a (fine) grid and then computing an estimated point density at each grid point. There are several ways in which the density can be estimated, but the most widely used within GIS and related packages is kernel estimation (discussed in some detail in Section 4.3.4). Once a density value has been assigned to each intersection or cell, the region taken for the NPCH may be determined by specifying a threshold density and using the grid directly, or by contouring the grid. The final result may then be used directly or converted to polygonal form, depending on the requirement.

4.2.13.3 Minimum Bounding Rectangles (MBRs)

One difficulty with convex hulls is that they are slightly awkward shapes to work with, and in many cases it is simpler to deal with a boundary that is a regular shape, such as a circle or rectangle. Rectangles that align with the coordinate system and enclose selected features compactly are known as Minimum Bounding Rectangles, or Minimum Enclosing (or Envelope) Rectangles (MBRs or MERs). It is very fast to determine whether the MBR for one feature or set of features is completely contained within a study region or within the MBR of another set of features. Identifying where the centre of an MBR lies is also very simple and thus they are useful for first-stage identification of simple overlays and membership. MBRs and convex hulls provide an indication of the limits to the area over which you can reliably interpolate values and many GIS packages take one or other of these boundaries as the limit for procedures such as interpolation, generation of contour lines or the creation of triangulated irregular networks (TINs).

Figure 4-21 shows the result of interpolating income per capita data to a fine grid using the census map shown earlier in Figure 4-6 (darker areas have higher per capita income). In this case the ArcGIS Spatial Analyst facility has been used, and as can be seen the interpolation process extends to the MBR of

census tract centres and no further. We can also see, in this example: (i) the interpolation process replaces census tracts with a grid surface of values, comprising around 250x500 cells, which we have classified into 9 groups for display purposes; (ii) the generated surface pattern is closely tied to "centroid" locations; and (iii) interpolation has occurred for our rogue locations, including those that should have been screened out as being lakes (no data) *prior* to interpolation, and two small polygons which are classified as land but for which no census data are provided.

Figure 4-21 Interpolation within "centroid" MBR

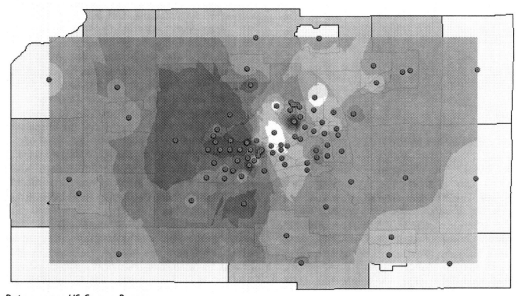

Data source: US Census Bureau

MBRs and other simple shapes are provided as options for feature enclosure in many GIS packages and in more specialised toolsets such as Crimestat (which is concerned with point events). Frequently the aim is to select a boundary which is reasonable and meaningful for subsequent analysis, often analyses that involve statistical calculations that rely on knowing the point density in advance. As boundaries are altered, the area, *A*, is changed and this can have a dramatic influence on the analytical results.

Another problem with MBRs and convex hulls is that by definition they enclose the points for which you have data, and you may be aware that other points which could or should be taken into consideration, lie outside this region. This may distort analysis, for example giving poor interpolation results near to the region boundary or under-estimating the average distances between sampled points. For example, the nearest tree to one on the boundary may actually be in the adjacent field and not within the pre-defined boundary (Figure 4-22).

Figure 4-22 Point locations inside and outside polygon

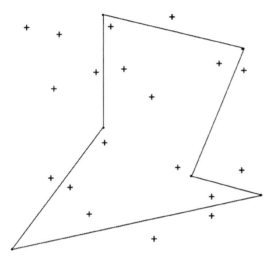

4.2.13.4 Fuzzy boundaries

A number of GIS packages include special facilities for identifying, selecting and analysing boundaries. These may be the borders of distinct zones or areas, within which values on one or more attributes are relatively homogeneous, and distinct from those in adjacent areas, or they may be zones of rapid spatial change in more continuously varying attribute data. Amongst the tools available for boundary determination are raster-based GIS products, like Idrisi, and the more specialised Terraseer family of spatial analysis tools (indeed the latter has a specific product, Boundaryseer, for boundary detection *and* analysis). The problem these products seek to address is that of identifying where and how boundaries between different zones should be drawn and interpreted when the underlying data varies continuously, without sharp breaks. For example, soil types may be classified by the proportion of sand, clay, organic matter and mineral content. Each of these variables will occur in a mix across a study region and may gradually merge from one recognised type to another.

In order to define boundaries between areas with, for example, different soil types, clear polygon-like boundaries are inappropriate. In such cases a widely used procedure is to allocate all grid cells to a set of *k* zones or classes based on *fuzzy membership* rather than binary or 0/1 membership. The idea here is to assign a number *between* 0 and 1 to indicate how much membership of the zone or class the cell is to be allocated, where 0 means not a member and 0.5 means the cell's *grade of membership* is 50% (but this is not a probabilistic measure). If a zone is mapped with a fuzzy boundary, the 50% or 0.5 set of cells is sometime regarded as the equivalent to the boundary or delimiter (cross-over point) in a crisp, discrete model situation. Using this notion, crisp polygon boundaries or identified zone edges in raster modelled data may be replaced with a band based on the degree of zone membership. Alternatively grid datasets may be subjected to fuzzy classification procedures whereby each grid cell acquires a membership value for one of *k* fuzzy classes, and a series of membership maps are generated, one for each class. Subsequent processing on these maps may then be used to locate boundaries.

Several fuzzy membership functions (MFs) have been developed to enable the assignment process to be automated and to reflect expert opinion on the nature and extent of transitional zones. Burrough and McDonnell (1998, Ch.11) provides an excellent discussion of fuzzy sets and their applications in spatial classification and boundary determination problems. The GIS product Idrisi, which utilises some of Burrough's work, provides four principal fuzzy MFs:

1. Sigmoidal or (double) s-shaped functions, which are produced by a combination of linear and $cos^2()$ functions in Idrisi's case, or as an algebraic expression of the form $m=1/(1+a(z-c)^2)$ where a is a parameter that varies the width of the function, z is the property being modelled (e.g. proportion of clay) and c is the value of this proportion at the function midpoint (see Figure 4-23; here membership values of >0.5 are regarded as being definitely members of the set A)

2. J-shaped functions, which are rather like the Sigmoidal MFs but with the rounded top sliced off flat over some distance, x (if $x=0$ then the two sides of the J meet at a point). The equation used in this case is of the form:
$m=1/(1+((z-a)/(a-c))^2)$

3. Linear functions, which are like the J-shaped function but with linear sides, like the slope of a pitched roof, and are thus simple to calculate and have a fixed and well-defined extent, and

4. User-defined functions, which are self-explanatory. In most applications MFs are symmetric, although monotonic increasing or decreasing options are provided in Idrisi

Figure 4-23 Sigmoidal fuzzy membership functions

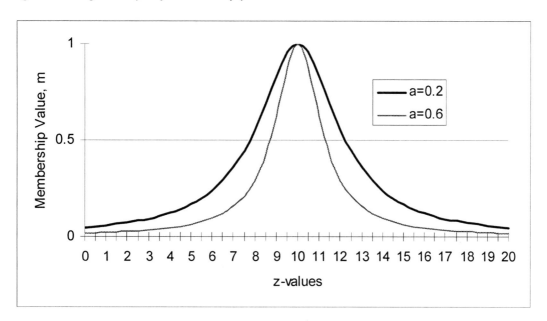

An alternative to using MFs as spreading functions is to apply a two-stage process: first, use a (fuzzy) classification procedure to assign a membership value (m_{ik}=0.0-1.0) to each grid cell, i, for each of k classes. This value is based on how similar the cell attributes are to each of k (pre-selected) attribute types or clusters. These may then be separately mapped, as discussed earlier. Having generated the membership assignments boundaries are then generated by a second algorithm. Boundaryseer provides three alternative procedures:

- Wombling (a family of procedures named after the author, Womble, 1951)

- Confusion Index (CI) measures, which are almost self-explanatory, and

- Classification Entropy (CE) based on information theoretic ideas

Wombling involves examining the gradient of the surface or surfaces under consideration in the immediate neighbourhood of each cell or point. Typically, with raster datasets, this process examines the four cells in Rook's or

Bishop's move position relative to the target cell or point — boundaries are identified based on the rate of change of a linear surface fitted through these four points, with high or sudden rates of change being the most significant. Wombling methods can be applied to vector and image datasets as well as raster data.

The Confusion Index (CI) works on the presumption that if we compute the ratio of the second highest membership value of a cell i, m_{i2}, to the highest, m_{i1}, then any values close to 1 indicate that the cell could realistically be assigned to either class, and hence is likely to be on the boundary between the two classes.

Finally, Classification Entropy, devised by Brown (1998), creates a similar value to the CI measure, again in the range [0,1], but this time using the normalised entropy measure:

$$CE = \frac{\sum_{i=1}^{k} m_{ij} \ln(m_{ij})}{\ln(k)}$$

where the summation extends over all the k-classes and the measure applies to each of the j cells or locations. This latter kind of measure is used in a number of spatial analysis applications (see further, Section 4.6).

Fuzzy procedures are by no means the only methods for detecting and mapping boundaries. Many other techniques and options exist. For example, the allocation of cells to one of k-classes can be undertaken using purely deterministic (spatially constrained) clustering procedures or probabilistic methods, including Bayesian and belief-based systems. In the latter methods cells are assigned probabilities of membership of the various classes. A fuller description of such methods is provided in the documentation for the Boundaryseer and Idrisi packages. As is apparent from this discussion, boundary detection is not only a complex and subtle process, but is also closely related to general-purpose classification and clustering methods,

spatially-constrained clustering methods and techniques applied in image processing.

Boundaries may also be subject to analysis, for example to test hypotheses such as: "is the overlap observed between these two boundaries significant, and if so in what way, or could it have occurred at random?"; "is the form of a boundary unusual or related to those of neighbouring areas?". Tools to answer such questions are provided in Boundaryseer but are otherwise not generally available.

4.2.13.5 Breaklines and natural boundaries

Natural boundaries, such as coastlines, rivers, lake shores, cliffs, man-made obstacles, geological intrusions and fault lines are widespread and are frequently significant in terms of understanding spatial processes. When conducting spatial analysis it is often important to examine datasets for the existence of such features, and to take explicit account of their impact where appropriate. A large number of structures and arrangements may be encountered, and the standard analytical tools provided within your preferred GIS package may not handle these adequately for your purposes. For example, simply masking out regions, such as lakes, areas of sea, regions where the underlying geology is not igneous etc., is often misleading, if all this procedure does is to hide those areas which are covered. Ideally the analytical procedure applied should be able to take account of the breakline or natural boundary in a meaningful manner. In the case of breaklines there are three different types that tend to be supported, principally in connection with surface analysis and modelling (see further, Chapter 6).

Terminology usage varies here, so we shall use the form provided within ArcGIS 3D Analyst for convenience:

- Case 1: hard breaklines, which are well-defined surface structures, whose shape and heights are known and fixed, and which mark a significant change of surface continuity — for example, indicating a road, a stream, a dam or a major change of slope which has been surveyed along its length (e.g. a ridge). Such breaklines are usually specified as a set of triples {x,y,z} and intermediate values are usually determined by simple linear interpolation

- Case 2: soft breaklines are similar to hard breaklines, but do not imply a change in surface continuity. They indicate that the known values along the breakline should be maintained, for example when creating a TIN of the surface, but no sharp change in the continuity of the underlying surface is implied

- Case 3: faults are more complex, in that they often have a spatial extent due to displacement and not just a simple linear form

The surface analysis package, Surfer, models breaklines as 3D polylines and faults as 2D polylines or polygons, reflecting their variable form. ArcGIS models breaklines as 3D polylines and faults as 2D polylines. Software packages treat these types of boundary in very different ways. Cases (1) and (2) take into account all data points provided, although in Case (1) z-values computed for the linear breakline take precedence over points that may lie on the other side of the line from the point being analysed. Case (3) is quite different, since fault lines and zones are treated as barriers. In ArcGIS data values beyond a barrier are not included in computations, where the notion of beyond is defined by visibility from the source point; in Surfer linear barriers are treated as obstacles, which may be skirted around (involving considerable additional distances and extra processing), whilst polygonal faults are treated more like self-contained regions, with interpolation inside and outside the polygonal regions being treated separately.

4.2.14 Tessellations and triangulations

A region may be divided into a set of non-overlapping areas in many ways. As we noted in Table 1.2 a (regular or irregular) tessellation

of a plane involves the subdivision of the plane into polygonal tiles that completely cover it. Within GIS these tiles are almost always either square or rectangular, and form a (continuous) grid structure. Regular triangular and hexagonal grids are also possible in the plane but are rarely implemented in software packages. GIS does make extensive use of irregular triangular tessellations, both for division of plane regions and as an efficient means of representing surfaces.

4.2.14.1 Delaunay Triangulation

Given a set of points in the plane, *P*, lines may be drawn between these points to create a complete set of non-overlapping triangles (a triangulation) with the outer boundary being the convex hull of the point set. It has long been known that triangulation provides a secure method for locating points on the Earth's surface by field survey. It has also been found that a desirable characteristic of such triangulations is that long thin triangles with very acute internal angles are to be avoided in order to provide the best framework for measurement and analysis. To ensure the triangulation will have the best chance of meeting the characteristics desired a construction rule was devised by the mathematician Delaunay. This states that three points form a Delaunay Triangulation if and only if (*iff*) a circle which passes through all three points contains no other points in the set. Figure 4-24 illustrates this process for a set of spot height locations in a small area south of Edinburgh in Scotland (GB Ordnance Survey tile NT04). Each location is represented by an (*x,y*) pair. Many programs support the creation of the triangulation required, for example the delaunay(x,y) function in MATLab where x, and y are *n*x1 vectors containing the coordinates of all the points, or (as here) the Grid Data operation in Surfer, using the option to export the triangulation used as a base map. The two circles illustrate how the circle circumscribing each triangle has no other points from the set within it. Worboys and Duckham (2004, pp. 202-207) provide a useful discussion of triangulation algorithms.

Figure 4-24 Delaunay triangulation of spot height locations

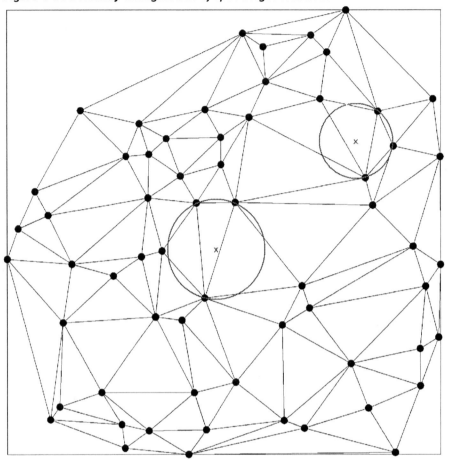

A number of other observations should be made about this process:

- this triangulation is unique in the sense that it is the only triangulation of the point set that satisfies the rule described above. However, with some point set arrangements (e.g. four points at the corners of a square or rectangle) multiple Delaunay triangulations are possible, hence to this extent it is not unique

- long thin triangles with acute angles do occur using this procedure, although the

number and severity of these is minimised compared to other triangulations

- the centre of the circumscribing circle (marked with an x here) may lie outside the triangle in question, as is the case with the smaller circle illustrated

- this procedure is by no means the only, or necessarily the best method of triangulation for all purposes. For example, when modelling a surface there may be important breaks of surface continuity or surveyed transects which it is highly desirable to include within the dataset, which are not captured by simple Delaunay-type procedures (as noted in Section 4.2.13.5)

If all points included in the Delaunay triangulation of a set *P* also lie on the convex hull of *P* the result is known as a furthest point Delaunay triangulation. The set of lines (edges) that belongs to all possible furthest site Delaunay triangulations is known as a *Delaunay diagram*, and has important applications in network analysis (see further, Chapter 7)

4.2.14.2 TINs — Triangulated irregular networks

Regular lattices such as grids provide a very inefficient means of storing details of surface topography. Areas that have similar slope could be better represented and stored using a series of inter-connected triangles forming an irregular mesh or network (a TIN). The nodes or vertices of these triangles can be placed closer together where landscape detail is more complex and further apart where surfaces are simpler. TINs may be designed or derived. Designed TINs arise in engineering and surveying models, whereas derived TINs are typically programmatically generated from grid-based datasets (digital elevation models or DEMs). Although such TINs could be generated by selecting vertices using key feature extraction and surface modelling, a simpler and faster approach is to systematically divide the entire region into more and more triangles based on simple subdivision rules. For example, if you start with a rectangular region and define vertices at each of the four corners, two triangulations are possible: (i) a line connecting the NE-SW corners creates one pair of triangles; or (ii) a line connecting the NW-SE corners creates a second pair. For each triangle a statistic can be computed, such as the maximum deviation in elevation of the triangle from the original DEM. The arrangement that produces the smallest maximum absolute deviation (a minimax criterion) can then be selected as the preferred triangulation. Each of the two initial triangles can then be subdivided further. For

example, a Steiner point can be placed within each triangle which creates a further three triangles, or 6 in total. The minimax criterion test can be applied for each of the subdivisions of the two original triangles and the process proceeds iteratively until the minimax criterion is less than a pre-defined threshold value (e.g. 5 metres) or the number of triangles generated has reached a maximum value, also pre-defined. This kind of procedure is used in many GIS packages — for example TNTMips describes this process as *adaptive densification*. TINs and DEMs are discussed in more detail in Chapter 6 (see for example, Figure 6-5).

4.2.14.3 Voronoi/Thiessen polygons

Given a set of points in the plane, there exists an associated set of regions surrounding these points, such that all locations within any given region are closer to one of the points than to any other point. These regions may be regarded as the *dual* of the point set, and are known as proximity polygons, Voronoi polygons or Thiessen regions. These various terms are equivalent, as is the term Dirichlet cell, all being derived from the names of their proponents: Dirichlet (in 1850); Voronoi (in 1909); and Thiessen (in 1912).

Figure 4-25A shows an example of Voronoi region creation, in this case using ArcGIS. The point set from which the map was generated is also shown and as can be seen, points are frequently close to polygon edges. Indeed, in some applications it may be desirable to exclude points that are closer together than some predefined tolerance level. Note that in this example Voronoi regions at the edges of the sample extend to infinity, and are shown as bounded (e.g. by the MBR of the point set plus 10%). Figure 4-25B shows the same point set plus the associated Delaunay triangulation once more, created within the MATLab mathematical modelling package. This enables us to see exactly how Voronoi regions are created. There are two steps: the first is the triangulation process, connecting the point set as described in subsection 4.2.14.1; and the second is bisection of the triangulation lines of adjacent points to define Voronoi polygon boundaries.

There are many reasons an analyst may wish to generate Voronoi polygons from a point set, but a common requirement is to assign values to regions or cells based on the attributes of the point set. The most common assignment is that of the individual point, or nearest-neighbour model. Other options are offered by some GIS packages in much the same way as is applied to grid and image files (in which each cell is surrounded by 4 or 8 immediate neighbours, depending on the model used). Note that the MATLab implementation does not extend to the MBR or other bounding region (e.g. rectangle) defined by the user.

Figure 4-25 Voronoi regions generated in ArcGIS and MATLab

A. ArcGIS

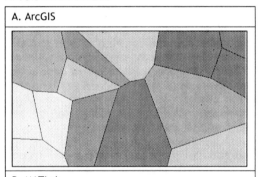

B. MATLab

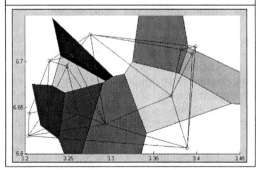

As an example the assignment facilities provided within ArcGIS include: local mean value – the cell is assigned the average value of the cell and its direct neighbours; and local

modal value – determined by computing the frequency distribution of all adjacent cells in 5 classes and then assigning the cell the class value from this set that is the most common amongst the cell and its immediate neighbours. These assignments represent local smoothing procedures. Local variation rather than values may also be computed. For example:

1. local entropy value (equivalent to Shannon's Landscape Diversity Index, SHDI – see Section 5.3.4) – in this model the adjacent cells are again grouped into 5 sets (smart quantiles or natural breaks classes) and the number in each class is represented as a proportion of the total, p_i, i=1-5, from which Shannon's entropy statistic is computed (Table 1.4) and this value is assigned to the cell – values will range from 0 to $\log_2(5)$; and

2. local standard deviation and local inter-quartile range – these statistics are computed for the cell and its immediate neighbours and the value assigned to the selected cell.

Voronoi polygons may also be generated in grid or raster datasets (perhaps better described as Voronoi patches). Since the definition of a Voronoi polygon in the plane is that it consists of all locations closer to selected points than to any other points, such regions can be generated by a form of spreading algorithm. In this case we start with a number of points (the source points, sometimes called the target points) and spread outwards in all directions at a constant rate, like ripples spreading across the surface of a pond when a stone has been thrown into the middle. The process stops when the ripple (spreading) from another point is encountered. The rectangular grid structure and the number of search directions (typically 8) affect the form of the resulting cells (Figure 4-26). With the raster representation Voronoi patch shape and boundary definition are not as clear cut as in vector operations.

Within GIS packages processing activities of this type are usually carried out using techniques described as cost distance, or accumulated cost distance tools (e.g. in ArcGIS

Spatial Analyst). Other packages have specific functions to generate regions of this kind (e.g. the Fence function in MFWorks). In image processing the procedures adopt a somewhat different set of algorithms, based on scanning rather than spreading, which are known as distance transforms. The latter may be exact (in Euclidean distance terms) or approximate (see further, Section 4.4.2). Furthermore, with both approaches the space across which the scan or spread procedure is conducted may be homogenous (i.e. treated as uniform in all directions and at all points) or non-homogenous, containing areas of varying difficulty to traverse, varying cost, and/or including obstacles.

Figure 4-26 Voronoi cells for a homogeneous grid

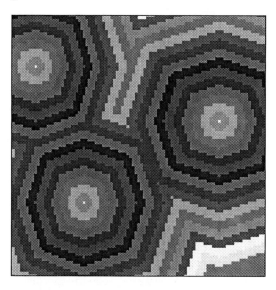

The most comprehensive book on this field is by Okabe, Boots and Sugihara (1992), a monumental work entitled "Spatial tessellations: Concepts and applications of Voronoi diagrams". For details on the huge range of applications and methods relating to Voronoi diagrams readers are recommended to dip into Okabe *et al.*'s work. Our discussion at various points in this Handbook will be restricted to a small subset of these applications.

A selection of (a) MATLab and (b) GRASS functions that support operations discussed in this section is shown in Table 4.7.

Table 4.7 Selected MATLab/GRASS planar geometric analysis functions

Function	Description
(a) convhull (b) v.hull	Compute the convex hull of a point set, optionally providing its area
(a) delaunay (b) v.delaunay	Delaunay triangulation
(a) dsearch	Nearest point search of Delaunay triangulation. Identifies the vertex of a Delaunay triangle that is closest to a specified point
(a) inpolygon	True (1) for points inside polygonal region, False (0) for points outside, and assigned 0.5 if points lie on the boundary
(a) polyarea (b) area_poly2.c	Area of a polygon
(a) rectint	Area of intersection for two or more rectangles
(a) tsearch	Closest triangle search. Identifies the Delaunay triangle that includes a specified point
(a) voronoi (b) v.voronoi	Voronoi diagram creation

4.3 Queries and Computations

4.3.1 Spatial selection and spatial queries

As with many other terms, the expression "spatial query" means different things to different suppliers. In the majority of cases it refers to the process of *selecting* information (features and associated attribute data in tables) by some form of query operation that includes a spatial constraint. For example: SELECT all features WITHIN a specified radius of a specified point; or SELECT all point objects that fall WITHIN a specific polygon. For most (all) GIS packages this selection process can be achieved by a point-and-click visual interface, subject to certain selection rules. In parallel there is the option to execute a single command or series of commands as a script or program, or as a form of extended SQL query. The result of such queries may be a new table and feature set, which can be viewed, analysed further or saved for additional analysis. Raster-based GIS packages do not generally support spatial query operations, although they may support standard SQL queries on attribute data.

The various approaches are best understood by describing a number of specific implementations:

1. ArcGIS provides full support for standard SQL SELECT queries, such as: SELECT <all polygons in a given layer> WHERE <attribute table field> EQUALS <value>. It does not appear to support other forms of SQL query, such as MAKE TABLE, UPDATE or DELETE. It supports JOIN and RELATE operations between attribute tables based on specific fields, but it also provides spatial extensions to the JOIN and RELATE process enabling tables to be joined and queried based on their relative locations. Where joins are based on spatial location, data from one set of features are incorporated into the attribute table of the other set of features according to selectable

rules. The ArcMAP visual interface provides a "Select by location" facility with selection qualified by a variety of spatial conditions (applicable to point, line and polygon features where meaningful): Intersect; Are within a distance of; Completely contain; Are completely within; Have their centre in; Share a line segment with; Touch the boundary of; Are identical to; Are crossed by the outline of; Contain; Are contained by. Each of the options may be further modified by selecting a tolerance level (a buffer — see further Section 4.4.4) around the points, lines or polygons in question

2. Manifold provides support for specific spatial extensions to the standard SQL syntax. This enables a very large range of operations to be carried out, including so-called Action Queries (e.g. queries that CREATE, UPDATE and DELETE table data). Example spatial extensions that result in true or false being returned include: Adjacent, Contains, Intersection, IsArea/IsLine/IsPoint and Touches. These operations may be combined with other spatial operations (such as distance and buffer functions) to select records, and may also be combined with operators that generate spatial objects (e.g. a triangulation or convex hull of a point set) rather than simply selecting or amending existing objects. As such many of Manifold's spatial query functions are an extension of its "Transform" facilities, performing geometrical processing on the selected data as well as simply selecting items

3. MapInfo, like ArcGIS, provides a range of SQL SELECT facilities and support for spatial joins. Spatial join operators supported include: Contains (A contains at least the other object, B's centroid), Contains Entire (A contains whole object B), Within (as per Contain, but A and B roles are reversed), Entirely within, and Intersects (A and B have at least one point in common). These operators enable attribute tables to be JOINed without sharing a common field (i.e. a spatial join only)

4.3.2 Simple calculations

Almost all GIS packages provide a wide range of calculation facilities. By simple calculations we mean arithmetic operations and standard functions applied to: (a) tabular data; (b) within queries and similar operations; and (c) to map layers (see further, Section 4.6). As with spatial queries, there are many different implementations of simple calculation operations and these are widely used, readily understandable, and applied as often to raster datasets as to vector data. A number of examples help to clarify the range of facilities provided:

(a) ArcGIS provides a field level calculator within attribute tables by right-clicking on any field (existing or newly added). This enables basic operations such as +, -, x and / to be performed, plus a range of standard functions such as cos(), exp(), log(), abs() etc. Operations are applied to one or more fields, e.g. current $field=(field1+field2*100)/field3$, and this process either updates the values for all records in the table selected or, if a subset have been pre-selected, updates records in just that subset. In addition an "Advanced" option is provided enabling Visual Basic for Application (VBA) commands to be entered and executed. Such commands are often needed in ArcGIS in order to add or update intrinsic data (as explicit fields in the attribute table) such as the area of a polygon or the (x,y) coordinates of a polygon centroid. Manifold implements such operations through its extended SQL query facilities and using the Query toolbar facility it provides.

(b) Simple calculations and expressions are extensively used in SQL queries and in association with spatial extensions to standard SQL. Again, the common arithmetic operators are used (extended in some cases to facilitate integer division, modulo arithmetic etc.) combined with comparison operators (e.g. =, >, < etc.) and Boolean operators (e.g. AND, OR, NOT, XOR). Since queries often involve non-numeric fields additional operators are provided to enable the use of strings, dates and times. Note that it is often possible, and sometimes simpler, to perform calculations outside of the GIS framework using standard tools, such as spreadsheet or database programs.

(c) Within raster-based GIS, and in the raster processing facilities of general purpose GIS, there is a very large set of simple calculation facilities. Such facilities are a standard part of the set of operations provided in specifically raster-based GIS packages (e.g. MapCalc, Idrisi, MFWorks, PCRaster), but are often additional components in general purpose (vector-oriented) GIS. For example in ArcGIS the majority of such tools are provided in the Spatial Analyst extension whilst in Manifold they are provided in the Surface Tools extension.

The term Map Algebra is often used to describe operations that combine two or more grid layers according to a simple algebraic expression, e.g. $A+2*B+C/100$ where A, B and C are matching grid layers (Figure 4-27a). This is commonly the case with multi-band remote sensing data (e.g. 7 bands representing different spectral components of a single image). Map Algebra may also be used where the grid resolutions do not match, in which case decisions have to be made (rules applied) on how resolution differences are to be accommodated. The example in Figure 4-27b show how grid C, being lower resolution, determines the resolution of the result grid. The general term for this kind of process is *resampling*.

However, Map Algebra refers to a far broader set of operations, extending to functions that operate on individual grid cells, blocks of grid cells or entire grid layers (often described as *local, focal or neighbourhood, zonal,* and *global* analysis) as we shall see in Section 4.6 and Chapter 6. In this Section we restrict our consideration to operations that apply on a cell-by-cell or point-by-point basis (the term *local* is generally used for such operations).

Figure 4-27 Cell-by-cell or Local operations

a. Matching grids

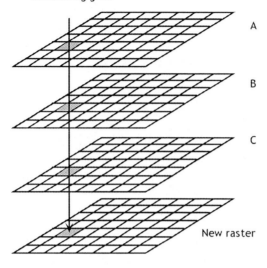

b. Grid resampling

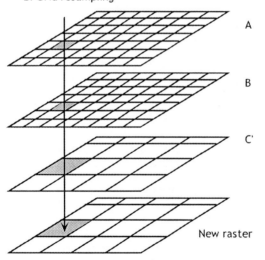

based on the position of the centre of the output grid relative to the centres of finer resolution input grids. Grids do not have to consist entirely of square zones, nor do they have to align precisely, but if they do not then resampling and some form of localised interpolation is inevitable.

In Idrisi, for example, pairs of images (grids of pixels) may be combined using the operators: +, -, x, /, ^ (power), maximum, minimum, and cover. These operators combine images *A* and *B* on a pixel by pixel basis, for example adding each pixel in image *B* to each matching pixel in image *A*. Operations such as maximum and minimum identify a value from corresponding positions in image *A* or *B* as appropriate, whilst the cover function creates an image just containing pixels from *A* except where these are zero, in which case the pixels of *B* are taken (showing through the "holes" in *A*). With the exception of + and -, all other operations are non-symmetric, i.e. *A*<operator>*B* is not equal to *B*<operator>*A*. An example application of operations of this kind would be the algebraic expression:

$$C=(A-B)/(A+B)$$

where *C* is the output raster, *A* and *B* are input raster maps (e.g. remote sensed images or image bands) of vegetation cover, with pixels in *C* being taken as a vegetation index (with values being in the range -1 to +1). Note that a check is required to ensure that if *A*=*B*=0 for a pixel location then the calculation gives the result 0.

Many GIS packages allow the creation and saving of macros or scripts that implement regularly used map algebra expressions. For example TNTMips and ENVI include numerous pre-built combination formulas of this type. TNTMips holds such formulas as scripts that may be modified or augmented. For example, its two-band transformed vegetation index (TVI) script is of the form:

If distinct raster layers have different resolutions (cell sizes) then one or more must be resampled to ensure their resolutions match. The normal practice is to adjust the layers of finest resolution to match the coarsest, and to use simple nearest-neighbour interpolation to estimate values. By nearest-neighbour is meant that values are assigned

$$C = 100\left(\sqrt{\frac{A-B}{A+B}}\right), A-B \geq 0, A+B \geq 0$$

$C = 0$ if $A - B < 0$

Here A might be the near infra-red band from a remote sensing image and B the red band image, with the result being an index raster providing a measure of vegetation vigour (Figure 4-28C). The basic index has the range [0,100], and as elsewhere, checks for division by 0 are required.

PCRaster, a non-commercial package designed specifically for manipulating raster data, provides a rich set of point operations, many of which have been derived from the work of Tomlin (1990). These include a similar set to those described above, i.e. arithmetic, trigonometric and logarithmic functions, coupled with Boolean and relational operators. Point operations are augmented by facilities for missing data detection and substitution, conditional expressions, data lookup and other facilities that mirror those provided in the extended SQL-type command structures described earlier.

Combination operations of this type are not confined to raster datasets. Polygon layers may be combined with raster layers, either by converting the polygon layers to rasters and then proceeding as above (the most common approach in GIS packages), or by direct support for such combinations within the GIS (as is the case with TNTMips for example).

Figure 4-28 Map algebra: Index creation

A. Near infra-red band

B. Red band

C. TVI

4.3.3 Ratios, indices, normalisation and standardisation

The vegetation index calculations we described in Section 4.3.2 were generated in a manner that ensured all output pixels (grid cells) would contain entries that lie in the range [-1,+1] or [-100,100]; an index value of 0 indicates that matching pixel locations have the same value. There are many reasons for changing attribute data values in this manner before mapping and/or as part of the analysis of spatial datasets. One of the principal reasons is that if you create a map showing zones (e.g. counties) for a *spatially extensive* attribute, such as total population, it is very misleading — obviously a single total value (indicated perhaps by a colour) does not apply throughout the zone in question. A more correct or meaningful view would be to divide by the area of the zone and show a *spatially intensive* attribute, the density of the population across the zone in this case. In this way zones of different size can be analysed and mapped in a consistent manner. The field of geodemographics, which involves the analysis of population demographics by where they live, involves the widespread use of index values — for a full discussion of this subject see Harris *et al.* (2005) and the "Geodemographics Knowledgebase" web site maintained by the Market Research Society (http://geodemographics.org.uk/).

In many instances, especially with data that are organised by area (e.g. census data, data provided by zip/post code, data by local authority or health districts etc.) production of ratios is vital. The aim is to adjust quantitative data that are provided as a simple count (number of people, sheep, cars etc.) or as a continuous variable or *weight*, by a population value for the zone, in a process known as *normalisation*. This term, rather confusingly, is used in several other spatial contexts including statistical analysis, in mathematical expressions and in connection with certain topological operations. The context should make clear the form of normalisation under consideration. Essentially, in the current context, this process involves dividing the count or weight data by a value (often another field in the attribute table) that will convert the information into a measure of intensity for the zone in question — e.g. cars owned per household, or trees per hectare. This process removes the effects associated with data drawn from zones of unequal area or variable "population" size and makes zone-based data comparable.

Three particular types of normalisation are widely used: (i) averages; (ii) percentages; and (iii) densities. In the case of averages, an attribute value in each zone (for example the number of children) is adjusted by dividing by, for example, the number of households in each zone, to provide the average number of children per household, which may then be stored in a new field and/or mapped. In the case of percentages, or conversion to a [0,1] range, a selected attribute value is divided by the total set from which it has been drawn. For example, the number of people in a zone registered as unemployed divided by the total population of working age (e.g. 18-65). The final example is that of densities, where the divisor is the area of the zone (which may be stored as an intrinsic attribute by some software packages rather than a field) in which the specified attribute is found. The result is a set of values representing population per unit area, e.g. dwellings per square kilometre, mature trees per hectare, tons of wheat per acre etc. — all of which may be described as density measures.

The term *standardisation* is closely related to the term normalisation and is also widely used in analysis. It tends to appear in two separate contexts. The first is as a means of ensuring that data from separate sources are comparable for the problem at hand. For example, suppose that a measure of the success rate of medical treatments by health district were being studied. For each health district the success rates as a percentage for each type of operation could be analysed, mapped and reported. However, variations in the demographic makeup of health districts have not been factored into the equation. If there is a very high proportion of elderly people in some districts the success rates may be lower for certain types of treatment than in other districts. Standardisation is the process whereby the rates are adjusted to take account of such district variations, for example in age, sex, social deprivation etc. Direct standardisation may be achieved by taking the district proportion in various age groups and comparing this to the regional or national proportions and adjusting the reported rates by using these values to create an age-weighted rate. Indirect standardisation involves computing expected values for each district based on the regional or national figures broken down by mix of age/sex/deprivation etc., and comparing these expected values with the actual rates. The latter approach is less susceptible to small sample sizes in the mix of standardising variables (i.e. when all are combined and coupled with specific treatments in specific districts).

The second use of the term standardisation involves adjusting the magnitude of a particular dataset to facilitate comparison or combination with other datasets. Two principal forms of such standardisation are commonly applied. The first is z-score standardisation (see Table 1.4). This involves computing the mean and standard deviation of each dataset, and altering all values to a z-score in which the mean is subtracted from the value and the result divided by the standard deviation (which must be non-zero). This generates values that have a mean of zero and a standard deviation of 1. This procedure is widely used in the statistical analysis of datasets. The second approach involves range-based standardisation. In this method the minimum and maximum of the data values are computed (thus the *range=max-min*), and each value has the min subtracted from it and is then divided by the range. To reduce the effect of extreme values range standardisation may omit a number or percentage of the smallest and largest values (e.g. the lowest and highest 10%), giving a form of trimmed range. Trimmed range-based standardisation is the method currently applied within the UK Office of National Statistics (ONS)

classification of census areas (wards) using census variables such as demographics, housing, socio-economic data and employment.

Another way of looking at the process of dividing a count for a zone, e.g. population, by the zone area is to imagine the zone is divided into very many small squares. Each square will contain a number of people registered as living in that square. The total for the zone is simply the sum of the counts in all the little squares. But this form of representation is exactly that of raster or grid GIS, and hence raster representations generally do not require area-based normalisation, since they are already in a suitable form. Of course, the process of generating a grid representation may have involved computation of values for each cell from vector data, either assuming a uniform model or some other (volume preserving model) as discussed in Section 4.2.10. Normalisation is used with raster data for a variety of reasons, for example: (i) to change to set of values recorded in a single map layer to a [0,1] or [-1,1] range (e.g. by dividing each cell value, z, by the theoretical or observed maximum absolute value for the entire grid:

$$z^*=z/z_{max}$$

or using the transformation

$$z^*=(z-z_{min})/(z_{max}-z_{min})$$

and (ii) to combine two or more grid layers or images into a single, normalised or "indexed" value (sometimes this process is described as indexed overlay). In this latter case the overlay process may be a weighted sum, which is then normalised by dividing the resulting total by the sum of the weights.

These various observations highlight a number of issues:

- the division process needs to have a way of dealing with cases where the divisor is 0

- missing data (as opposed to values that are 0) need to be handled

- if an attribute to be analysed and/or mapped has already been normalised (e.g. is a percentage, a density, an average or a median) then it should not be further normalised as this will generally provide a meaningless result

- highly variable data can lead to confusing or misleading conclusions. For example, suppose one zone has a population of 10,000 children, and the incidence of childhood leukaemia recorded for this zone over a 10 year period is 7 cases. The normalised rate is thus 0.7 per 1000 children. In an adjacent zone with only 500 children one case has been recorded. The rate is thus 2 per 1000 children, apparently far higher. Effects of this type, which are sometimes described as *variance instability*, are commonplace and cause particular problems when trying to determine whether rare illnesses or events occur with unexpectedly high frequencies in particular areas

- the result of normalisation should be meaningful and defensible in terms of the problem at hand and the input datasets — using an inappropriate divisor or source data which are of dubious value or relevance will inevitably lead to unacceptable results

- obtaining an appropriate divisor may be problematic — for example if one is looking at the pattern of disease incidence in a particular year, census data (which in the UK are only collected every 10 years) may be an inadequate source of information on the size of the population at risk

GIS packages rarely emphasise the need for normalisation. One exception is ArcGIS, where the Properties form for map layers includes a normalisation field when quantitative data are to be classified and mapped (Figure 4-29). In

this example we are normalising the number of people who are categorised as owner occupiers (home owners, OWN_OCC) by a field called TOTPOP (total population). The normalisation prompt is provided for graduated colours, graduated symbols and proportional symbols, but not for dot density representation as this already provides a spatially intensive representation of the data.

Figure 4-29 Normalisation within ArcGIS

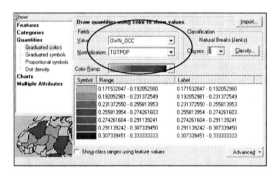

The result of normalisation in such cases is a simple ratio. Assuming the field being normalised is a subset of the divisor (which it should be) the ratio will not exceed 1.0.

Specialised spatial analysis software, like GeoDa, may provide a more extensive and powerful set of pre-defined normalisation tools (see further, Section 5.2.1), although in this case the software does not support built-in user-definable calculation facilities. Many of GeoDa's facilities are designed to address problems associated with simple normalisation, or in ESDA terminology, the computation of *raw rates*. The example of OWN_OCC/TOTPOP above would be a raw rate, which could be plotted as a percentile map in order to identify which zones have relatively high or low rates of owner occupancy, unemployment or some other variable of interest.

As an illustration of this procedure, Figure 4-30 shows 100 counties in North Carolina, USA where the incidence of sudden infant death syndrome (SIDS, also known as "cot death") in the period 1st July 1974 to end June 1978 is mapped. This dataset has been normalised to form a *raw rate* by dividing the number of cases in each zone by the births recorded for the same period (BIR74). The resulting raw rates have been sorted, from the lowest to the highest, and divided into 7 groups (or quantiles), each containing approximately the same number of zones. The uppermost range is comprised of the zones in which the relative incidence or *rate* of SIDS is highest.

Figure 4-30 Quantile map of normalised SIDS data

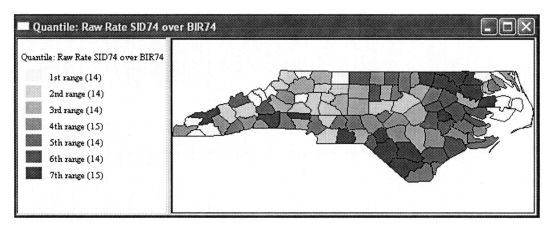

However, even raw rates can be misleading, as noted earlier, and various techniques have been devised for analysing and mapping such data to bring out the real points of interest. One example will suffice to explain the types of procedure available. The ratio SID74/BIR74 is normally computed for every zone. However, one could compute this ratio for all zones first, based on the sum of the selected fields. This would give an average ratio or rate for the entire study area. This value could be regarded as the *expected rate* for each zone, E, say. We can then calculate the *expected number* of SIDS cases for each zone by multiplying the average rate E by the relevant population in each zone, i:

$$E_i = E * BIR74_i$$

Now consider the ratio: $R_i = SID74_i / E_i$

This is a ratio with a value from $[0,n]$ that highlights those areas with higher than expected or lower than expected rates (Figure 4-31). As can be seen, one zone is picked out in red/dark grey in the lower centre of the mapped zones as appearing to have an unusually high SIDS rate, being at least 4 times the average for the region as a whole. This technique is known as *rate smoothing* and is one of a series of such procedures that help to focus attention on interesting data variations rather than artefacts of the way the zones are laid out or their size. When attempting to develop explanatory models of such data (mortality data, disease incidence data, crime data etc.) it is the widely accepted practice to utilise the expected rate based on overall regional or national statistics as an essential component of the analysis. This issue is discussed further in Section 5.6.4 where the data illustrated here are re-visited – see also, Cressie and Chan (1989); and Berke (2004).

Figure 4-31 Excess risk rate map for SIDS data

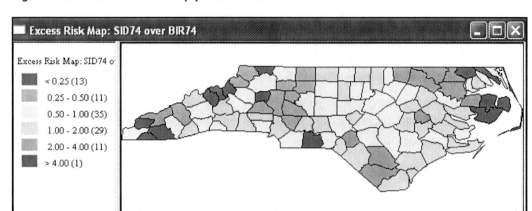

Single ratios provide the standard form of normalisation, as described above, but more complex combinations of attributes, from one or more sets of spatial data, may be combined in a variety of ways in order to generate *index values*. One advantage of such procedures as that index values may be used as a guide for planning policy or as comparative measures, for example as a means of comparing crime rates and clear-up rates between different police forces. As the need to incorporate several factors in the construction of such indices becomes more important, simple weighting and ratio procedures will be found wanting, and more sophisticated methods (such as various forms of cluster analysis) will be required.

4.3.4   Density, kernels and occupancy

4.3.4.1   Point density

In Section 4.3.3 we noted that spatially extensive variables, such as total population, should not be plotted directly for zones under normal circumstances, but standardised in some manner. For variables that are a subset of a total it is often useful to compute and map these data as a proportion of this total, but for totals themselves we will need to compute a density measure — dividing the total population, N, by the zone area, A. This yields a density value, N/A, for each zone, but

assumes that population density is constant through the zone and then may suddenly change to some other value at the zone boundaries. In our discussions on boundaries and on areal interpolation we have already alluded to the inadequacies of simple density calculation, but it remains in widespread use. Conversion of the source dataset to a raster model using some form of intelligent interpolation may well provide a more satisfactory result. A good introduction to mapping density is provided in Mitchell (1999, Chapter 4).

If the dataset we have is not zone-based but point- or line-based, alternative methods of determining density are required. The simplest measures apply some form of zoning and count the number of events in each zone, returning us to a zone-based dataset. For example, with data on thefts from cars (TFC events), each event may be geocoded and assigned to a unique zone such as a pre-defined 100mx100m grid, or to city blocks, census blocks, or policing districts. The density of TFC events is then simply the event count divided by the area of the associated zone in which they are reported to have occurred. A difficulty with this approach is that densities will vary substantially depending on how the zoning or grid is selected, and on how variable are the size and associated attributes of each zone (e.g. variations in car ownership and usage patterns). Overall density calculation for all zones (e.g. a city-wide figure for the density of TFC events) suffers similar problems, especially with regard to definition of the city boundary. Use of such information to compare crime event levels within or across cities is fraught with difficulties.

An alternative approach to density computation for two-dimensional point-sets is based on techniques developed in one-dimensional (univariate) statistical analysis. This is simplest to understand by looking at the example in Figure 4-32. We have a set of 5 events marked by crosses. They occur at positions 7, 8, 9, 12 and 14 along the line. We could argue that the point density across the entire 20-wide line length is 5/20=0.25 points per unit length and assign a value of 0.25 to each section, as shown on the grey line. We might equally well argue that if we divide the overall line into two halves, the density over the first half should be 0.3 per unit length and over the second, 0.2 per unit length, to reflect the variation in positioning of the 5 events.

There is clearly no single right answer or single method to assign the points to the line's entire length, so the method we choose will depend on the application we are considering. Important observations to notice about this problem include: the length of the line we start with seems to have an important effect on the density values we obtain, and since this may be arbitrary, some method of removing dependence on the line length is desirable; if the line is partitioned into discrete chunks a sudden break in density occurs where the partition boundaries occur; depending on the number of partitions and distribution of points, areas may contain zero density, even if this is not the kind of spread we are seeking or regard as meaningful; the line is assumed to be continuous, and allocation of density values to every part is valid; and finally, if we have too many partitions all sections will only contain values of 1 or 0, which is essentially back to where we started from.

Figure 4-32 Point data

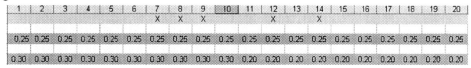

Figure 4-33 Simple linear (box or uniform) kernel smoothing

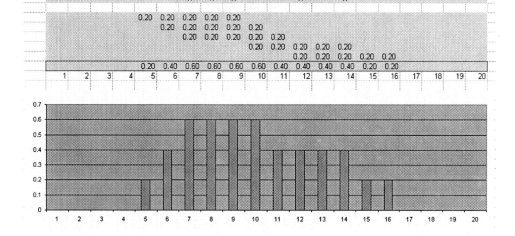

These observations can be dealt with by treating each point in the original set as if it was spread over a range, then adding together overlapping zones and checking that the total adds up to the original value. For example, choosing each point and smoothing it over 5 units in a uniform symmetric manner, we obtain the result shown in Figure 4-33. The rows in the diagram show the spreading of each of the 5 original points, with the total row showing the sums (densities) assigned to each unit segment. These add up to 5, as they should, and a chart showing this distribution confirms the pattern of spread. This method still leaves us with some difficulties: there are no density values towards the edges of our linear region; density values still jump abruptly from one value to the next; and values are evenly spread around the initial points, whereas it might be more realistic to have a greater weighting of importance towards the centre of each point.

All of these concerns can be addressed by selecting a well-defined, smooth and optionally unbounded function, known as a kernel, and using this to spread the values. The function often used is a Normal distribution, which is a bell-shaped curve extending to infinity in each direction, but with a finite (unit) area contained underneath the bell (Figure 4-34). In this diagram, for each point (7,8,9,12 and 14) we have provided a Normal distribution curve with central value (the mean) at the point in question and with an average spread (standard deviation) of one unit. We can then add the areas under each of these curves together to obtain the brown (cumulative) upper curve with two peaks, and then divide this curve by 5 if we want to adjust the area under the curve back to 1 (effectively a form of normalisation of the distribution, a term not to be confused with the Normal distribution itself). When adjusted in this way the values are often described as probability densities, and when extended to two dimensions, the resulting surface is described as a probability density surface, rather than a density surface.

Figure 4-34 Univariate Normal kernel smoothing and cumulative (Cum) densities

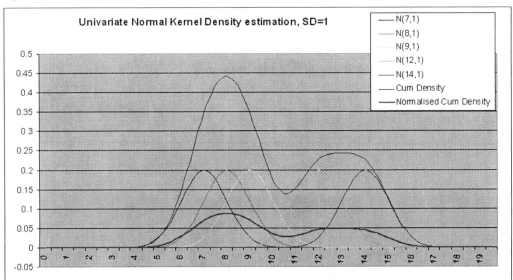

Figure 4-35 Alternative univariate kernel density functions

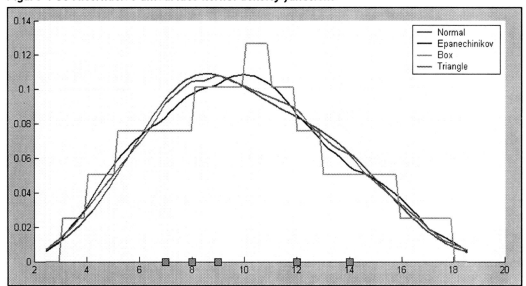

We now have a density value for every position along the original line, with smooth transitions between the values, which is exactly what we were trying to achieve. There still remain some questions: why should we use the Normal distribution? could we not use almost any symmetric function with a finite area under it? and why did we choose a value of 1 unit for the average spread? The answer to these questions is that the specific selections made are a matter of choice and experience, although in some instances a symmetric distribution with a finite extent (e.g. a box or triangular function) may be regarded as more suitable than one with an infinite possible extent. Figure 4-35 shows a selection of commonly used functions plotted for the same point set, using the MATLab Statistics Toolbox function ksdensity(). For further details on these and other functions used for smoothing and density estimation see Bowman and Azzalini (1997) and Silverman (1986). As may be guessed from examining the various curves shown in Figure 4-35 the exact form of the kernel function does not tend to have a major impact on the set of density values assigned across the linear segment (or area in 2D applications). Of much greater impact is the choice of the spread parameter, or bandwidth.

All of this discussion addresses problems in one dimension (univariate smoothing). We now need to extend the process to two dimensions, which turns out to be simply a matter of taking the univariate procedures and adding a second dimension (effectively rotating the function about each point). If we were to use the Normal distribution again as our smoothing function it would have a two-dimensional bell-shaped form over every point (Figure 4-36). As before, we place the kernel function over each point in our study region and calculate the value contributed by that point over a finely drawn grid. The grid resolution does not affect the resulting surface form to any great degree, but if possible should be set to be meaningful within the context of the dataset being analysed including any known spatial errors or rounding that may have been applied, and making allowance for any areas that should be omitted from the computations (e.g. industrial zones, water, parks etc. when considering residential-based data). Values for all points at every grid intersection or for every grid cell are then computed and added together to give a composite density surface. This may then be plotted in 2D (e.g. as density contours) or as a 3D surface.

Figure 4-36 2D Normal Kernel

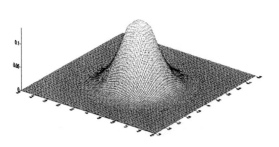

Figure 4-37 Normal kernel density map, lung cancer cases

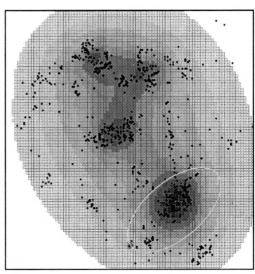

The resulting grid values may be provided as: (i) relative densities — these provide values in events per unit area (i.e. they are adjusted by the grid size, giving a figure as events per square metre or per hectare) — this is the default or only option in many GIS packages, including ArcGIS; (ii) absolute densities — these provide values in terms of events per grid cell, and hence are not adjusted by cell size. The sum of the values across all cells should equal the number of events used in the analysis; (iii) probabilities — as per (ii) but divided by the total number of events. Crimestat supports all three variants.

In Figure 4-37 the kernel density procedure has been applied to a dataset of reported cases of lung cancer in part of Lancashire, England. Cases are shown as points in this map, with areas of higher kernel density being shown in darker tones. The highlighted point in the lower left of the map is the location of a disused incinerator (the white oval is a hypothetical plume extent based on the prevailing wind direction). This example is discussed further in Section 5.4.3, and fully in Diggle (1990).

The software used in this instance was Crimestat, with a Normal kernel function and average spread (often called the bandwidth) determined from the point pattern itself. Map data were output in ESRI shape file format (SHP) and mapped in ArcGIS. Other GIS packages support a variety of kernel functions and procedures. ArcGIS Spatial Analyst provides kernel density estimation for point and line objects, but only supports one kernel function, which it describes as a quadratic kernel (a bounded kernel) but which is often described as an Epanechnikov kernel (see further, Table 4.8). MapInfo's grid analysis add-on package, Vertical Mapper, includes kernel density mapping as does the spatial statistics package SPLANCS and the crime analysis add-on for MapInfo, Hot Spot Detective (in the latter two cases based on quartic kernels). TransCAD/Maptitude supports what it describes as density grid creation with the option of count (simple), quartic, triangular or uniform kernels. Crimestat supports four alternative kernels to the Normal, all of which have finite extent (i.e. typically are defined to have a value of 0 beyond a specified distance). These are known as the quartic, exponential, triangular and uniform kernels.

The details of each of the main kernel functions used in various GIS packages are as shown in Table 4.8. The value at grid location g_j, at a distance d_{ij} from an event point i, is obtained as the sum of individual applications of the kernel function over all event points in the source dataset. The table shows normalised functions, where the distances d_{ij} have been divided by the kernel bandwidth, h, i.e. $t=d_{ij}/h$. Graphs of these functions are shown in Figure 4-38, where each has been normalised such that the area under the graph sums to 1.

Whether the kernel for a particular event point contributes to the value at a grid point depends on: (i) the type of kernel function (bounded or not); (ii) the parameter, k, if applicable, which may be user defined or determined automatically in some instances; and (iii) and the bandwidth, h, that is selected (a larger bandwidth spreads the influence of event points over a greater distance, but is also more likely to experience edge effects close to the study region boundaries). Event point sets may be weighted resulting in some event points having greater influence than others.

Bandwidth selection is often more of an art than a science, but it may be subject to formal analysis and estimation, for example by applying kernel density estimation (KDE) procedures to sets of data where actual densities are known. An alternative to fixed bandwidth selection is adaptive selection, whereby the user specifies the selection criterion, for example defining the number of event points to include within a circle centred on each event point, and taking the radius of this circle as the bandwidth around that point.

Table 4.8 Widely used univariate kernel density functions

Kernel	Formula	Comments. Note $t=d_{ij}/h$, h is the bandwidth
Normal (or Gaussian)	$\dfrac{1}{2k}e^{-\frac{t^2}{2}}$	Unbounded, hence defined for all t. The standard kernel in Crimestat; bandwidth h is the standard deviation (and may be fixed or adaptive)
Quartic (spherical)	$\dfrac{3}{k}\left(1-t^2\right)^2$, $t\leq1$ $=0$ $t>1$	Bounded. Approximates the Normal. k is a constant
(Negative) Exponential	$Ae^{-k\lvert t\rvert}$, $\lvert t\rvert\leq1$ $=0$ $t>1$ (optional)	Optionally bounded. A is a constant (e.g. $A=3/2$) and k is a parameter (e.g. $k=3$). Weights more heavily to the central point than other kernels
Triangular (conic)	$1-\lvert t\rvert$, $\lvert t\rvert\leq1$ $=0$ $t>1$	Bounded. Very simple linear decay with distance.
Uniform (flat)	k, $\lvert t\rvert\leq1$ $=0$ $t>1$	Bounded. k=a constant. No central weighting so function is like a uniform disk placed over each event point
Epanechnikov (paraboloid/quadratic)	$\dfrac{3}{4}\left(1-t^2\right)$, $t\leq1$ $=0$ $t>1$	Bounded; optimal smoothing function for some statistical applications; used as the smoothing function in the Geographical Analysis Machine (GAM) and in ArcGIS

Figure 4-38 Univariate kernel density functions, unit bandwidth

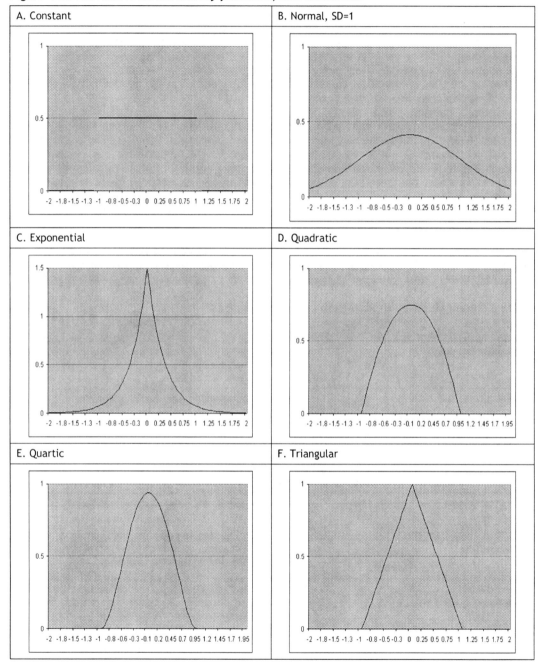

Kernel smoothing methods (KDE methods) of the type described have a variety of applications: point data smoothing; creation of continuous surfaces from point data in order to combine these with other datasets that are continuous/in raster form; probability distribution estimation; interpolation (although this terminology is confusing and not recommended — Crimestat and SPANS are amongst packages that use this terminology, which is essentially incorrect); and hot spot detection. KDE can also be used in visualising and analysing temporal patterns, for example crime events at different times of day and/or over different periods, with the objective of understanding and potentially predicting event patterns. KDE can also be applied with more than one point set, for example a set of cases and a set of controls. The output of such "dual" dataset analysis is normally a ratio of the primary set to the secondary set, with the objective being to analyse the primary pattern with background effects being removed or minimised. Care must be taken in such cases that the results are not subject to distortion by very low or zero values in the second density surface. Crimestat provides support for 6 alternative dual density outputs: simple ratio; log ratio; difference in densities (two variants, with and without standardisation); and sum of densities (again, with and without standardisation). Levine (2005, Chapter 8) provides an excellent discussion of the various techniques and options, including alternative methods for bandwidth selection, together with examples from crime analysis, health research, urban development and ecology.

The use of kernel functions as a form of point weighting enables the creation of local weighted means and variances. The locally weighted (kernel) statistics supported within the GWR software are defined as follows:

$$\bar{x}(u) = \frac{\sum_{i=1}^{n} w(u)_i x_i}{\sum_{i=1}^{n} w(u)_i}$$

and

$$\sigma^2(u) = \frac{\sum_{i=1}^{n} w(u)_i (x_i - \bar{x}(u))^2}{\sum_{i=1}^{n} w(u)_i}$$

These are local statistics, based on locations, u, and weighting function w() defined by a fixed or adaptive kernel function. The bandwidth in this case must be user-defined, and might be chosen on the basis of experience or could use an adaptively derived bandwidth (e.g. from an associated regression study) for comparison purposes.

4.3.4.2 Line and intersection densities

Density methods may be applied to linear forms as well as point forms. Some statistics of interest include:

- the number of line segments per unit area (sometimes called the *line frequency*, λ, rather than the density)

- the length of line segments (by type) per unit area. The equation $L_a = L_0 e^{-r/b}$ has been found to provide a good model of the distribution of urban road length with radial distance, r, from many city centres. L_0=the road density at the city centre, which for London is c.19.5 kms/km^2; b is a parameter to be fitted from sample data. The inverse of this density measure is the average area associated with a unit of path length, sometimes called the *constant of path maintenance*

- the number of *line intersections* per unit area

- the number of line intersections per unit of length (the latter is known as the *intersection density*, k). This is a particular case of the density of points on lines, which may be subject to kernel density smoothing in the manner discussed earlier in this Section. It can be shown that if λ is the line density (frequency) then for a random network of straight lines there will be an average of $k=\lambda^2/\pi$ intersections per unit length. $K=k/\lambda^2$ is sometimes called the *crossings factor*, and experimental evidence suggests that the random model provides an upper limit to this measure, so K has the range $[0, 1/\pi]$.

All these measures provide information that may be useful for analysis, including inter-regional comparisons. Applications include hydrology, ecology and road traffic analysis. For further discussion of applications in these areas and statistical measures see Haggett and Chorley (1969, Chapter 2.II) and Vaughan (1987, Chapter 3).

A number of GIS packages provide facilities for kernel density mapping of lines as well as points (e.g. ArcGIS Spatial Analyst, Density operations). Essentially these work in a similar manner to point operations, spreading the linear form symmetrically according to a specified kernel function. ArcGIS provides two sets of point density and line density functions (for SIMPLE or KERNEL density operations), one set that results in a grid being generated and the other for use in map algebra environments (i.e. where a grid already exists). SIMPLE density operations generate a set of grid values based on a neighbourhood shape (e.g. circular, annular, rectangular, or polygonal) and size (e.g. radius).

Similar facilities do not appear to be available with most other mainstream GIS packages. The GRASS OpenSource package include a vector processing function v.kernel (formerly s.kernel) that now includes linear kernel generation (i.e. a 1D kernel) in addition to its generation of 2D raster grids from vector point sets with a Normal (Gaussian) kernel.

4.4 Distance Operations

Fundamental to all spatial analysis is the use of distance calculations. If our model of the world (mapped region under investigation) is regarded as a flat plane (2D Euclidean space) or a perfect sphere, then we can use the simple distance formulas d_E and d_S provided in Section 4.2 to compute distances between point pairs. These provide the basic building blocks for many forms of spatial analysis within most GIS packages. However, these two formulas are too limiting for certain forms of analysis and additional distance measures are

required. In some cases these alternative measures (or metrics) can be computed directly from the coordinates of input point pairs, but in others computation is an incremental process. In fact incremental methods that compute length along a designated path (between a series of closely spaced point pairs) by summing the lengths of path segments, are the most general way of determining distance, i.e. the notions of distance and path are inextricably linked. Figure 4-39 provides a simple cross-sectional illustration of the kind of issues that arise. We wish to determine the distance separating points A and B.

Figure 4-39 Alternative measures of terrain distance

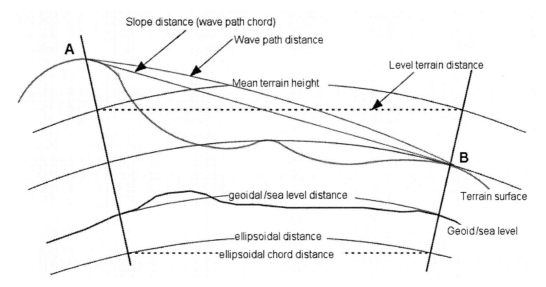

If A and B are not too far apart (e.g. less than 10 kms) we could use a high precision laser rangefinder to establish the *slope distance* between A and B, assuming there is no atmospheric distortion. In practice there will be some distortion and the laser wave path will need to be adjusted in order to provide an estimated slope path distance. This in turn will require further adjustment if it is to be referenced to a common datum or a level terrain surface. In each case the distance recorded between A and B will be different. From the diagram it is clear that none of these

distances corresponds to the actual distance across the terrain surface along a fixed transect, nor to a distance adjusted or computed to reflect the particular model of the Earth or region of the globe we are using. In some cases these differences will be small, whilst in other they may be highly significant. In general simple Euclidean 2D (incremental) computations are satisfactory for many problems involving relatively smooth terrains extending over an area of perhaps 20km by 20km, even using suitably defined longitude and latitude coordinates (assuming these latter coordinates are not inside the Arctic or Antarctic regions). Closer to the poles the distortion of lines of latitude results in increasing errors in this formula and a variant known as the Polar Coordinate Flat-Earth formula may be used as an improved approximation:

$$d = R\sqrt{A - B}$$

where

$$A = (\pi / 2 - \phi_1)^2 + (\pi / 2 - \phi_2)^2$$
$$B = 2(\pi / 2 - \phi_1)(\pi / 2 - \phi_2)\cos(\lambda_2 - \lambda_1)$$

In this formula we are using polar coordinates (ϕ, λ) to represent latitude and longitude expressed in radians, and R = the radius of the terrestrial sphere. With this formula the error is at most 20 metres over 20 kms even at 88°N or S. For example, the formula evaluates the distance between A(70°N,30°N) and B(71°N,31°N) as 117.6 kms (using the WGS84 ellipsoid), 112.8 kms 10° further north, and 111.4 kms at 87°N/88°N. One degree separation (ignoring longitude differences) would be 111.3 kms.

However, for: (a) larger regions, or problems where *surface distance* is important; (b) problems involving networks; and (c) for problems involving variable costs; alternative procedures are required. Within mainstream GIS packages a range of tools exist to support such analysis, including the provision of support for Euclidean and Spherical distance (but not always ellipsoidal distance), network distance and cost distance (where cost is a generalised concept as we discuss in the subsections that follow).

Most software packages offer at least three categories of distance measure: (i) coordinate-based distance measures in the plane or on a sphere (which may be adjusted to a more close approximation of the Earth's shape); (ii) network-based measures determined by summing the stored length or related measure of a single link or multiple set of links along a specified route (e.g. shortest, quickest); and (iii) polyline-based measures, determined by

summing the length of stored straight line segments (sequences of closely spaced plane coordinates) representing a given feature, such as a boundary. The first and third of these distance measures utilise the simple and familiar formulas for distance in uniform 2D and 3D Euclidean space and on the surface of a perfect sphere or spheroid. These formulas make two fundamental assumptions: (i) the distance to be measured is to be calculated along the shortest physical path in the selected space — this is defined to be the shortest straight line between the selected points (Euclidean straight lines or great circle arcs on the sphere); and (ii) that the space is completely uniform — there are no variations in terms of direction or location. These assumptions facilitate the use of expressions that only require knowledge of the coordinates of the initial and final locations, and thereby avoid the difficult question of how you actually get from A to B.

In most cases measured terrestrial distances are reduced to a common base, such as the WGS84 reference ellipsoid. Note that the reference ellipsoid may be above or below the geoidal surface, and neither distance corresponds to distance across the landscape. Laser-based measurement is rarely used for distances of 50km or more, and increasingly DGPS measurements of location followed by direct coordinate-based calculations avoid the requirement for the wave path adjustments discussed above.

Whilst coordinate-based formulation is very convenient, it generally means that a GIS will not provide true distances between locations across the physical surface. It is possible for such systems to provide slope distance, and potentially surface distance along a selected transect or profile, but such measures are not provided in most packages. A number of packages do provide the option to specify transects, as lines or polylines, and then to provide a range of information relating to these transects. Examples include the Cross Section facility in MapInfo's Vertical Mapper, the Grid|Slice command procedures in Surfer, the Profiles and Elevations facilities in

Manifold (Surface Tools) and the Profile facility in ArcGIS 3D Analyst. In addition to path profiles (elevations, lengths and related transect statistics) such paths may be linked with volumetric cut-and-fill analyses (see further Sections 6.2.3 and 6.2.6).

For problems that are not constrained to lie on networks it is possible to derive distances from the surface representation within the GIS. For example this can be achieved using a regular or irregular lattice representation along whose edges a set of links may be accumulated to form a (shortest) path between the origin and destination points. For the subset of problems concerned with existing networks, such questions could be restricted to locations that lie on these networks (or at network nodes) where path link lengths have been previously obtained. Both of these approaches immediately raise a series of additional questions: how accurate is the representation being used and does it distort the resulting path? which path through the network or surface representation should be used (e.g. shortest length, least effort, most scenic)? how are intermediate start and end points to be handled? are selected paths symmetric (in terms of physical distance, time, cost, effort etc.)? what constraints on the path are being applied (e.g. in terms of gradient, curvature, restriction to the existing physical surface etc.)? is the measure returned understandable and meaningful in terms of the requirements? how are multi-point/multi-path problems to be handled? Answers to some of these questions are discussed in Section 4.4.2, but we first discuss the notion of a metric.

4.4.1 Metrics

4.4.1.1 Introduction

The term metric in connection with spatial analysis normally refers to the form of distance calculation applied, although it has a number of other meanings (e.g. the metric system of weights and measures; in the Manifold GIS package it means the sequence of coordinates that defines a vector object). Here we stick to its use in connection with distance computation. The essential features of the notion of distance are:

- a finite or infinite set of distinct, definable objects, and

- a rule or set of rules for determining the degree of separation of object pairs (a "measure").

From a mathematical perspective a metric d_{ij}, denoting a measure of separation between points i and j, should satisfy a number of specific conditions. These are:

- $d_{ij}>0$ if $i \neq j$ (distinction/separation)

- $d_{ij}=0$ if $i=j$ (co-location/equivalence)

- $d_{ij}+d_{jk} \geq d_{ik}$ (triangle inequality)

- $d_{ij}=d_{ji}$ (symmetry)

The standard formulas d_E and d_S we discussed earlier in this Guide meet all four of these requirements, but there are many situations in which one or more of these conditions fails to be met. Particularly common is cases in which either the triangle inequality or symmetry conditions fail. Typically these occur in transport networks, where one-way routing systems apply or where there are variable tariffs or traffic flow asymmetries. Circular routes, such as the inner and outer circuits of Glasgow's "Clockwork Orange" underground system (Figure 4-40) provide an example of asymmetry. This service operates one clockwise and one anti-clockwise route, each of length 10.4 kms and transit time of 24 minutes. On a given circuit, the journey-time from Kelvinbridge to Hillhead, for example, which is 2 to 3 minutes, is clearly not the same as Hillhead to Kelvinbridge (over 20 minutes), and between stations it is not possible to change circuits.

Figure 4-40 Glasgow's Clockwork Orange Underground

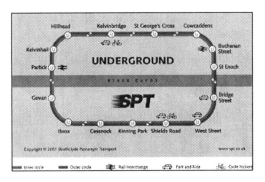

Asymmetry is also a characteristic of a number of telecommunications networks, where some modem standards and some of the more recent broadband internet technologies (e.g. ADSL) operate in this way – information is generally received at a faster rate than it can be transmitted. A similar situation applies in many hybrid broadcast/telecommunications systems, such as satellite, coax or fibre-optic downstream and copper/cable upstream, and in some mobile network architectures. Pure broadcast technologies, such as terrestrial television and radio, are strictly one-way and only facilitate two-way communication via completely separate media (such as post, telephone or email). This kind of asymmetry also reflects a fundamental human characteristic – our ability to receive and process information is far greater than our ability to respond to it in an active manner.

If the triangularity constraint is not met we have the possibility that $d_{ij}+d_{jk}<d_{ik}$ which offends our notion of "between-ness" and thus of conventional notions of distance and space. For example, this would suggest that the "distance" from London to Madrid plus the "distance" from Madrid to Lima might be *less than* (cheaper than, faster than) the direct "distance" from London to Lima. A similar pattern can be found with railway and other transport timetables, where express trains and stopping trains in combination may reach a destination faster than a direct stopping train, even though the rail distance travelled may be greater. It may not be possible to draw a time- or space-consistent map of such data, but this does not mean that the measure used is not valuable. Indeed, in such cases non-cartographic representations (e.g. GIS internal representation, to/from matrices) of the data are more meaningful and useful in optimisation and routing problems, and as such are explicitly supported in GIS-T packages such as TransCAD.

4.4.1.2 Terrestrial distances

Most GIS software uses the standard Euclidean formula, d_E, when computing distances between plane coordinate pairs and for calculation of "local" distances, as in the case of the buffer operator (see further, Section 4.4.4). Some, such as MapInfo, provide the option of using spherical distance, d_S, as a default. Manifold computes distances either using Euclidean or Ellipsoidal methods (supported via its functions Distance() and DistanceEarth(), and in its Tracker facility). The details of the latter computations are not provided. The MATLab Mapping Toolbox provides a range of distance and associated mapping functions, which help to illustrate some of the differences that apply on larger scale maps. Figure 4-41 shows a map of the North Atlantic on a Mercator projection, with the great circle path and line of constant bearing, or rhumb line between Boston, Mass. and Bristol, England. The MATLab function distance() will compute the length of each path, taking the latitude and longitude of Boston and Bristol as inputs, and optionally a geoid parameter. The GRASS module geodist.c provides similar functionality. In the example shown the great circle distance is computed as 5105.6 kms based on its default terrestrial radius of 6371 kms; the Haversine formula we gave earlier in this Guide produces the same result, whilst the rhumb line distance is 5283.4 kms.

Figure 4-41 Great circle and constant bearing paths, Boston to Bristol

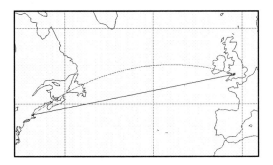

The geoid parameter, rather confusingly named, provides the option to select a model of the Earth which has an ellipsoidal form, specified by a two element vector: (i) the length of the major semi-axis of the ellipsoid and; (ii) its eccentricity or flattening coefficient. If these parameters are not specified it is assumed that the two semi-axes are equal, and thus provide a spherical or great circle measure. If geoid (ellipsoid) values are provided then an approximation method is used to calculate distances. It first treats the two terrestrial coordinates as if they were Cartesian coordinates in three dimensions and calculates the Euclidean distance between them. This value, which is the chord length of an elliptical section, is then adjusted by a correction factor to produce the final estimate of ellipsoidal distance. In our example this gives a revised distance of 5110.4 kms, a few kilometres longer than the great circle distance, as expected. Finally, yet another alternative is to use the more accurate Vincenty (1975) algorithm to obtain the ellipsoidal distance, which gives a figure of 5119.7 kms for the matching ellipsoid in this example.

4.4.1.3 Extended Euclidean and L_p-metric distances

Euclidean distance is the most widely used GIS metric. In 2D space (planar or projected space) the Euclidean distance between points $a(x_1, y_1)$ and $b(x_2, y_2)$ is a special case of the n-dimensional Euclidean metric. Within GIS this extended Euclidean metric is used to determine 3D distances (across surfaces and within solid structures) and in multi-dimensional analyses such as clustering. For 3D problems using $\{x, y, z\}$ as our coordinate representation the measure is simply:

$$d(a,b) = \left(\left(x_1 - x_2\right)^2 + \left(y_1 - y_2\right)^2 + \left(z_1 - z_2\right)^2 \right)^{1/2}$$

More generally we may represent our points a and b using subscripts to represent the dimensions $1, 2, \ldots n$. Thus we have $a = (a_1, a_2)$ and $b = (b_1, b_2)$ for the 2D case. The general form of the Euclidean metric in n-space then becomes:

$$d(a,b) = \left(\sum_{i=1}^{n} (a_i - b_i)^2 \right)^{1/2}$$

This formula is processor-intensive to compute for very large number of point pairs and dimensions, mainly due to the square root operation. For some applications, such as K-means clustering (see Section 4.2.12) and line generalisation (see Section 4.2.4) the square root is not taken, which breaks its metric behaviour, but remains suitable for these specific purposes. In other areas, such as incremental grid operations, approximations to the Euclidean metric are taken to avoid such problems (see further, Section 4.4.2.2).

The formula above also arises, with some modifications, in the analysis of statistical datasets. The standard Euclidean metric in n-space takes no account of the average variation of the data points. However, it can be normalised (made scale free) by calculating the variance of the point set components, σ_i^2, and using these as a divisor:

$$d(a,b) = \left(\sum_{i=1}^{n} \frac{(a_i - b_i)^2}{\sigma_i^2} \right)^{1/2}$$

This is known as the *normalised* Euclidean metric (see also, Section 4.3.3) and is essentially the same as the Euclidean metric if the variances of point set components (the average variation of values or attributes of each dimensional component are equal. Pre-normalising the attributes of input datasets (e.g. raster files) and then computing conventional n-dimensional Euclidean distances will produce the same result. This expression is also closely related to a more general measure, introduced by P C Mahalanobis in 1936 and named after him, which is used to determine the similarity of an unknown sample set to a known one. In addition to the variances of the components *Mahalanobis distance* takes into account the covariances that may exist. This more general measure is used in a number of GIS packages in connection with the analysis of remote sensing data.

For applications that require network distance measures between locations, Euclidean distance is generally inadequate and computing large numbers of (shortest) network distances may be impractical or inappropriate. In order to obtain more useful measures of inter-location distance many authors have suggested that the standard Euclidean distance formula be amended in various ways. The most frequently used modifications are based on replacing the use of the square and square root operations — i.e. replacing the powers of 2 and 1/2 — with other values, typically p and $1/p$, estimated from sampling the network to be analysed. We shall refer to such modifications as the family of L_p metrics or p-norms, with L_2 being the standard Euclidean measure. They are amongst a range of alternative metrics supported by the Locational Analysis package, LOLA, which provides output in ArcGIS shape file format (SHP) as an option (see further, Section 7.4).

The distance between two points in the plane: $a(x_1,y_1)$ and $b(x_2,y_2)$ with this metric is defined as $d_p(a,b)$, where:

$$d_p(a,b) = \left(\left| x_1 - x_2 \right|^p + \left| y_1 - y_2 \right|^p \right)^{1/p}$$

Figure 4-42 plots "circles" using various values for p, with the case $p=2$ (the Euclidean metric) giving a perfect circle — the only value which provides a rotationally invariant distance measure. The lines show the position of a point at a fixed distance from the centre using the simple p^{th} power/p^{th} root model.

Figure 4-42 p-metric circles

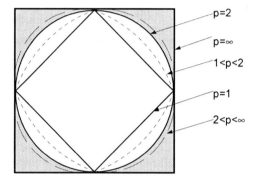

In geographic literature the case $p=1$ is variously called the Manhattan, Taxicab, Rectilinear or City block metric, and is supported in a number of software packages, including LOLA and the Crimestat package (in the latter instance with both Plane and Spherical computations). With $p=2$ we have the standard Euclidean metric and with $p=\infty$ we have the so-called minimax or Chebyshev metric used in a number of optimisation problems. Note that the shape of the "circle" for $p=1$ is a square at a 45 degree angle, and this is the same as the shape for $p=\infty$.

Values for p of approximately 1.7 have been found by many studies to approximate road distances within and between towns, and such estimates may be used in place of direct-line distance (where $p=2$) in problems such as locational analysis and the modelling of travel patterns — see for example, Love *et al.* (1972) and (1988). An alternative is to use a simple *route factor* (typically around 1.3) as a multiplier and to continue using Euclidean distances, but inflated by this route factor. Note that if $p<1$ the triangle inequality no longer holds, i.e. it is possible for the distance (or time) from *A* to *B* plus the distance from B to *C* to be *less than* the distance (or time) from *A* to *C*. This characteristic is quite common in transport infrastructures, where one-way systems, timetabled routes and traffic congestion alter the expected paths significantly. Numerous variations on the simple L_p metric model have been studied, including using a mixture of powers for the *x*- and *y*-directions (*p* and *q*), adding weighting factors for the two directions, including a different power, *s*, for the root etc. Most of these variations provide modest improvements in estimation of transport route distances at the expense of increased complexity in determination of the parameters. A summary of the possible values for *p* in the L_p metrics and their interpretation is provided in Table 4.9.

One of the most striking features of the L_p metric diagram (Figure 4-42) is that for every value of *p* except 2 the "circles" show rotational bias — maximised at 45 degrees. Similar bias exists if other variants of the L_p metric family are plotted. This highlights the fact that a distance value obtained by measurement and application of any of these modified formulas will change if the reference frame is rotated. If this seems a bit obscure, imagine measuring the distance between two points randomly located on a rectangular street network oriented north-south/east-west. In general this will involve finding the length of a route in the north-south direction and adding the length of a route in the east-west direction. If, however, we could ignore the physical network and rotate and translate the pattern of streets about one of the points until the two points both lay along one road (a rotated north-south route), the length measured would be shorter. This comment applies equally to grid files, where distances (and many other measures) are often calculated in an incremental manner which is affected by the grid orientation.

Table 4.9 Interpretation of p-values

0<*p*<1	The L_p formula does not yield a metric because the triangle inequality no longer holds — for example, construction of minimal spanning trees on the basis of progressively connecting nearest-neighbours is not valid. Routes in cities may have values of *p*<1 indicating that routes are complex — a value of *p*=2/3 indicates that distances are effectively twice as far as direct line estimates would suggest. The locus of points in this case is concave.
1<*p*<2	The distance, *d*, is a metric which is intermediate between the Manhattan grid-like street pattern and the normal Euclidean metric, and is thus applicable to more complex city street networks.
p>2	As *p* increases greater weight is given to the largest components of distance in the network. Consider the distance from (0,0) to (1,1-*E*) where 0<*E*<1. For *p*=1 *d*=2-*E*; for *p*=2 and *E*<<1 $d \approx \sqrt{2}\sqrt{(1-E)}$ and as *p* increases, *d*→1. In all cases as *p*→∞ then *d*= max($|x_1-x_2|$, $|y_1-y_2|$)

The above observations on rotational relationships can be used in problem solving — for example, in some cases it may be possible to solve selected problems by applying the sequence of actions:

- rotate the problem coordinate frame

- solve the problem using the metric appropriate to the rotated frame, and then

- rotate the frame back to its original position.

For example, Okabe *et al.* (2000, Section 3.7) show that in order to determine the Voronoi diagram for a set of planar points under the L_∞ metric, the set of points may be rotated by an angle of +π/4 radians and the regions computed using the L_1 metric, with the resulting diagram rotated back by -π/4 radians to produce the final solution.

Although the *p*-metric family provides real value in analytic studies, this lack of rotational invariance is undesirable in the context of distance measurement. Clearly it would be simple for suppliers to include such distance expressions in their GIS packages, but the estimation of the parameters and the interpretation of their results require considerable care. In practice the advent of more powerful hardware and software, and greatly improved network datasets, has meant that many problems can be tackled by explicit network analysis rather than the application of alternative metrics (see further, Chapter 7).

4.4.2 Cost distance

The term cost distance is widely used within GIS and in this context has a dual meaning: (i) the notion of an alternative family of distance metrics; and (ii) a procedure for determining least cost paths across continuous surfaces, typically using grid representations. Within GIS the term is distinct from the computation of varying costs or times of travel across a network. The term cost in this connection is also a generic idea, not necessarily implying financial cost, but some composite measure that varies (i.e. is not constant) across the study region, and which needs to be taken into account when computing paths and distances. An example serves to clarify this notion.

Suppose that we wish to construct a section of road between two points, $A(0,2)$ and $B(4,2)$, where costs are constant over the region of interest. In this case the distance is simply 4 units and the path involved is a straight line in the plane (Figure 4-43, bold line). Construction costs will simply be $4k$ units, where k is some constant (the cost per unit length of path construction). Now suppose that construction costs also depend on the cost of land acquisition, and that land costs vary in our study region in a simple manner — greater to the "north" of our region than to the "south". In this situation it is cheaper to construct our road with a curve in it, taking advantage of the fact that land costs are lower to the south (nearer the *x*-axis in our model).

Figure 4-43 Cost distance model

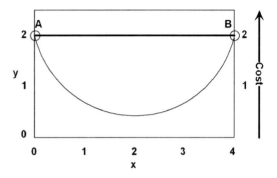

If costs increase in a precisely linear manner, the optimal (least cost) path will have a smooth curve form of the kind shown by the thinner line (a shape like a chain hanging between two posts). This case turns out to be solvable mathematically, as are cases where costs vary in radially symmetric manner (e.g. around a city centre), whereas almost every other example of varying costs cannot be solved precisely and we need to resort to computational procedures to find the least cost path.

4.4.2.1 Accumulated cost surfaces and least cost paths

Within GIS packages solutions to such problems are found using a family of algorithms known as Accumulated Cost Surface (ACS) methods — for a seminal discussion of ACS methods and representational accuracy see Douglas (1994). The ACS procedure must be applied to raster datasets, in which the primary input surface is a complete grid of generalised costs, i.e. every cell is assigned an absolute or relative cost measure, where costs must be (positive) ratio-scaled variables. In Figure 4-44 we show a hypothetical cost grid for the same example as shown in Figure 4-43. Values increase linearly away from the x-axis, from 0.5 to 2.5. This essentially a much simplified representation of the function $f(x,y)=y$, so $f(x,y)=0$ when $y=0$ for all x and increases steadily so at $y=2$ $f(x,y)=2$. In practice cost functions should rarely if ever have zero-valued sections, so this example should be considered illustrative!

Providing a gridded representation of the cost surface highlights the importance of using a sufficiently fine representation of the surface, since the solution path must be computed by adding up the costs along alternative paths. If the grid resolution is quite coarse, as in Figure 4-44, the solution path will not have the curved form of the optimal path, but will simply be a very blocky stepped line of cells approximating the curve. Amending the grid resolution (halving the cell size to 0.25 units) as shown in Figure 4-45A improves both the cost surface representation and the possible solution paths that are considered.

Figure 4-44 Cost surface as grid

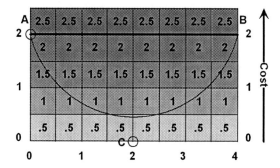

If we look at a path that starts at $A(0,2)$ and goes diagonally down towards $C(2,0)$ and then back up to $B(4,2)$ we have cell costs of 2+1.5+1+.5+.5+1+1.5+2=10 units (times cell size of 0.5 gives a cost distance of 5 units). But this path is longer than the sum of the steps — all the diagonal steps are $\sqrt{2}$ (1.414... units) longer than horizontal and vertical steps, so the total cost must be increased for such steps by this factor, giving a final cost distance of 6.85 units. This is less than the cost of 8 units that would be incurred for a straight line path from A to B. But can we be sure this is the best route possible for this problem and is the calculation "correct"?

Figure 4-45 Grid resolution and cost distance

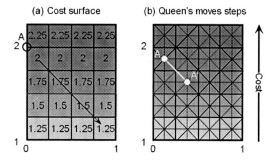

ACS provides a systematic procedure for answering this question. With this algorithm we start at A and sum up the costs in all directions, step by step, choosing the least cost increment at each step. Eventually, when we arrive at B, we will have the overall least cost of getting to B, and if we have kept track of our route (the steps we have taken) we will have the least cost path from A to B. Thus ACS is a form of spread algorithm, with values increasing in all directions depending on the grid resolution, cost surface and algorithmic implementation. Almost all GIS implementations of ACS use immediate neighbours in calculating values (i.e. queen's moves, to the 8 adjacent cells) but some facilitate calculation to neighbours further away. GRASS, for example, offers the option in the r.cost function, of using knight's move neighbours (one step up and one diagonal) in its calculations, and distance transform methods (see further Section 4.4.2.2) utilise a much wider range of neighbours, and hence are not restricted to 8-directional cost surfaces and path alignments.

As we start to consider the ACS idea more closely we see that a number of additional issues need to be addressed. The first is that in a GIS-enabled grid structure cells (or more specifically cell centres) are frequently used as references rather than grid lines. Thus our starting point A cannot be (0,2) exactly but must be a cell centre and our end points will also all be cell centres. For example, see Figure 4-45B where $A(0,2)$ is approximated on the grid by the point $A(0.125, 1.875)$. For the best results we should estimate the value of the cost function at cell centres if possible, rather than at cell edges or grid lines. In the example we are using we have ignored this latter factor, which will often be the case since finer detail for within cell estimation may not be available (although could be obtained by interpolation).

A second issue is that if we take grid cell centres as our reference point, the start point will be half way across the first cell, and the next point will be half way across an adjacent cell, so it can be argued that the cell values (costs) should be allocated 50:50 over this one step, and likewise for steps 2, 3 etc. This can be seen in Figure 4-45B where the starting point for A is shown as a selected cell centre and the step to A' in the diagonally adjacent cell lies 50% in a cell with cost 2 and 50% in a cell with cost 1.75. This 50:50 division of costs is the normal approach used in GIS packages. The averaged cost from A to A' is then $(2+1.75)/2=3.75/2$ and thus the incremental cost distance is $\sqrt{2}*0.25*3.75/2$ units (adjusting for the diagonal path using $\sqrt{2}$ and for the cell size of 0.25 units in this example).

We now have all the basic elements we need in order to compute accumulated costs and identify the least cost path from A to B. Using our example cost grid, with a cell size of 0.25 units, we need to carry out the following operations using a typical GIS:

- Create a source grid that contains a 0 in the cell in which our starting point A lies, and some other value (e.g. -1, 9999) for all other grid cells. This can be done with a text editor and converted to a GIS compatible grid format if necessary

- Create a cost grid in the same manner, exactly the same size and ideally the same resolution as the source grid, but containing the costs associated with each cell position (i.e. similar to the cells in Figure 4-45a)

- Run the cost distance facility within the GIS on these two input grids, requesting the output as a cost distance (ACS) grid, and also requesting (if necessary) that a direction or tracking dataset is generated that can be subsequently used to identify least cost paths

- Finally, to create the least cost paths, a target grid is required, which contains the target or destination points (again as 0s or a similar cell identifier for the target points). The least cost path facility in the GIS is then run using the target grid, ACS grid and tracking grid.

A sample result for our test problem is shown in Figure 4-46. Each grid cell has been separately coloured and a series of additional destination points (cells) have been included in the least cost path finding stage. The least cost path from A to B, corresponding to our original problem, hugs the bottom of the diagram as it seeks to take advantage of the low cost first row (cells with a value of only 0.25 units). Routes to other cells may share common paths for parts of their route and may not be unique, as indicated by the two routes to the cell diagonally below point A to the right.

Figure 4-46 Accumulated cost surface and least cost paths

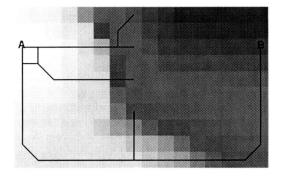

If the cost surface had been flatter, for example $f(x,y)=y/4+1$, the ACS model with the cell resolution illustrated would fail to select the correct path, but would show a horizontal path from A to B. To obtain a more accurate estimate of the correct path and cost distance a much finer grid would be required. For this latter example the true path is a gentle curve with a minimum value at around y=1.65 and total cost of 5.88 units.

This simple example provides a flavour of cost distance computation, but this procedure (or family of procedures) has far more to offer. The first, and perhaps most important point to note is that the source and target features do not need to be single points. They can be multiple distinct points (cells), lines (linear groups of cells), or regions (clusters of adjacent cells). In all such cases the ACS procedure computes the lowest total distance to a target cell from its nearest source cell. If requested the GIS may also produce an allocation raster, which identifies which source each cell in the grid is closest to (typically by giving each cell a unique integer identifier). Obstacles (e.g. lakes, cliffs etc.) may be included in these models by setting cells in the cost raster to very high values or by the GIS explicitly supporting unavailable (masked) cells.

The input cost raster may be, and often is, a composite grid created from a number of other (normalised) input grids by simple map algebraic operations. The final cost grid may also be normalised to provide an indication for each cell of the relative cost, difficulty or *friction* of crossing that cell. Figure 4-47 illustrates the result of such a process, which has been applied to determining possible alternative routes for a main road in part of South-East England. The map shows the existing road (the A259T) from Hastings in the south-west to Brenzett in the north-east. Three alternative routes, identified here by arrows, were generated by ACS methods. These routes were generated from cost surfaces that included different weighted values for a series of inputs: land use (grade and type of land); SSSIs (Sites of special scientific interest); areas of designated Ancient Woodland; the slope of the land across the study area (derived from a DEM); existing

(buffered) built-up areas (many of which are conservation areas) and ancient monuments; National Trust land; and existing (buffered) road and rail routes. The final route alternatives were then computed from the composite ACS surface and smoothed to reduce the effects of the gridding of all the datasets and output paths. This information was then provided as input to a public consultation exercise.

Figure 4-47 Alternative route selection by ACS

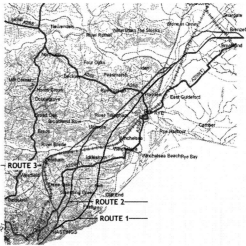

© Crown Copyright Data

The above examples are based on the procedures utilised in ArcGIS, Spatial Analyst. Several other packages offer similar facilities. For example, Idrisi offers two alternative algorithms: one, called the *pushbroom* algorithm, is similar to that described above and is fast but not suitable for use with more complex problems (for example truly implementing barriers rather than setting selected cells to high values, and supporting street patterns as well as open regions); a second Idrisi procedure, *costgrow*, is much slower but does facilitate complex friction patterns. Idrisi also allows for anisotropy in the cost modelling process, with some directions (e.g. downhill) being preferable to others (e.g. uphill). MapCalc, GRASS and PCRaster provide similar functions, with inputs optionally including a friction surface (with relative friction values and absolute barriers) and a physical surface, with movements restricted to uphill, downhill and either stopping at or crossing flat areas. Such procedures are clearly targeted at hydrological and related studies, so whilst being well suited to these environments are not readily generalised. Many such programs, particularly those with an environmental sciences and remote sensing background, perform the path finding stage using a form of steepest descent procedure (often described as a *drainage* procedure) applied to the ACS. This approach is not guaranteed to provide the correct shortest path (see for example Figure 4-48).

Figure 4-48 Steepest path vs tracked path

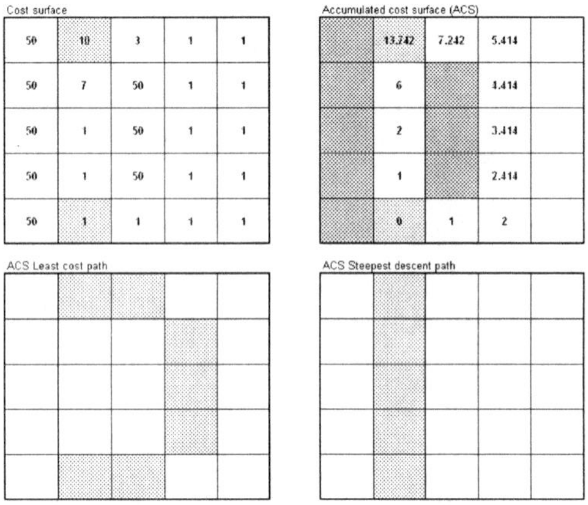

These procedures are subject to a number of other problems, for example: becoming trapped in localised plateaux or pits; inadequate handling of multiple equivalent paths; or failing when input and/or output datasets are specified to operate in integer rather than float mode.

A number of GIS packages incorporate extensions to the standard cost distance/ACS model. For example, ArcGIS provides a variant of the ACS procedure which it describes as Path Distance. This is essentially the same as ACS but includes options such as inclusion of

surface distance (using a surface raster or DEM), and vertical and horizontal movement rasters that are utilised in adjusting the optimum path computation. TNTMips also includes a path analysis procedure that takes into account the physical surface form, but the authors are cautious about its optimality.

Idrisi provides a similar facility, Disperse, designed to provide a simple force-based model of anisotropic object dispersal. In this case the inputs are a set of four grid files: (i) the location of the target and source features (non-zero entries are targets, zeros are sources); (ii) and (iii) two grids, one containing the magnitude and the other containing the direction (azimuth) of the anisotropic force field, i.e. a vector field modelled using a pair of grids; and (iv) an output grid. The notion of the magnitude of the force field in this implementation is as the inverse of a friction field, F, force=$1/F$ where F in this case is taken to be greater than 1. The anisotropic dispersal model is either user specified or utilises an inbuilt cosine function that modifies the stated friction field:

$$F'=F*f, \text{ where } f = \frac{1}{\cos^k \alpha}$$

α is an angular difference between the angle being considered and the stated friction angle, and k is a parameter ($k=2$ is the default). The process may be modelled as a series of steps (time periods) by pre-specifying a maximum dispersal distance at each stage.

4.4.2.2 Distance transforms

In the image processing world a procedure similar to ACS has been developed in connection with work on pattern recognition. The procedure is known as the Distance Transform (DT) process, and is typically implemented on binary images in a scanning manner, starting from the top left of the image, proceeding to the bottom right, and then scanning back to the top left. Extremely efficient algorithms have been developed to generate exact Euclidean DTs and excellent approximations to exact Euclidean DTs on large images, including implementation within hardware for real-time applications. GIS vendors do not currently utilise DT methods, but they are included in many image processing toolsets and packages, including the MATLab Image Processing Toolkit — see the bwdist() function, which provides a variety of metrics for black and white images, including City Block and exact Euclidean. Another example with more direct links with GIS is the Leica product suite, ERDAS Imagine.

The basic DT procedure has been extended and developed by Ratti *et al.* (2001, 2004) and de Smith (2004, 2006) to provide a new and fast family of ACS-like procedures for solving many grid-related distance problems. Details of the procedures and sample code are provided in the references cited, so we limit discussion here to a small sample of applications. The first deals with radially symmetric models of traffic in cities (see Angel and Hyman, Urban Fields (1976) for a full discussion of the models involved).

Figure 4-49A shows the result of applying a least cost distance transform (LCDT) to a city with traffic speeds varying from almost 0kph in the centre (where the crossed lines meet) to much higher levels (e.g. 50kph) at the city edge. The widening concentric circles in the diagram represent the cost field, in this case being traffic speeds (a velocity field) and the curved lines (blue/dark grey and light grey) show contours of equal travel time (isochrones) to/from a point due South of the city centre. Shortest paths (minimum travel time paths) across this transformed surface may be plotted by following routes to or from the source point in any given direction, remaining at right angles (orthogonal) to the equal time contours, or by using vector tracking of the computed shortest paths (Figure 4-49B). Angel and Hyman applied models of this type to the analysis of traffic flows in Manchester, UK. Distance transform methods enable such models to be calibrated against known analytical results and then applied to more realistic complex velocity and cost fields, for example to examine the

possible effects of introducing central zone congestion charging.

Figure 4-49 Urban traffic modelling

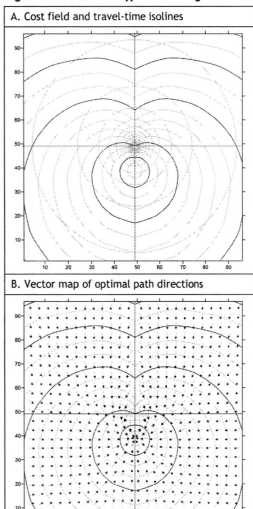

A. Cost field and travel-time isolines

B. Vector map of optimal path directions

A second example of this procedure is shown in Figure 4-50. This shows a section of the Notting Hill area of central London, with the annual Carnival route shown in white and areas in dark grey being masked out (buildings etc.). Colours indicate increasing distances (in metres) from the Carnival route along nearby streets. Although this example has been created by a DT scanning procedure the result is sometimes described as a "burning fire" analogy, with the fire spreading from the source locations (in this case a route) outwards along adjacent streets and squares — for true fire spread models see www.fire.org and Stratton (2004). The same techniques can be applied in public open spaces and even within building complexes such as shopping malls, galleries and airports. This illustrates the ability of DT methods to operate in what might be described as complex friction environments.

Figure 4-50 Notting Hill Carnival routes

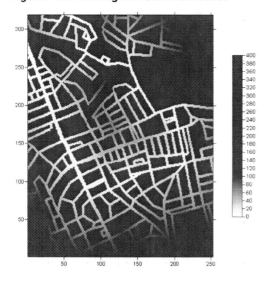

Distance transforms are very flexible in their ability to incorporate problem specific constraints. For example, in a typical ACS procedure variable slopes are modelled by regarding steeper slopes as higher cost, rather than relating the spreading process to slope directly (other than some uphill/downhill constrained implementations). The route for a road, railway or pipeline is affected by slope, but primarily in the direction of the path rather than the maximum slope of the physical surface. Figure 4-51 illustrates the application of a gradient-constrained DT (GCDT) to the selection of road routes through a hilly

landscape. The most northerly pair of these routes was constrained to a 1:12 and 1:11 gradient (before engineering grading) whilst the most southerly pair represent routes found using less restrictive 1:9 and 1:8 gradient constraints.

Figure 4-51 Alternative routes selected by gradient constrained DT

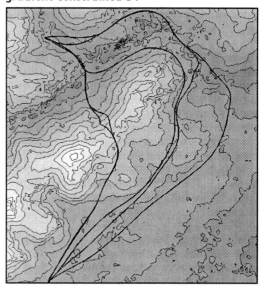

With minor modifications the same procedure has been utilised with a very fine resolution (10metre) DEM to locate pipeline routes for an 80MW geothermal power station in Iceland by Kristinsson, Jonsson and Jonsdottir (2005). In this case additional constraints were applied giving the maximum and minimum heights permitted for the route and differential directional constraints that permitted limited flow uphill, which is possible for these kind of two-phase pipelines (Figure 4-52A,B). Ratti and Richens (2004) and Ratti, Baker and Steemes (2004) have applied similar ideas to urban landscapes, deriving measures of accessibility, noise diffusion, wind flow and solar factors (shadows, energy, light etc.) from DEMs utilising DTs to determine distances within the built environment as part of LT (lighting and thermal) modelling.

Figure 4-52 Hellisheiði power plant pipeline route selection

A. Overview of power plant area

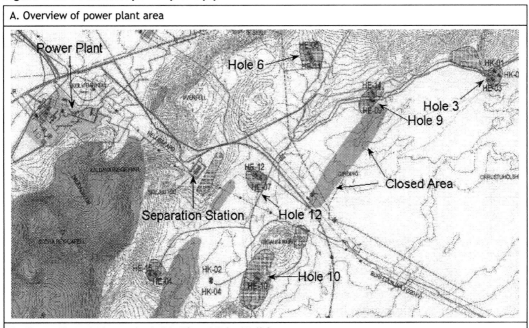

B. Alternative solution paths A, B1, B2 and C for Well 3

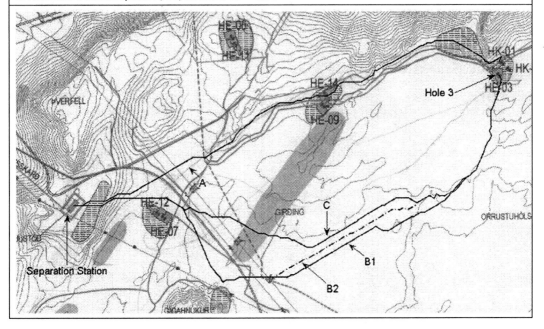

4.4.3 Network distance

Network distance, as the name suggests, is the distance (or cost, time etc.) involved in traversing a network. In most instances the term refers to the lowest value distance measure between pairs or lists of points (generally network nodes or points upstream and/or downstream in hydrological applications). Typically the network is a logical collection of linked polylines, joined at network nodes (or vertices), and possessing a network topology (sometimes called a geometry). Network links and nodes may have additional explicit attributes, such as type of link, weights applied to links and nodes, rules governing behaviour of flows (e.g. link directionality, permitted turns, barriers) etc., and intrinsic attributes (e.g. position and segment lengths). When network distances are computed by a GIS the procedure takes into account the various explicit and intrinsic attributes of the network and the rules applied for the problem selected. Thus network distance will vary in meaning according to the problem at hand. A variety of hybrid concepts exist, which may be modelled within some GIS packages, but generally only in a limited manner. These include multi-modal network (thus seeking routes that may combine road, rail and bus for example), and mixed network/cost distance situations, in which sources or targets may not lie on the network, but for which sections of optimal routes (or network building programme) need to utilise existing network infrastructures.

Most GIS packages do not include extensive networking tools as part of their standard offering, but may provide limited functionality in this area (e.g. ArcGIS Utility Network Analyst and Linear Referencing tools). Manifold does include a range of network analysis tools via its Business Tools extension, and ArcGIS V9.1 now ships with the new Network Analyst option as part of the main distribution kit. Similarly, TNTMips now includes a network analysis module with a wide range of facilities. The OpenSource GIS, GRASS, includes a range of network analysis tools in its core modules for vector processing, with functions such as v.net.path generating shortest paths and outputting associated attribute data such as distances and times in linked attribute tables.

TransCAD differs from most GIS products as it is effectively a GIS for transportation (GIS-T) offering. It is built on a general purpose GIS engine, Maptitude. Figure 4-53 illustrates the use of network distance within the TransCAD shortest path facility, utilising least distance (heavy line shown in blue/dark grey) and least time (heavy line shown in green/light grey) with a single intermediate node. Such algorithms are widely used in web-based GIS providing routing and traffic advice (see for example, www.getmethere.com). Output from these systems is in a variety of forms, including tabulation of section-by-section network distance/time and cumulative distance and/or time.

Figure 4-53 Shortest and least time paths

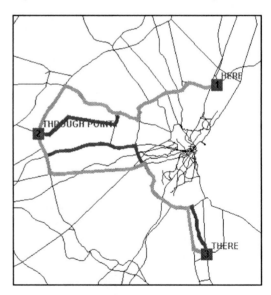

Depending on the GIS package, for large networks with potentially complex routing it is often advisable to compute link lengths as an explicit field and work with this data rather than expecting the GIS to re-compute such information dynamically from an intrinsic field. More details on this topic are provided in Chapter 7.

4.4.4 Buffering

4.4.4.1 Vector buffering

Buffering is the process of creating one or more zones around selected features, within a pre-specified distance (almost always Euclidean distance) from these features. The resulting buffer is an area object or series of area objects (typically polygons) that may be automatically merged (dissolved) where they overlap. In many cases the buffer regions are created from a combination of straight lines and circular arcs centred on the vertices of the convex hull. In Figure 4-54A simple symmetric buffers generated using the GRASS v.buffer operation are shown, whilst in Figure 4-54B the result of applying a 100metre buffer to sections of road network is shown, without buffer segment merging (selected segment highlighted). Buffers may be applied to:

- point, line or polygon objects, or to cells in a grid dataset

- they may be symmetric, i.e. spreading outwards from lines and polygon boundaries in both directions (left and right or inside and out), or

- they may be asymmetric, spreading inwards or outwards only (see for example the buffering used in polygon skeletonisation, Figure 4-11)

- buffers may also be applied differentially, for example larger buffers applied to major roads and smaller buffers applied to minor roads, as part of a pollution analysis exercise (air-born, noise, vibration etc.)

- buffers may be applied incrementally, progressively increasing and/or reducing in size

Figure 4-54 Simple buffering

A. v.buffer in GRASS

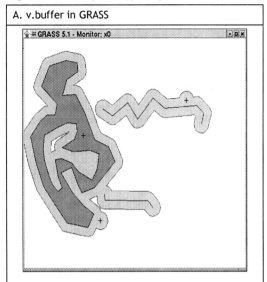

http://grass.itc.it/grass60/screenshots/vector.php

B. Buffer tool, line element, ArcToolbox

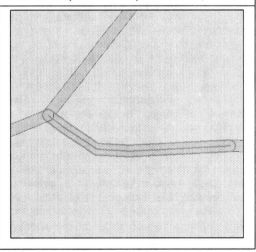

Although the idea of buffering appears very straightforward, and has many applications, different GIS packages implement the facility in a variety of ways. In ArcGIS extensive support is provide in ArcToolbox for buffer creation of selected features, with facilities to select buffering: to the left, right or both sides of polylines and polygons; to select whether linear feature buffers are terminated with rounded or squared-off ends; and to select features which may have their buffered regions combined (dissolved together) and optionally a selected attribute (field) to be used for determining whether dissolve operations should be performed. An additional buffer tool is provided to generate *multiple buffer rings* in a single operation. MapInfo provides a similar facility, which it calls "Concentric ring buffers".

Manifold's implementation is rather different. Its principal buffering facility is invoked via the Transform toolbar, and the standard Buffers operation when applied to a polygonal area provides a buffer that extends outside the polygon plus the entire polygon internal area, effectively growing the entire polygon. To limit the buffer zone to the boundary of a polygon the Border buffer zone operation must be used (Figure 4-55B). An additional operation, Inner buffer zone is provided which selects the inner part of a polygon, comprising all areas at least the buffer distance from the boundary (Figure 4-55C). Repeated buffering operations may be applied, either to the original feature or to the resulting buffered entity itself. Buffer features may also by modified, e.g. by clipping operations or intersection, to generate new or modified zones, as required. The Buffers operation in Manifold creates distinct buffer zones (Figure 4-56A), whilst an alternative Transform function, Common buffers, creates merged zones (Figure 4-56B). In this example buffers have been created around a selection of features from the hydrographic features layer of the Bay Area sample map. Once a buffer has been created, or as part of the creation process, a new map layer containing the buffer polygons will be generated. This map layer will have an associated attribute table, and if not already generated, attributes such as polygon areas can be computed. The new layer may, in turn be combined with other layers (e.g. via an Intersect overlay) to analyse attribute data within the buffer zone.

Figure 4-55 Manifold: Buffer operations

A. Selected polygon

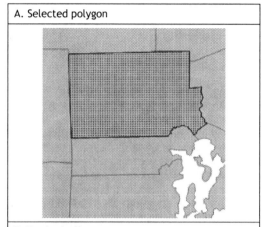

B. Border buffer

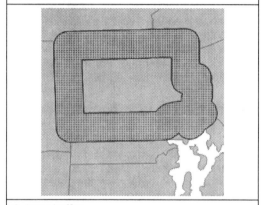

C. Inner buffer

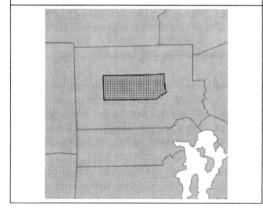

Figure 4-56 Manifold: Buffer and Common buffer

A. Buffer

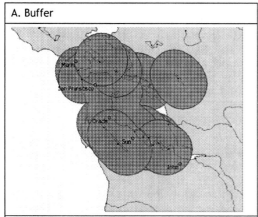

B. Common buffer

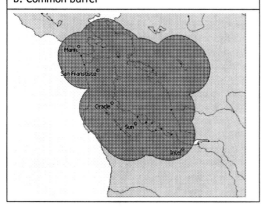

4.4.4.2 Raster buffering

Buffering may also be applied in raster GIS, whereby cells are classified according to their distance from one or more target cells (grid distance, Euclidean distance or a more complex, cost-related distance). For example, in Idrisi by default target cells are coded as 1, buffer cells as 2 and other cells as 0, and the buffer width to apply is specified in reference system units (based on true Euclidean distances or grid-based cost distances). The process of generating a buffer in this environment is very similar to performing a standard image processing task, that of generating a distance transform of the (binary) input image, and then selecting from the resulting output image all those cells that have entries (distance values) less than or equal to the buffer width specified. These selected cells are then re-classified (RECLASS in Idrisi) as required (e.g. to a value of 2; see also, Section 4.4.2.2).

4.4.4.3 Hybrid buffering

Buffer operations within vector GIS generate regions, and distances within those regions meet the overall buffering criterion set. The use of multiple buffering does enable a degree of gradation within such buffer regions, but for more detail some form of hybrid buffering may be more appropriate. In this approach either hybrid buffering is directly supported by the GIS tool (e.g. as with MapInfo's Vertical Mapper, Grid buffer facility) or the relevant vector objects will require conversion to grid format and then grid buffering applied. The latter may be achieved using cost distance (ACS) methods or distance transforms (DTs), as described in Section 4.4.2.

4.4.5 Network buffering

A simple buffer operation applied to a vector map of roads will create a set of merged or unmerged regions either side of the selected roads. An alternative form of buffering for such networks is a time or distance buffer from a point source or sources with respect to the network. In the case of roads, this form of buffer generates travel-distance or drive-time zones. The Manifold GIS Business Tools option provides for such buffering. In their model drive times are computed to all network points in one or more time zones, plus a distance buffer (speed related) for off-road travel (e.g. walking). This is a very processor intensive operation. TransCAD and ArcGIS Network Analyst provide similar functionality via their "drive time" analysis facilities.

4.4.6 Distance decay models

Almost by definition, spatial modelling makes extensive use of distance measures. In many instances a suitable metric is selected and used directly, but for many other situations some generalised function of distance is used. Typically these models involve the assumption that relationships between locations, or the effect one location has upon another, diminishes as separation increases. There are a wide range of GIS analysis techniques that make use of this concept — we describe a number of these briefly below, and address them separately in a number of the later chapters of this Guide.

The most widely used distance decay models are those in which distance is introduced as an inverse function to some power, typically 1 or 2. Thus the value of some variable of interest, z, at location j, z_j, is modelled as a combination of attribute values, z_i, associated with other locations, i, weighted by the inverse of the distance separating locations i and j, d_{ij} raised to a power, β:

$$z_j = \frac{f(\{z_i\})}{d_{ij}{}^{\beta}}, \beta \geq 0$$

The exponent, β, has the effect of reducing the influence of other locations as the distance to these increases. With $\beta=0$ distance has no effect, whilst with $\beta=1$ the impact is linear. Values of $\beta>>1$ rapidly diminish the contribution to the expression from locations that are more remote.

One of the commonest applications of this kind of model is in surface interpolation. Most GIS and surface analysis packages provide support for such operations (see further, Section 6.6). However, inverse distance weighting (IDW) is used in many other situations. An example is the study by Draper *et al.* (2005) referred to in Section 3.2.1. The distance of cancer cases and controls from overhead power lines was calculated and the inverse and inverse squared distances were then used as weights in modelling the observations. The idea behind this approach was to take into account the fact that electromagnetic radiation intensity diminishes with distance from the source.

Inverse distance modelling is one of the main techniques used in the fields of transportation, travel demand modelling and trade area analysis. The propensity to travel between pairs of locations is often assumed to be related to the separation between the locations and some measure of the size of the source and destination zones. The general form of such models is:

$$T_{ij} = A_i B_j O_i D_j f(d_{ij})$$

where T_{ij} is the number of trips between zones i and j, O_i is the size of the origin location i (e.g. the total number of trips to work/commuters from zone i), D_j is the size of the destination zone (e.g. the total number of work trip destinations/work places in zone j), and $f(d_{ij})$ is some appropriate function of the distance or degree of separation of zones i and j. A_i and B_j are balancing factors to ensure that the modelled row and column totals add up to the total number of trips from and to each zone (i.e. a so-called doubly-constrained model, which assumes that the totals are correct and not subject to sampling error) — see further, Wilson (1967). A common variant of this type of model is the origin-constrained version, in which case only the A_i are determined by balancing. Typical models for $f(d_{ij})$ include: simple inverse distance (as above); negative exponential models; and combined or Gamma models. All three types (plus others) are supported by TransCAD as impedance functions within its trip distribution modelling facility. Crimestat incorporates similar functions within its trip distribution modelling component.

The statistical models within GeoBUGS provide for a generalised powered exponential family of distance decay models of the form:

$$f(d) = e^{-(\phi d)^k}, k \in [0,2], \phi > 0$$

where ϕ is the principal decay parameter relating the decay pattern to the effective range of correlations that are meaningful. Note that the parameter k is effectively a smoothing factor and with $k=2$ this is essentially a Gaussian distance decay function. Geographically Weighted Regression (GWR) utilises similar decay functions (the authors describe these as kernel functions — see further, Section 5.6.3), this time of the form:

$$f(d) = e^{-d^2/2h^2}, or$$

$$f(d) = e^{-d/h}, or$$

$$f(d) = \left(1 - \frac{d^2}{h^2}\right)^2, d < r$$

$$f(d) = 0 \text{ otherwise}$$

In these functions the parameter, h, also known as the *bandwidth*. A small bandwidth results in very rapid distance decay, whereas a larger value will result in a smoother weighting scheme.

The Gaussian model is an unbounded function, and as an alternative GeoBUGS provides the following bounded or disk model of distance decay:

$$f(d) = \frac{2}{\pi}\left[\cos^{-1}\left(\frac{d}{\alpha}\right) - \left\{\left(\frac{d}{\alpha}\right)\left(1-\left(\frac{d}{\alpha}\right)^2\right)\right\}^{1/2}\right], d < \alpha$$

$$f(d) = 0 \text{ otherwise}$$

This model decays slightly more rapidly than a straight line over the interval $[0,\alpha)$, especially over the initial part of the interval. The parameter, α, here has a similar role to that of the bandwidth parameter in the GWR expressions above.

All such models may suffer from problems associated with clustering — for example where a number of alternative (competing) destinations are equidistant from an origin zone, but may be located either in approximately the same direction or in very different directions. Such clustering has implications for modelling, since competition and even interaction *between* destination zones is inadequately represented. Most GIS

and related software packages do not provide direct support for analysis of such factors, and it is up to the user to specify and implement an appropriate modelling framework to address such issues. In the case of trade area modelling, so-called gravity models or Huff models (which are origin constrained) are often used. The former typically use inverse distance power models coupled with measures of location attraction, such as retail square footage. Vertical Mapper, for the MapInfo product, is an example of GIS software that provides such facilities. For a fuller discussion and background to such modelling, see the Crimestat manual, Chapter 14, Miller and Shaw (2001, Ch.8), Wilson (1967) and Haggett (1965, 2.II).

These are all models based on a mathematical function of distance. In addition to such models other models or measurements of impedance are extensively used, typically in the form of origin-destination (O-D) matrices. The latter may be derived from measurements or from alternative distance-related models such as intervening opportunities, see Stouffler (1940), or lagged adjacencies (e.g. see the spatial weights modelling options in GeoDa, and Section 5.5.2.3). Distance, d, in this instance is a free-space or network-related measure of impedance to interaction (e.g. network distance, time, cost or adjacency) and most models implement distance impedance expressions of the form:

$$\alpha/d^\beta, \; \alpha e^{-\beta d} \text{ and } \alpha e^{-\gamma d}/d^\beta$$

where α, β and γ are parameters to be selected or estimated. Figure 4-57 illustrates the first two of these models, in each case standardised with an α value to provide a value of 10 at $d=0.1$ units and a range of values for β. The Gamma model provides curves that are intermediate between the two sets shown, depending on the specific values of γ and β chosen or fitted from sample data. Pure inverse distance models (and the Gamma model) have a (serious) problem when distances are small, since the expression will tend to infinity as d tends to 0. A partial solution to this problem is to include a small distance adjustment to such formulas, such that $d=d+\varepsilon$ where $\varepsilon>0$, or to exclude distances that are below a given threshold from the inverse distance calculations (possibly defining a fixed value for the function in such cases). These approaches stabilise computations but are essentially arbitrary. The exponential model has the advantage that it tends to 1 as d tends to 0, so may be preferable if observations are known to be very closely placed or even coincident.

Models of this kind may be constrained in various ways: by limiting the maximum range to be included in computations; by limiting the maximum number of locations to be considered; by restricting the range of values the parameters may take; and by explicitly taking account of barriers (e.g. inaccessible regions). More sophisticated analysis would take into account aspects of the dynamics and uncertainty associated with the interactions or flows being modelled.

Somewhat different models of distance decay are incorporated into radial basis interpolation and geostatistical analysis (see further, Sections 6.6.4 and 6.7). The analysis of data variation by distance or distance bands (e.g. variograms and correlogram analysis) is designed to identify overall patterns of variation in a measured variable with linear (Euclidean) distance. In such instances a maximum distance or range is identified if possible, and the pattern of variation in the measured variable is modelled to reflect the diminishing associations observed as distance increases. Here the models effectively treat distance as an independent variable and the observations (generally averaged by distance band) as a form of dependent variable.

Figure 4-57 Distance decay models

A. Inverse distance decay, α/d^{β}

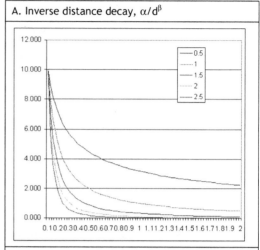

B. Exponential distance decay, $\alpha e^{-\beta d}$

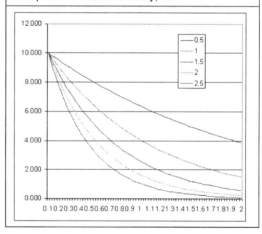

4.5 Directional Operations

4.5.1 Directional analysis - overview

Directional analysis is perhaps less well known and well used than other forms of spatial analysis. It primarily deals with the analysis of lines, points sets and surface orientation, sometimes from different time periods, examining patterns in order to identify and utilise information on specific directional trends. Researchers In the fields of geology, sedimentology and geomorphology utilise directional analysis to examine fault lines and fracture systems, glacial striations, particulate distributions and lithography patterns; in ecology it is used to study patterns of wildlife movement and plant dispersal; in hydrology and other forms of flow analysis directional considerations are paramount; and directional analysis can be used when studying patterns of disease incidence or the distribution of particular crime events. If a particular dataset exhibits no discernable variation in direction across a study region it is described as *isotropic*. Whilst such situations may be regarded as unusual in the real world, it is frequently the starting assumption for many forms of analysis. Variations that have a particular directional bias are known as *anisotropic*. In some areas of GIS analysis, for example geostatistics, examination of anisotropy is a fundamental part of the modelling and prediction process (see further, Section 6.7).

It should be noted that when comparing and analysing directional datasets computed with projected (plane) coordinate systems, there may be introduced directional bias. Comparisons should be made between datasets that have been generated using the same projections and approximate scale, or by utilising spherical coordinates where necessary. For small areas (under 100kmx100km) the problems for most applications should be minor.

4.5.2 Directional analysis of linear datasets

Directional analysis applied to linear forms (line segments, polylines, or smooth curves) is sometimes known as alignment or lineament analysis. Line elements within a GIS are normally directed, in the sense that they have a start and end point determined during data capture. This directed aspect of the line may or may not be relevant, and if relevant may be incorrect for the analysis you are considering. Hence all such datasets need to be examined carefully before proceeding with analysis.

In addition to line-based data from which orientation is derived, the same types of analysis may be applied to point datasets with an associated directional attribute or attributes. For example, these might be: recordings of the initial migration direction of birds of various types from a small area of woodland; the dispersal pattern of seeds from a tree; the wind direction recorded at a series of meteorological stations at hourly intervals; or the direction of a particular type of crime event in a city with respect to a given reference point. As such this is simply a particular application of such statistics to a defined dataset. However, if such data are stored within a GIS as attributes in a table associated with a point set, then conventional descriptive statistics such as mean and variance should not be used. Similar concerns apply to any form of modulo or cyclic data, for example fields containing temporal data recording the time of day in minutes from midnight or numeric day of week associated with events recorded over a number of weeks.

Data with directional information is essentially of two types: directed (or vector) data, where the direction is unique and specified as an angle in the range $[0°, 360°]$; and oriented or bi-directional data, often referred to as *axial* data, where only the orientation is defined. Axial data are normally doubled, and converted to a 360° range using modulo arithmetic, processed and the results are then re-converted to a $[0°, 180°]$ range.

Determination and processing of line direction within a GIS is problematic for a number of reasons:

- the way in which a line is represented, the level of detail it portrays, and the extent to which real world features are generalised

- the data capture process and composite form of line representation (polylines) — where is the true start and end of the line for directional analysis purposes?

- the nature of circular (or cyclic) measure: the orientation of an undirected line at 90 degrees to the vertical is indistinguishable from one at 270°; furthermore the difference between 280° and 90° is not taken to be 280°-90°=190°, but 360°-280°+90°=170°.

Similarly the mean direction of 3 lines at 280° (or -80°), 90° and 90° is not sensibly calculated as (280°+90°+90°)/3=186.7° or as (-80°+90°+90°)/3=16.7°, but a consistent and meaningful definition of such an average is needed. With two angles, 350° and 10° from due north this is even more obvious, as a mean direction of 180° (i.e. due South) does not make sense

We treat each of these issues in turn:

If the source data of forms are suspected of being fractal in nature, across a very wide range of scales, then in theory they may exhibit no reliable tangents (directions) at any point. In practice the process of data capture accepts that our model within a GIS has involved simplification of the real world, and that the data may be represented as a collection of non-fractal discrete elements (polylines or pixels). Assuming this representation is acceptable and meaningful for the problem being investigated, analysis of the captured data can proceed. Particular attention should be given to the location of the initial and final points (end nodes) in such cases, as these may offer more acceptable point pairs for the analysis of broader directional patterns than any of the intervening line segments.

More problematic is the issue of generalisation. Subsequent to data capture, feature representation within the GIS may be subject to some form of generalisation as discussed earlier (Section 4.2.3). These processes significantly alter the direction of linear forms, especially the component parts (segments) of polylines. For these reasons, when performing directional analysis on polylines, a range of alternative ways of describing the line direction may be needed. Examples include: end node to end node; linear best fit to all nodes; disaggregated analysis (treating all line segments as separate elements); and weighted analysis (treating a polyline as a weighted average of the directions of its component segments, e.g. weighted by segment length). Smooth curves (e.g. contours) may be treated in a similar manner, approximating these by polylines and applying the same concepts as for standard polylines. A GIS package that supports such selection facilities is TNTMips — most other packages have more limited facilities.

These approaches address the first and second bullet points above. One way to address the third bullet point is to treat the lines or line segments as vectors (i.e. having an origin, a magnitude and a direction with respect to that origin) and then to use trigonometric functions of the directions rather than the directions themselves to perform the computations. This will give us the direction (and optionally magnitude) of the *resultant vector*, r (the mean effect of each vector "pulling" in its own direction). For example:

- let the set of $N=i+1$ points determining a polyline define a set of i directions $\{\theta_i\}$ from a given origin with respect to a predefined direction (e.g. grid north)

- compute the two vector components (northing and easting): $V_n=\Sigma\cos\theta_i$ and $V_e=\Sigma\sin\theta_i$

- the resultant vector, **r**, has mean or preferred direction $\tan^{-1}(V_e/V_n)$. e.g. using our earlier example with three vectors at -80°, 90° and 90° from horizontal the resultant mean direction is +80.3°; with the two vectors 350° and 10° the resultant is 0° (due north). The directional mean by itself is of limited value unless the underlying data demonstrate a consistent directional pattern that one wishes to summarise. If a set of vectors show an arbitrary or random pattern the mean direction could be almost any value, which is of little use

- the length or *magnitude* of the resultant vector if all N component vectors have unit magnitude or are simply provided as angular measurements is simply:

$$|r| = \sqrt{V_n^2 + V_e^2}$$

If $\{\theta_i\}=0$ for all i, all the V_n components will be 0 and all the V_e components will be 1, hence $|r|=N$ where $|r|$ denotes the vector magnitude of **r**. Dividing through by N standardises $|r|$ such that the length of the mean vector, $r^*=|r^*|$, lies in the range [0,1]. Larger values *suggest* greater clustering of the sample vectors around the mean.

- the circular variance is simply var=1-$|r|/N$, or var=1-r^*, and again lies in the range [0,1]

- the circular standard deviation is defined as

$$sd = \sqrt{-2\ln(r^*)}$$

- if the point set for analysis includes the vector length, v_i, then the expression for the directional mean can be generalised (weighted by length) as $V_n=\Sigma v_i\cos\theta_i$ and $V_e=\Sigma v_i\sin\theta_i$. The resultant vector magnitude in this case will no longer lie in

the range [0,1] if normalised by the number of vectors

Assuming that we have a suitably encoded dataset consisting of distinct or linked polylines (e.g. a stream network), we can explore directional trends in the entire dataset or subsections of the data. This process is illustrated in Figure 4-58 for a sample stream network using tools from the TNTMips package. The two direction rose diagrams show the number of stream segments having orientations in 10 degree groupings (i.e. rather like a circular frequency diagram or histogram). The pattern is completely symmetric since in this case lines are regarded as undirected.

Any given section of the rose diagram can be selected, and the number of components (frequency of the sample) will be displayed. Higher frequencies are shown as longer segments, although some software provides the option of using areas to indicate frequency variations rather than lengths on rose diagrams. Furthermore the line elements that contribute to this total can be identified on the associated map (a form of linking, familiar to most users of GIS packages in object selection procedures). In this example we have selected those lines in the 90 degree segment of the upper rose diagram (85°-95°) highlighted here in blue/dark grey, and separately those in the 350° segment (345°-355°), highlighted here in green/very pale grey. The directional data for all lines in this case were treated as end node to end node. The lower rose diagram shows the results when all component parts of polylines are included in the analysis, showing a much more even spread of orientations. As with almost all forms of GIS analysis, it is important to observe and take account of the effects of the region selection process, and as can be seen here, many of the streams are arbitrarily dissected by the boundary, which may impact the interpretation of results.

Figure 4-58 Directional analysis of streams

A. Hydrology, Crow Butte region

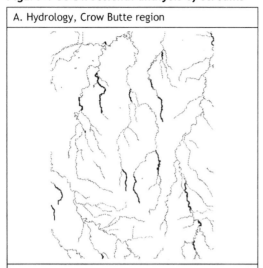

B. Direction roses: end point and segment versions

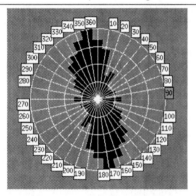

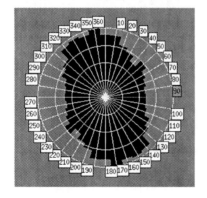

Similar functionality is provided within ArcGIS, in its Linear Directional Mean facility. Whilst this does provide a choice of metric (Euclidean or Manhattan) and selection of directed or non-directed computation, selection of alternative modes of interpreting polylines (as described above) is not provided so polylines would need to be segmented (or combined) as required prior to analysis.

For a more complete set of tools, including distributional analysis, specialised software such as Oriana from Kovach Computer Services is required. This facilitates analysis of point-like datasets or datasets that have been extracted from a GIS in this form. Input is in tabular form (text, csv, Excel etc.), and graphing and distributional analysis of single or multiple datasets is supported. Oriana may be unable to process large datasets (such as grid outputs in column format), in which case the use of generic programs such as MATLab and the Surfer-related program Grapher may be more effective. The formulas used by Oriana follow those provided in Fisher (1993) and Mardia and Jupp (1999). Measures are provided which compare the observed distribution to a uniform distribution and/or the distribution due to von Mises that is the circular equivalent of the Normal distribution. The general form of the von Mises distribution is:

$$p(x) = \frac{e^{\kappa \cos(x-\alpha)}}{2\pi I_0(\kappa)}$$

where α is the mean direction of the distribution and κ is a shape parameter known as the concentration. $I_0(\kappa)$ is the modified Bessel function of the first kind, of order 0. The circular variance of this distribution is:

$$\sigma^2 = 1 - \frac{I_1(\kappa)}{I_0(\kappa)}$$

This modified Bessel function is relatively straightforward to compute for integer orders, although it requires summation of an infinite series:

$$I_0(\kappa) = \sum_{i=0}^{\infty} \frac{(-1)^i (\kappa/2)^{2i}}{(i!)^2} \text{, and}$$

$$I_1(\kappa) = \frac{\kappa}{2} \sum_{i=0}^{\infty} \frac{(-1)^i (\kappa/2)^{2i+1}}{i!(i+1)!}$$

Several types of spatial dataset consist of multiple components that incorporate directional information. For example: wind direction and speed over time for a specific location or set of locations; single grid files of derived gradient data (slope, or gradient magnitude and aspect, or direction of maximum gradient — see further, Section 6.2); dual grid files containing either direction and magnitude data (polar information) in separate grids, or Cartesian data (x and y components) in each grid, for example simple optimal path tracking data (see further, Section 4.4.2.2).

Plotting such data may utilise a variety of display methods.

For temporal data a variant of the rose diagram can be utilised that provides stacked histogram functionality in the radial direction in addition to the circular histogram facility provided by a standard rose diagram. The Grapher program, available from the same providers as Surfer, is an example of a package that supports such displays, as does Oriana, but many statistical graphing packages provide similar functionality. This kind of diagram is illustrated in Figure 4-59, where the radial extent shows the frequency of measured wind speeds (hourly intervals) separated into speed categories in knots (with zero speed values removed). The black arrow shows the mean vector (direction and speed) for the entire dataset of almost 4500 records.

Figure 4-59 Two-variable wind rose

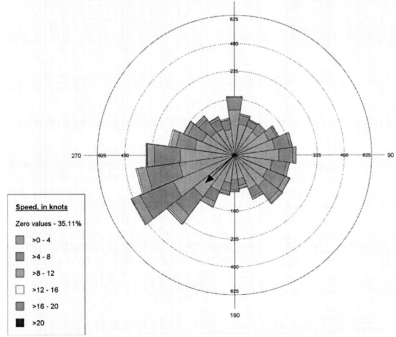

For vector-like grid data (single or dual-grid source data) a vector plot may be informative. An example of such a plot has been utilised in Figure 4-49B. A more familiar variant is shown in Figure 4-60 where gradient vectors have been plotted using Surfer for a sample grid file of the Mt St Helens volcano. In this case colour/grey scale has been used to indicate vectors of greater magnitude (from white through blue to red/light to dark grey).

Figure 4-60 Slope and aspect plot, Mt St Helens data, USA

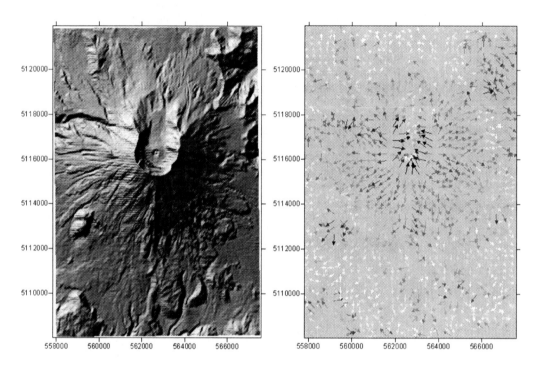

4.5.3 Directional analysis of point datasets

A somewhat different form of directional analysis is used within many packages to summarise the distribution of point sets around a mean centre. The average variation in the distance of points around the mean can be viewed as a circle or set of circles at a set of standard distances (like standard deviations in univariate statistics). In addition, separate variations in *x*- and *y*-coordinate values may be used to generate a standard distance ellipse, with major and minor axes reflecting the directional variation of the point pattern. Figure 4-61 illustrates this for the location of churches on Romney Marsh in SE England. The two standard deviational ellipses are set at one and two standard deviations respectively.

Determination of the standard deviations in *x* and *y* are not made using the original data, but using transformed coordinates selected by minimising the squared deviations from the point set to transformed (rotated) *x* and *y* axes. This is similar to the process of computing a linear regression on the (*x,y*) coordinate pairs and using the slope of the best fit line in angular measure to calculate the adjusted standard deviations.

Figure 4-61 Standard distance circle and ellipses

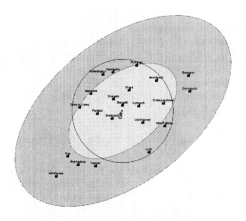

The formula for the angle of (clockwise) rotation of the *y*-axis is somewhat lengthy:

$$\theta = \tan^{-1}\left(\frac{\sum(x_i - \bar{x})^2 - \sum(y_i - \bar{y})^2 + \sqrt{C}}{2\sum(x_i - \bar{x})(y_i - \bar{y})}\right)$$

where

$$C = \left(\sum(x_i - \bar{x})^2 - \sum(y_i - \bar{y})^2\right)^2 + 4\sum\left((x_i - \bar{x})(y_i - \bar{y})\right)^2$$

Using this value the two standard deviations can then be computed in much the same way as for conventional standard deviations, but with the degrees of freedom in this case being *n*-2, where *n* is the number of points in the sample (for *n* large the divisor will be close to *n*, as per the common formulation of standard distance). The formulas cited are those used within Crimestat, which have been adjusted to ensure the ellipse axes are the correct length and the formula is consistent if the standard deviations in *x*- and *y*- are equal:

$$SD_x = \sqrt{\frac{2\sum\left((x_i - \bar{x})\cos\theta - (y_i - \bar{y})\sin\theta\right)^2}{n-2}}$$

$$SD_y = \sqrt{\frac{2\sum\left((x_i - \bar{x})\sin\theta - (y_i - \bar{y})\cos\theta\right)^2}{n-2}}$$

Software packages may provide information on the rotational angle and other attributes of the standard deviational ellipse. For example, Crimestat includes the angle of rotation and the individual standard deviation values, together with the area of the standard deviational ellipse (which is simply $A = SD_x SD_y \pi$; the lengths of the two ellipse axes are $2SD_x$ and $2SD_y$). For the data illustrated in Figure 4-61 the clockwise rotation of the *y*-axis is 53.2 degrees and the ratio of major to minor axis length, or *coefficient of eccentricity*, is 1.76.

Crimestat includes directional data in another technique, Correlated Walk Analysis (CWA). CWA is an analytical procedure based on examining a sequence of events (e.g. crimes) in terms of location, bearing (direction from a given point) and time lapse. Its objective is to predict where subsequent events, e.g. burglaries, may occur. Random walks, of various types (see further below), may be used as part of such analysis. For more details of these methods and their applications see the Crimestat manual, Chapters 4 and 9. An essential part of such analyses is the examination of patterns in the distribution of point data as the distance and direction of examination is varied.

A simple Correlated Random Walk (CRW) simulation facility is provided with Hawth's Tools for ArcGIS that enables model paths to be generated and saved as features (Figure 4-62). This facility highlights the impact of model parameters on the resulting path form (in this example using CRW as the basic model, but other models such as pure random/Brownian motion and so-called persistent random walks are also supported).

Figure 4-62 Correlated Random Walk simulation

A. 500 step CRW, variable (random uniform) step length, directional model N(0,1) degrees

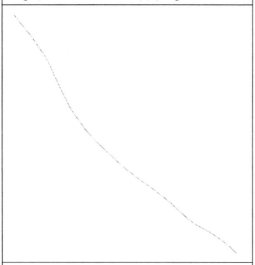

B. 500 step CRW, variable (random uniform) step length, directional model N(30,15) degrees

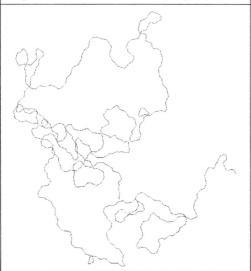

Graphs that plot the average difference in values for pairs of points in a region, over given blocks of distance, optionally grouped into different blocks of directions, are known as correlograms or variograms. If the region is isotropic there will be no significant differences between the variograms drawn for one direction (e.g. 0-30 degrees) or from another (e.g. 60-90 degrees), whereas with significant anisotropy the shapes of these variograms will differ. Hence if modelling is to utilise such information, different models or model parameters may be required to accommodate these directional differences (these issues are covered in greater detail in Section 6.7.1.11).

4.5.4 Directional analysis of surfaces

There are many applications of directional analysis relating to surfaces. These include: analysing and mapping surface aspect (e.g. the d.rast.arrow function in GRASS, the Aspect generation tools of Landserf, TAS, ArcGIS and TNTMips), which is defined as the direction of the surface gradient (as opposed to the magnitude of the gradient – see Figure 4-60); analysing and modelling flow directions and *friction* as part of hydrological and similar studies (e.g. local drainage directions such as the lddcreate function in PCRaster, the Flow function in Idrisi, and the FlowDirection function in the ArcGIS Spatial Analyst hydrology toolset); and analyses relating to the lighting of surfaces, as part of solar energy analysis, lighting for visualisation, and for visibility analysis (e.g. in the ArcGIS 3D Analyst and Spatial Analyst toolsets). Each of these topics is discussed in some detail in Chapter 6. In general the representation of direction in such instances is strongly affected by the underlying model or representation of the surface — for example, with grid (raster) surfaces there is often directional bias as a result of the limited range of directions available in immediate neighbourhoods (e.g. only 8 directions in a 3x3 neighbourhood).

4.6 Grid Operations and Map Algebra

In this Section we provide an introduction to grid analysis methods, including an overview of the kind of facilities available in raster analysis packages and toolsets. Many of the facilities relate to operations on surfaces (e.g. computation of gradient and aspect) and field-like datasets (e.g. interpolation and prediction techniques). These topics are covered in detail in Section 6.

4.6.1 Operations on single and multiple grids

Grid or raster data files often have a single attribute associated with each cell position. Datasets of this type can be thought of as triples (x,y,z) where x and y are the cell coordinates and z the attribute values. z-values might be: a colour code in the case of image data; a binary value, for example in the case of converted vector data (without additional attributes); a cell classification number, such as the type of vegetation cover associated with that location; or a measured or calculated data value, such as an interpolated figure for soil heavy metal content in parts per million (ppm) or the height of a surface in metres. In all instances the set of values observed in a single grid of size M-rows and N-columns is limited, either by the coding or classification procedure adopted, or by the fact that ultimately only MxN values can possibly be recorded.

This set of values may be summarised and analysed using basic univariate statistics (non-spatial analysis) and many GIS packages provide a wide number of simple single-grid measures. These include many of the basic statistics described in Table 1.4, including the arithmetic mean, max, min, count, sum, variety, minority, majority, range, mid-range, variance and standard deviation. Each of these operations applies to the set of individual cells of a single grid and may be described as *local* statistics.

Simple statistical operations may be applied to groups of cells, for example the immediate 8 neighbourhood (or window) of each cell or to individual cells across multiple grids (see also, Section 5.3). Where regular neighbourhoods of cells are involved such operations are termed *focal*, or if the neighbourhoods are defined using pre-defined (irregular) zones, then the operations are described as *zonal*. MFWorks, for example, includes a range of focal operations within its SCAN facility based on a sample window size and shape (square or circular). It allows for windows of varying sizes specified in terms of grid cells or measured distance, with facilities to include masked-off regions and annular-shaped windows. A matching set of facilities is provided via its SCORE option for zonal analysis.

There are a large number of basic grid operations supported for image and general raster files. These include local, focal and zonal operations depending on the scope of the operation. Such operations may be applied to a single grid, or to a number of input grids, depending on the operation in question. The set of possible operations of this type are often referred to as Map Algebra. Originally this term was introduced by Tomlin (1983) as the process of map combination for co-registered layers with rasters of identical size and resolution. Combinations involved arithmetic and Boolean operations. However the term is now used more widely by many suppliers. For example, ArcGIS describes the set of all operations performed using its Spatial Analyst option as "Map Algebra". More specifically it divides such functions into five main categories, the three above plus Global and Application-specific:

- Local functions, which include mathematical and statistical functions, reclassification, and selection operations

- Focal functions, which provide tools for neighbourhood analysis

- Zonal functions, which provide tools for zonal analysis and the calculation of zonal statistics

- Global functions, which provide tools for full raster layer or raster dataset analysis, for example the generation of Cost Distance rasters

- Application functions, specific to tasks such as hydrology and geometric transformation

MFWorks, a software package that is a direct descendent of Tomlin's original Map Analysis Package (MAP) software, includes a similarly broad range of grid functions. Its Operations menu includes the following:

- Classification: tools to re-code, slice (classify) single input layers, and to combine multiple layers using a range of local and focal operations

- Statistical: a number of statistical tools for grid processing, including focal (scan) and zonal (score) statistical calculations

- Proximity: tools to identify clumps of similar cells, to generate grid Voronoi polygons, and to generate cost distance grids

- Surface: computation of surface characteristics, such as gradient and orientation (focal functions)

- Network: tools to compute the number and pattern of stream networks

- Zone/area: tools to measure areas, boundaries of areas and to facilitate the creation of partitions

- Flow characterisation: hydrographic tools, such as identifying flow directions, drainage patterns, flows and accumulations across the surface

- Transformation: facilities to re-sample and/or adjust (warp) input grids

- Surface processing: procedures for filling-in pits and defined areas

- Interpolation: a range of different interpolation methods

- Visualise: image processing functions, such as merging input bands and applying filters, such as image sharpening

Similar facilities are provided in many of the mainstream raster-based GIS packages, such as PCRaster, TNTMips and Idrisi, and in the raster-processing toolsets of more generic packages such as GRASS and the Vertical Mapper facility for MapInfo. Many of these are described in greater detail in Chapter 6.

4.6.2 Linear spatial filtering

The term spatial filtering is principally associated with digital image processing, although such methods may be applied to almost any type of grid or image. The term is also used, in a related manner, in the area of spatial statistics (see further, Section 5.6.5).

Most GIS packages provide simple grid filtering functions, whilst those that have a bias towards image processing and grid file handling (e.g. TNTMips, Surfer, Idrisi) tend to provide a broader range of functions. The most commonly provided functions are so-called low-pass and high-pass spatial filters. These are focal functions whose operation is determined by a kernel or neighbourhood of NxN cells around each pixel or grid position. Grid cells "covered" by the kernel are multiplied by the matching kernel entry and then the weighted average is calculated and assigned as the value for the central cell, G. A 3x3 symmetric kernel might look something like:

$$\begin{bmatrix} a & a & a \\ a & b & a \\ a & a & a \end{bmatrix} \text{ or perhaps } \begin{bmatrix} c & a & c \\ a & b & a \\ c & a & c \end{bmatrix}$$

In the first of these examples a and b are positive numbers, typically integers. If $a=b=1$ then the kernel provides a simple averaging or smoothing operation, whereas if $a=1$ and $b=4$ more weight would be applied to the grid cell being filtered. In both cases the weighted average is divided by the sum of the kernel elements. No divisor is applied in cases where the kernel sums to 0 or 1.

Filters of this type are sometimes described as low-pass spatial filters.

For example, consider the filter

$$\begin{bmatrix} 1 & 1 & 1 \\ 1 & 4 & 1 \\ 1 & 1 & 1 \end{bmatrix} \text{ and grid section } \begin{bmatrix} 98 & 176 & 200 \\ 90 & 187 & 188 \\ 124 & 170 & 175 \end{bmatrix}.$$

The value at the central grid position $G(2,2)=187$ is replaced by the weighted average of itself (assigned a weight of 4) and the surrounding 8 cells (each assigned a weight of 1). The adjusted value is 164 after rounding. In the second example if a, b and c are positive with $b>a>c$ the kernel is described as Gaussian, being symmetric but centre-weighted, for example:

$$\begin{bmatrix} 1 & 2 & 1 \\ 2 & 4 & 2 \\ 1 & 2 & 1 \end{bmatrix}$$

With the same data grid as above the filtered value of $G=162$.

More generally, if $\{C_i\}$ is the set of coefficients of an $m=N\times N$ kernel matrix and $\{P_i\}$ is the set of target grid values within this kernel neighbourhood, then the filtered grid value, G, is defined as:

$$G = \sum_{i=1}^{m} C_i P_i \Bigg/ \sum_{i=1}^{m} C_i + B$$

The extra term, B, here is often set to 0. It is a bias factor that increases or decreases the resulting filter value by a fixed amount.

High-pass filters emphasise the difference between the central point of the kernel and the values in its immediate neighbourhood. Typically the entries in such kernels have a mix of positive and negative values, but the entries still add up to 1:

$$\begin{bmatrix} -a & -a & -a \\ -a & +b & -a \\ -a & -a & -a \end{bmatrix}$$

Values are often used to provide image sharpening prior to further processing. Table 4.10 provides examples and a summary of typical linear spatial filters used in GIS and image analysis. Basic linear image filtering operations fall into a number of categories:

- Sharpening — for which the kernel matrix elements sum to 1 and the matrix is symmetric. Matrix elements are a mix of positive, negative and zero entries

- Blurring (also known as a smoothing, averaging or low-pass filtering, since this reduces or removes extreme values). In this case the kernel matrix elements sum to >1, entries are normally all positive and the matrix is symmetric

- Edge detection — for which the kernel matrix elements sum to 0 and the matrix is normally asymmetric. As with sharpening, matrix elements are a mix of positive, negative and zero entries

- Embossing — for which the kernel matrix elements sum to 0 as per edge detection, but the matrix is asymmetric. It is usual to compute the embossed version of an image on a greyscale version, which is obtained by finding the average value of the bands, e.g. red (r), green (g) and blue (b) components, as $x=(r+g+b)/3$ for each pixel, and assigning the resulting values in the range [0,255] to a grey scale (where 0=black and 255=white)

A number of general characteristics of this type of filtering should be noted:

- if the range of values permitted within the grid are limited (e.g. RGB values in the range [0,255]) then computed values that fall outside of this range must be truncated to the range limits. This generates some computational (as opposed to user defined) bias in the processing

- multi-band image data filters are applied to each band separately. Hence a sharpening filter applied to an RGB image, for example, would be applied to the R, G and B components

- the kernel size may be any square arrangement, but odd squares of 3x3 and 5x5 cells are the most commonly used. Even-sided square arrangements are occasionally supported, but in this instance the central part of the kernel is a 2x2 grid rather than a single cell. Rectangular kernels are also possible but are less frequently applied

- filtering may be conducted more than once (single pass or multi-pass). Each pass re-applies the same filter or filters to the target grid or image (layer or band)

- at image or grid edges it is not possible to apply the kernel symmetrically. This effect is more pronounced for larger kernels, but applies in every instance. There are many options for resolving such effects. Common procedures include: leaving the edge pixels unaltered; blanking edge pixels (which causes the grid to shrink); using an asymmetric (i.e. partial) kernel, which can generate artefacts; and mirroring, where grid values inside the kernel are used as surrogates for the missing values outside the kernel

- user-defined kernels are widely supported, both in image processing packages and within GIS software. Users may be able to select a pre-defined filter and modify this to suit their requirements, or to create their own NxN kernel

- all of the examples described involve simple linear operations, essentially weighted averaging (sometimes described as *linear convolution*). Non-linear filtering is also widely available and is described in more detail in Section 4.6.3

- in addition to filters that operate in the spatial domain, pure frequency domain filters may also be applied. These operate on the set of grid values within an image without reference to the spatial pattern. Frequency domain filtering (for example Fast Fourier Transforms or FFTs) is provided in some GIS packages such as TNTMips and Idrisi, but such procedures are not described in this Guide. FFTs are used, for example, to removing striping effects from remote-sensed imagery

Table 4.10 Linear spatial filters

Filter type	Filter	Sample 3x3 kernels	Description
Low-pass (symmetric)	Averaging	$\begin{bmatrix} a & a & a \\ a & b & a \\ a & a & a \end{bmatrix}, \begin{bmatrix} 0 & a & 0 \\ a & b & a \\ 0 & a & 0 \end{bmatrix}$	Smoothing, noise reduction or blurring filter (focal mean)
	Gaussian	$\begin{bmatrix} c & a & c \\ a & b & a \\ c & a & c \end{bmatrix}$	Smoothing, noise reduction or blurring filter (focal weighted mean)
High-pass (symmetric)	Sharpening	$\begin{bmatrix} -a & -a & -a \\ -a & +b & -a \\ -a & -a & -a \end{bmatrix},$ $\begin{bmatrix} 0 & -a & 0 \\ -a & +b & -a \\ 0 & -a & 0 \end{bmatrix}$	Mean effect removal/sharpening filter (focal sum). Provides limited edge detection. Typically entries sum to 1 but may be greater. 3x3 Laplacian kernels typical add to 1. Larger Laplacian kernels (e.g. 7x7) may be more complex and sum to >1
Gradient (asymmetric)	Edge detection	$\begin{bmatrix} a & b & a \\ 0 & 0 & 0 \\ -a & -b & -a \end{bmatrix}, \begin{bmatrix} -a & 0 & a \\ -b & 0 & b \\ -a & 0 & a \end{bmatrix}$	Applied singly or as a two-pass process. These kernels highlight vertical and horizontal edges. When used in combination they are known as Gradient or Order 1 derivative filters. Typically $a=1$ and $b=1$ or 2 and entries sum to 0. A variant known as the Roberts method has row 3 all zero and column 1 zero for the first pass and column 3 zero for the second pass (i.e. a 2x2 filter)
	Embossing	$\begin{bmatrix} 0 & +a & +a \\ -a & +a & +a \\ -a & -a & 0 \end{bmatrix}$	Edge detecting filters that enhance edges in a selected compass direction to provide an embossed effect. The example here shows a sample north-east kernel
	Directional	$\begin{bmatrix} -1 & 1 & 1 \\ -1 & -2 & 1 \\ -1 & 1 & 1 \end{bmatrix}, \begin{bmatrix} 1 & 1 & 1 \\ 1 & -2 & 1 \\ -1 & -1 & -1 \end{bmatrix}$ $\begin{bmatrix} 1 & 2 & 1 \\ 0 & 0 & 0 \\ -1 & -2 & -1 \end{bmatrix}$	Simple computation of gradient in one of 8 compass directions. east and north directional derivatives are illustrated in the first two examples here. Note that the third example shown corresponds to Horn's method of derivative calculation for the north/south component (dz/dy) as described in Section 6.1.3. Compares with Edge detection filtering. A non-linear variant of this procedure is to compute all 8 values and select the maximum absolute value of these as the output

4.6.3 Non-linear spatial filtering

With linear filtering the value at each grid position, *G*, is altered using a simple kernel as a weights matrix. With non-linear filtering kernels are not generally used. Instead the value at *G* is determined by some procedure other than weighted averaging. For example, *G* may be determined by a statistical measure (other than the mean) based on an *NxN* neighbourhood of *G*. *NxN* is typically 5x5 or greater, and almost any basic statistic (as listed in Table 1.4) may be used. Typical examples might be the minimum value in the neighbourhood, or the cell value that is most common (occurs in the majority) within the neighbourhood. Other measures commonly supported included variance, standard deviation, coefficient of variation and inter-quartile range. In each case the computations are based on the defined neighbourhood.

ArcGIS describes such operations as Neighbourhood operations involving Focal or Block statistics, with neighbourhoods defined more generally than in typical filtering software (e.g. as rectangles, rings or wedges). Most other packages stick with the image processing terminology of filtering and rectangular neighbourhoods. ArcGIS also provides a special function to handle the Majority case (MajorityFilter) with integer-valued grids that incorporates an additional spatial contiguity constraint. The rule applied is that either half or at least half of the cell values in the 3x3 neighbourhood must be the same as the value at *G*, and these must satisfy a rook's move (4-way) or queen's move (8-way) contiguity constraint. Multiple passes of this filter can be made and will stabilise to a fixed pattern.

Several non-linear filters examine local deviations in grid values. In addition to standard statistical measures, median deviation and threshold averaging may be applied (see also, Section 6.7.2.6 on median-polishing). In median deviation filtering the value at *G* is replaced by *G-B*, where *B* is the local median. This emphasises outliers. With thresholding averaging the value at *G* is

retained if it is within a defined absolute value of the local mean (calculated excluding the value at *G* itself), otherwise it is replaced with this local average. The effect is to remove outliers. This is a form of "noise reduction", a procedure that is supported in many image processing packages and in some GIS software. TNTMips, for example, provides 6 noise reduction filters, all of which utilise local statistics to adjust the resulting dataset or image. These are principally based on the local median or mode value (variously computed), but also include a form of trimmed mean (the Olympic filter) in which 1 or more values at the upper and lower end of the local ranked set of values is dropped and the mean of the subset calculated as the selected output value.

Simple hillshading may be achieved using a form of non-linear filter. In this case horizontal, vertical or diagonal difference filtering is applied. The procedure subtracts all the values in row 1 say, from row 2 and assigns a value of -1, 0 or 1 for example to the outcome, depending on the sign of the result. A colour-graded map of the output grid provides the hillshading effect.

4.6.4 Erosion and dilation

Erosion and *dilation*, in the context of grid operations, refer to morphological operations applied to binary and greyscale images. These are terms commonly used within image processing but only arise in a small number of GIS packages in this context. Erosion involves the removal (alteration) of pixels at the edges of regions, for example changing binary 1 values to 0, whilst dilation is the reverse process with regions growing out from their boundaries. Buffering is a form of dilation.

These two processes are often carried out using a form of kernel known as a structural element. A structural element is an *NxN* kernel with entries classified according to a binary scheme, typically as 0 or 1. If all entries are coded 1 then the structural element is a solid square block, the centre of which is laid over each pixel in the source image in turn. Pixels that are coded as 1 in the structural element and extend beyond the boundary of a shape in the source image result in that element being extended or dilated. The shape of the structural element may vary, for example as a vertical bar, horizontal bar, cross shape or a user-defined pattern.

To illustrate this process we have taken a small section of a USGS DEM file and then applied dilation and erosion processes to this source. The results are shown in Figure 4-63. If dilation is followed by erosion the process is described as a *Closing* operation, whilst Erosion followed by dilation is known as *Opening*. These processes are not symmetric, and thus are generally not reversible. Opening eliminates small and thinner features, resulting in smoother edged regions, whilst Closing also smoothes shapes but makes thin narrow features larger and eliminates small holes and narrow gaps. These changes are the result of deliberate morphological modification of the source data files. However, similar effects may be observed as a result of transformation operations performed manually or automatically for other purposes — for example resampling (changing resolution), rotation, multiple overlay, or Map Algebra operations.

Figure 4-63 Dilation and erosion operations

A. Source DEM section

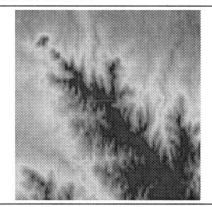

B. Dilation: 9x9 structural element, 9x3 vertical bar

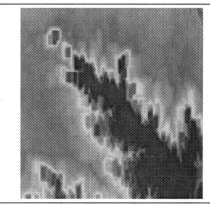

C. Erosion: 9x9 structural element, 9x3 vertical bar

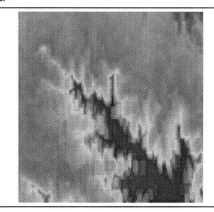

5 Data Exploration and Spatial Statistics

5.1 Statistical Methods and Spatial Data

This chapter of the Guide focuses on a range of exploratory data analysis techniques and statistical methods that have been implemented in widely available GIS and GIS-related software packages. Initially we describe a range of simple statistical facilities that are provided within GIS packages, notably those supporting descriptive statistical measures and special facilities relating to spatial sampling.

Section 5.2 addresses the question of exploratory data analysis in a explicitly spatial context and describes a range of tools currently available that support such analysis as a precursor to further investigation and modelling. Aspects of these methods have been touched upon in elsewhere in this Guide, notably in Section 4.3.

Many of the techniques of spatial analysis described in this Guide make use of statistical measures and methods. These include the contextual discussion in Section 2.4, listing of univariate statistical measures in Table 1.4, and directional analysis in Section 4.5. Statistical procedures with specific spatial extensions are described in this chapter and in Section 6.7. In each case ideas and methods from mainstream statistics have been extended and developed in order to address the specific needs of spatial datasets. The nature of these extensions is different from the ways in which multivariate statistics are derived from their univariate counterparts because of the ways in which they depend upon the fundamental organising concepts of distance, direction, contiguity and scale. In many instances classical hypothesis testing and inferential procedures cannot be applied to spatial problems (or at least, not without

reservations and/or the use of specialised modelling techniques). This may be because the datasets do not satisfy classical independence or distributional requirements and/or because the sampling framework is unknown or unsuitable.

The result of research in this field over the past 50 years or so has been the development of a collection of core statistical procedures for spatial analysis. Many of these procedures are descriptive and/or exploratory. Such methods are not without the support of strong statistical foundations, but recognise that the requirements and assumptions of classical statistics are often not strictly met. However, recent advances in spatial modelling and estimation procedures, inferential methods, and associated software tools mean that it is now possible to model relatively complex, fine structured, large spatial datasets using a wide range of methods.

Cressie (1993) proposed a taxonomy of spatial statistics based on the underlying model of the dataset being considered. The three main topics he identified using this approach are:

- **point pattern analysis** — corresponding to a location-specific view of the data (discussed here in Section 5.4) (vector data)

- **lattice or regional analysis** — corresponding to zonal models of space, notably planar enforced sets of regions (such as administrative or census districts — discussed here in Sections 5.2, 5.5 and 5.6), and

- **geostatistical modelling** — applying to a continuous field view of the underlying dataset — core issues relating to spatial modelling are discussed in Section 5.5, and geostatistical methods as applied to interpolation of field data in Section 6.7

Cressie's basic taxonomy has been developed further by Anselin (2002, p.14). He summarises key aspects associated with the object/field distinction in spatial data models and their implications as follows (Table 5.1):

Table 5.1 Implications of Data Models

	Object	Field
GIS	vector	raster
Spatial Data	points, lines, polygons	surfaces
Location	discrete	continuous
Observations	process realisation	sample
Spatial Arrangement	spatial weights	distance function
Statistical Analysis	lattice	geostatistics
Prediction	extrapolation	interpolation
Models	lag and error	error
Asymptotics	expanding domain	infill

The above analysis highlights not only the more obvious differences between the two main data models, but also the implications of these in terms of the modelling approaches implied. It also identifies the focus on interpolation and infill in the case of fields, versus extrapolation and domain expansion (spatial and temporal prediction) in the case of vector structures. The distinctions between the various groupings are not always clear, and methods applied in one area are often carried across in part to others. For example, a grid may represent a field view, a lattice view or an aggregated point set view of underlying data. Some aspects of this issue are described in Section 5.3, which deals with grid-based statistical analysis.

Finally, in Section 5.6, we make a brief foray into the field of statistical modelling using regression techniques, describing some of the main approaches that have been developed to tackle the specific difficulties that arise with spatial datasets.

5.1.1 Descriptive statistics

Almost all GIS packages provide a range of facilities for computing simple univariate statistical measures on tabular attribute data associated with vector objects. Many of the basic statistics listed in Table 1.4 are provided, either directly by opening an attribute table and selecting a column for analysis, or by executing some form of simple calculation facility. In addition to these basic measures, tools are often provided that display frequency histograms of data, with or without data transformation. Such facilities may be stand-alone or may be embedded in related functionality, such as map classification and symbology tools.

For image or grid datasets many packages provide a wide range of non-spatial and spatial statistical facilities. Purely non-spatial facilities include the same type of statistical measures as for vector objects, where the attributes are grid cell values rather than attribute table columns. In addition, many packages provide a range of statistical tools designed for multiple grid analysis. A number of these go beyond data description to provide facilities such as data reduction (examining data redundancy) and modelling (e.g. simple regression techniques). Some of these are described in the following subsections.

Much of the subject commonly described as "Spatial Statistics" deals with vector datasets, and many of the tools and techniques that have been developed apply (directly or indirectly) to point rather than line or area-based data. Historically much analysis of zone or area-based data has relied on converting the data to point sets (e.g. assigning attributes to a polygon centroid). Zone-based data have also often been analysed by simple classification and mapping techniques. Increasingly, however, tools (and data) are becoming available that facilitate the direct exploration, analysis and modelling of small area (local) patterns.

5.1.2 Spatial sampling

Principles and methods of spatial sampling have been described in Section 2.3.8. By definition, spatial data must be collected and stored prior to analysis, and for this reason GIS toolsets and related software incorporate few facilities that directly address issues of sampling and sample design. Most commonly the terms sampling and resampling in GIS are used to refer to the frequency with which an existing dataset (raster image or in some cases, vector object) is sampled for simple display or processing purposes (e.g. when overlaying multiple data layers, or computing surface transects). These operations are not directly related to questions of statistical sampling.

Two aspects of statistical sampling are explicitly supported within several GIS packages. These are: (i) the selection of specific point or grid cell locations within an existing dataset for separate analysis; and (ii) removing spatial bias from collected datasets, using a procedure known as declustering. We describe each of these in subsections 5.1.2.1 and 5.1.2.2.

5.1.2.1 Sampling frameworks

A number of software packages, such as TNTMips, ENVI, Idrisi and GRASS provide tools to assist in the selection of sample points, grid cells or regions of interest (ROI) from input datasets. Often these datasets are remote-sensing images, which may or may not have been subjected to some form of initial classification procedure. Examples of the facilities provided are listed below:

ENVI — takes a raster image file as input and provides three types of sampling, which it describes as:

- Stratified random sampling, which may be proportionate or disproportionate. In the former case random samples are made from each class or ROI in proportion to the class or region size. Disproportionate

sampling essentially requires users to specify the sample size, although the elements will still be randomly selected from each class or ROI

- Equalised random sampling, which selects an equal number of observations at random from each class or ROI

- Random sampling, which ignores classes or ROIs and simply selects a predefined number of cells or points at random

The selected points or cells (which may be output as a separate georeferenced list or table) are then used in post-classification analysis — comparing classifications with *ground truth* in each case (obtained from field survey or other independent data sources).

Idrisi offers similar facilities to ENVI via its SAMPLE function, providing random, systematic or stratified random point (cell) sampling from an input image (grid). The selection process for stratified random simply involves regarding the input image as being constructed from rectangular blocks of cells, and then sampling random cells within these larger blocks.

GRASS provides simple random sampling which may be combined with masking to create forms of stratified random samples. This facility may be somewhat cumbersome to implement. GRASS also provides a facility to generate random sets of cells that are at least D units apart, where D is a user-specified buffer distance. This can result in a more stratified than random sample and it is suggested that D should be derived with reference to observed levels of spatial autocorrelation (cf. Sections 5.5 and 6.7).

TNTMips supports a range of point sampling facilities to be used within vector polygons (e.g. field boundaries). These provide the more familiar form of statistical sampling frameworks that would precede field studies, and have application for research into soil composition (e.g. for precision farming), groundwater analysis, geological studies, or ecological research. However, they could equally well be applied to urban environments, as a precursor to environmental monitoring or even household surveys. The software provides for a two-stage process: (i) the creation of grids within the polygonal regions to be studied; and (ii) the selection of points within these grid structures. Grids are of user-definable size (edge length or area), shape (triangular, hexagonal, square, linear strips or random rectangles), and orientation (angle of rotation). Sample generated grids are shown in Figure 5-1. In this instance each grid cell in A and B is 0.25 hectares in area. In case C random orthogonal grid lines have been generated with a given *mean* separation (i.e. grid line separation is not fixed) and fixed or random rotation from the vertical.

Where a regular cell framework has been generated the software then supports creation of sampling points *within* each cell (Figure 5-2A) — single points in this example. The methods supported are regular (centre of cell); systematic unaligned (Figure 5-2B) the first cell point is selected with random *x,y* coordinates and subsequent points are selected using the same *x* or *y* coordinate as the previous cell, but with one of the two coordinates selected at random, alternating on a column-by-column basis); and random (Figure 5-2C). In the latter two cases a weighting factor is provided that biases selection towards the centre of the cell (100= no bias, 1=maximum bias). Selected sample points that nominally fall inside a cell but outside of the polygon boundary are excluded.

Figure 5-1 Grid generation examples

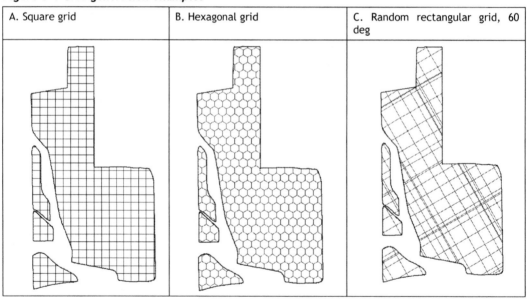

A. Square grid	B. Hexagonal grid	C. Random rectangular grid, 60 deg

Figure 5-2 Grid sampling examples within hexagonal grid, 1 hectare area

A. Regular (cell centre)	B. Systematic (random offset)	C. Random, no centre bias

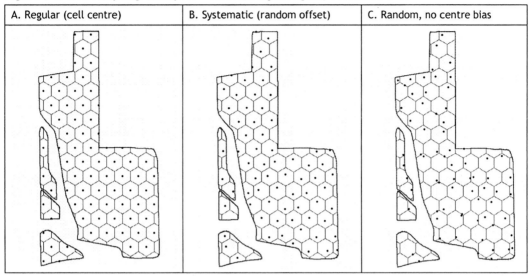

Figure 5-3 Random point generation examples — ArcGIS

A. Random point generation — selection form	B. Random sample points, 5 per county

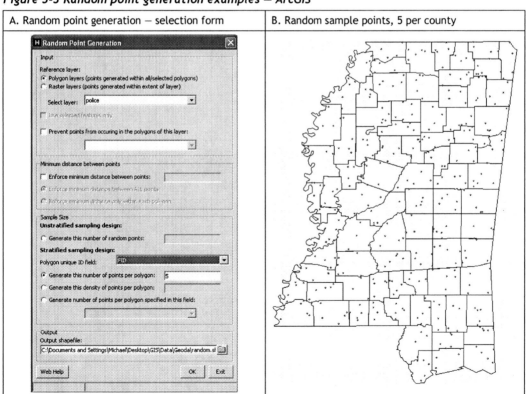

With general purpose GIS packages it is straightforward to generate a random, regular or partially randomised point set (within or externally to the GIS), and then to compute the intersection of this set with pre-defined polygons or grid cells. With this approach simple point-sampling schemes may be created, although precise matching to polygon forms, sample numbers or attribute weightings may be difficult. Purpose-built add-ins, such as Hawth's Tools for ArcGIS, provide a range of tailored sampling facilities. These include: (i) generation of random points, with a range of selection options (including use of raster or polygon reference layers — see Figure 5-3); (ii) random selection from an existing feature set (points, lines, polygons — see Figure 5-4); and (iii) conditional point sampling, designed for case-control analysis and similar applications. The latter facilitates a variety of random point generation methods in a region surrounding specified source points.

The term *quadrat sampling* is applied to schemes in which information on all static point data (e.g. trees, birds' nests etc.) is collected using an overlay of regular form (e.g. a square or hexagonal grid). Collected data are then aggregated to the level of the quadrat, whose size, orientation and internal variability will all affect the resultant figures (e.g. counts). Very small quadrats will ultimately contain 0 or 1 point objects, whilst very large quadrats will contain almost all the observations and hence will be of little value in understanding the variability of the data over space. An alternative to procedures based on lattices of quadrats is to "drop" quadrats onto the study area at random. Such quadrats may be of any size or shape, but circular forms have the advantage of being directionally invariant. A disadvantage with this approach is that some areas may be repeat sampled unless precautions are taken to exclude areas once sampled.

Figure 5-4 Random point sampling examples — ArcGIS

A. 10% random sample from existing point set	B. Stratified random selection, 30% of each stratum
800 radio-activity monitoring sites in Germany. Random sample of 80 (red/large dots)	200 radio-activity monitoring sites in Germany. Random sample of 30 (red/large dots)<100 units of radiation and 30 (crosses)>=100 units of radiation

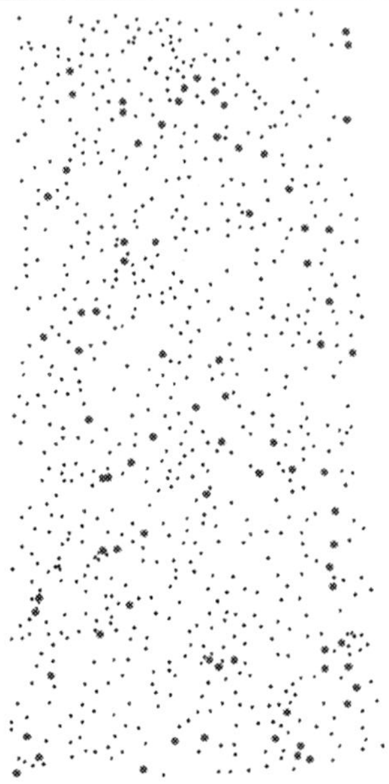

5.1.2.2 Declustering

Point samples may be unduly clustered spatially, for a variety of reasons. For example, samples from boreholes and wells may provide the basis for a chemical analysis of groundwater supplies, and the distribution of these may be clustered. Geological and subsea surveys frequently involve intensive data collection in localised areas, with sparsely sampled areas elsewhere. Practical constraints, such as access in built-up or industrialised zones, may also dictate sampling schemes that exhibit strong clustering. And of course, there may be clustering as a feature of the sampling design (e.g. stratified sampling, repeat sampling in small areas to obtain a representative measure of selected attributes). The latter may have been designed to ensure that different regions of interest (ROIs) are represented adequately, or that suspected areas of greater local variation are sampled in more detail than areas that are suspected of being more uniform.

Measured attributes in such instances may not be representative of population (whole region) attributes because observations in close proximity to one another may exhibit strong positive spatial autocorrelation — neighbouring measurements often have very similar attributes (see further Section 5.5.1). This results in attributes within these regions having undue weight in subsequent calculations. In the extreme, almost all observations may have been taken in a small region with consistently high or low attribute values, whilst very few have been taken from all remaining parts of the study area. Assuming spatial autocorrelation is present, clustering has the effect that measures such as the calculation of mean values, the estimation of regression parameters, or the determination of confidence intervals may be substantially biased.

A partial solution to problems of this kind is known as spatial declustering. Essentially this involves removing or reducing the known or estimated adverse effects of clustering in order to obtain a more representative picture

of the underlying population data and/or to ensure techniques such as feature extraction and surface modelling operate in an acceptable and useful manner. There are several approaches that may be adopted, each of which involves adjusting the sample values prior to further analysis. One of the simplest procedures for declustering involves defining a regular grid over sampled points (rather as per the grid generation procedure described in the subsection 5.1.2.1). The grid cell size is selected such that it is meaningful for the problem at hand (e.g. feature extraction) and/or ensures that the average number of points falling in a grid cell is 1 (typically). Cells which contain many sample points may then be regarded as clustered or possibly over-sampled, and a statistic such as the median value of the measured attribute(s) across all sampled points in that cell may then be used as the single assigned cell (centre) value. Another commonly provided declustering technique based on this grid-overlay approach is to use the density of points as a weighting function. For example, cells with 0 points have zero weight, cells with 1 point have a weight of 1, and cells with n points have each point weighted $1/n$ (hence in effect this is a simple averaging procedure). In reality both of these procedures amount to a kind of stratification of already sampled locations subsequent to their selection. It is important to note that procedures of this kind present no substitute for randomness in the selection of locations to be sampled and can amount to very dubious practice if the intention is subsequently to build an inferential statistical model using the observations that are retained.

In a similar vein, and as an alternative to count-based weighting, area-based weighting is provided as an option in several packages. This involves generating a set of Voronoi regions around each sample point, which results in small areas for closely spaced points and large areas for sparsely arranged sample points. The weights applied are then directly related to these areas. This method is simple but needs to have some justification and/or validation in terms of the problem under consideration, and may suffer from serious edge-effect problems, depending on how the Voronoi regions are computed (e.g. to the edge of the mapped region, or to the MBR or convex hull of the sample point set). Hybridised variants of area-based weighting (e.g. by adjusting the weights using known physical boundaries and/or nearest neighbour distances) have been shown to substantially reduce mean absolute error (MAE) and RMSE in some instances, e.g. see Dubois and Saisana (2002). Revised point-weighting schemes of this kind can be generated within GIS packages and then applied to the target attributes prior to further analysis. The scheme proposed by Dubois and Saisana, for example, which they tested on DEM data for Switzerland, was of the form:

$$w_i = \frac{s_i}{s_m} \times \frac{d_i^2}{s_m}$$

where w_i is the weight applied to the ith sample point, $i=1,2,...n$; s_i is the area of the ith Voronoi region; s_m is the average area of all the Voronoi regions (i.e. study area/n); and d_i^2 is the squared distance of the ith sample point to its nearest neighbour. Models of this type do not have universal application, and selection of appropriate declustering procedures requires careful analysis of the sample data, sub-sampling and cross-validating against some form of ground truth where necessary, and then applying adjustments in a manner appropriate to the problem and dataset to hand.

5.2 Exploratory Spatial Data Analysis

5.2.1 EDA, ESDA and ESTDA

Much of the groundwork in spatial statistics is concerned with the description and exploration of spatial datasets. The generic term for such methods is exploratory data analysis (EDA), or in the context of spatial and spatio-temporal analysis, ESDA and ESTDA respectively. Such methods are by no means exclusively statistical in nature, and for ESDA special forms of data mapping (i.e. visualisation) are of considerable importance — commercially available EDA or *data mining* tools do not generally provide spatial visualisation. This section provides a brief introduction to some of the methods that are specifically spatial in nature (ESDA) and which are supported in readily available software products.

The simplest form of EDA involves the computation of basic statistics, and in the context of spatial data, statistical summaries of attribute tables and grid values. A useful online reference on EDA is the NIST e-Handbook, referred to in the "Suggested reading" section of this Guide. The simplest forms of graphical analysis of such data tend to be histograms, pie charts, box plots and/or scatter plots. None of these provides an explicitly spatial perspective on the data. However, where such facilities are dynamically linked to mapped and tabular views of the data they can provide a powerful toolset for ESDA purposes. The selection of objects through such linking may be programmatically defined (e.g. all values lying more than 2 standard deviations from the mean) or user defined, often by graphical selection. The latter is known as *brushing*, and generally involves selection of a number of objects (e.g. points) from a graphical or mapped representation using a dragged region, generally of rectangular shape. Facilities of this type are implemented in a number of GIS packages, notably in ArcGIS V9 (with a range of

tools for different data types, but limited to 300 points for selected ESDA tools such as semivariance analysis) and in the stand-alone package, GeoDa. The latter has been built using ArcGIS objects and reads and writes ArcGIS shape files. It limits its attention to *lattice data*, by which is meant discrete spatial units (zones/areas) rather than point sets or point samples from a continuous surface.

Extensions to a number of these techniques to the spatio-temporal domain (ESTDA) have recently been made available in a number of software packages. These include the STARS OpenSource project and BioMedware's (commercial) Space-time Intelligence System (STIS) and the National Cancer Institute's SaTScan software, available free of charge, from http://www.satscan.org/.

5.2.2 Outlier detection

One of the areas ESDA tools focus on is outlier detection, as there are many instances in which so-called outliers are of great interest. In the present context these are spatial objects whose value on one or more attributes is markedly different from others in the set under consideration. The data in question may be correct or may be the result of some form of error (measurement, coding, representation etc.). Such data are of interest since they may represent the most important items in an investigation (e.g. mineral concentrations, a pollution source, an unexpectedly high incidence of a particular disease). Or they might represent data that need to be removed or adjusted (e.g. smoothed) if either the information is known or suspected to be incorrect, or if its retention will adversely affect the results obtained from the application of a particular analytical technique.

5.2.2.1 Mapped histograms

One of the simplest methods of highlighting possible outliers is to create a histogram of the data, typically using a fine class division, and then to examine the extreme classes. Where this facility is linked to a map of the data, the location of the object(s) may be identified and examined (Figure 5-5).

Figure 5-5 Histogram linkage

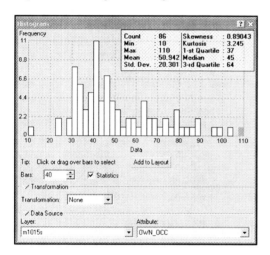

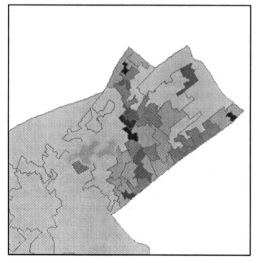

Source: UK 2001 Census Test Output Areas (OAs)

The upper figure shows the histogram and basic statistics for the attribute OWN_OCC (the number of Owner Occupiers, i.e. property owners) within 86 census districts (test census Output Areas, OAs, for part of Manchester, UK). The districts with the highest data values have been selected on the histogram window,

and are simultaneously highlighted in the map window (lower figure). The same approach may be applied for other vector object types, such as point data.

Data items that lie at the upper or lower limits of a dataset range may be described as *global outliers*. This term refers to values that are extreme compared to the dataset as a whole. However, within the dataset there may be values that are "relatively extreme" and these are referred to as *local outliers*. A local outlier is a value that is markedly different from (spatially) neighbouring values. An example of this might be a set of measurements taken along a transect, with a value part of the way along the transect that is very different from those immediately before or after, but still well within the overall range of the data recorded on the entire transect. Some ESDA software packages, such as ArcGIS Geostatistical Analyst, provide tools for displaying local as well as global outliers for selected data types.

5.2.2.2 Box plots

Box plots (or box-whisker plots) are a form of EDA provided in many data analysis and graphing packages (e.g. SPSS, STATA, Grapher, WinBUGS). Together with distribution plots and scatter plots they provide one of the three main ways in which statistical data are examined graphically. Because box plots are less familiar to many, and of particular use in examining outliers, we describe them in some detail (see Figure 5-6).

A box plot consists of a number of distinct elements. The example in Figure 5-6 was generated using MATLab Statistics Toolbox and we provide definitions below that apply to this particular implementation:

- The lower and upper lines of the "box" in the centre of the plot window are the 25th and 75th percentiles of the sample. The distance between the top and bottom of the box is the inter-quartile range (IQR)

- The line in the middle of the box is the sample median. If the median is not centred in the box it is an indication of skewness

- The *whiskers* are lines extending above and below the box. They show the extent of the rest of the sample (unless there are outliers). Assuming no outliers, the maximum of the sample is the top of the upper whisker. The minimum of the sample is the bottom of the lower whisker. By default, an outlier is a value that is more than 1.5 times the IQR away from the top or bottom of the box (a *hinge* value of 1.5), so with outliers the whiskers show a form of trimmed range, i.e. excluding the outliers (nb. the term *hinge* is also used in statistics to refer to locations within the main data range, in some instances matching the upper and lower quartile values)

- A symbol, e.g. a small circle, at the top and/or bottom of the plot is an indication of an outlier in the data. This point may be the result of a data entry error, a poor measurement or perhaps a highly significant observation

- The notches in the box are a graphic confidence interval about the median of a sample. A side-by-side comparison of two notched box plots is sometimes described as the graphical equivalent of a *t*-test. Box plots do not have notches by default

The box plots in Figure 5-6 are for a set of radioactivity observations made at 1008 sites in Germany on one day in 2004. The plot on the left (Sample 1) consists of 200 of the records, with whiskers extending to 1.5 times the IQR. Some packages allow user specification of the hinges, or provide an alternative set value (e.g. 3 times IQR — GeoDa; 2.5% and 97.5% limits — WinBUGS). The plot on the right shows a further 808 locations

and their readings (see also Figure 5-9). Three values in sample 1 were deliberately altered for this plot, e.g. simulating measurement or coding errors. One, for example, involved recording a measured value of 106.0 as 16.0. The plot picks out each of these outliers.

Figure 5-6 Simple box plot

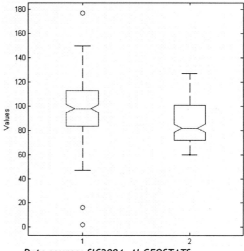

Data source: SIC2004, AI-GEOSTATS

Box plots with a link to zone-based spatial datasets are supported within GeoDa. Figure 5-7 illustrates the technique, again using the Manchester area test census Output Areas described earlier in this section. The census variable Owner Occupier has been selected for mapping, and a conventional box plot of the data is also illustrated. In the GeoDa implementation each data item that lies outside the box but within the whiskers is shown with a *, and the sole "true" outlier OA appears at the very top of the box plot above the upper whisker. The mapped box plot shows the OAs that fall into the various data quartiles (and the number of OAs in each), plus upper and lower outlier OAs — in this case just the one upper OA of rather complex shape, the same as that identified in Figure 5-5.

Figure 5-7 Mapped box plot

5.2.3 Cross tabulations and conditional choropleth plots

Mapped zonal data typically consist of a single variable, or a ratio of two variables, one of which is acting as a normalisation factor, e.g. mapping the ratio

r=persons_in_employment/total_population

Separate maps may be created for each variable or ratio of interest, but these are typically independent entities that may be difficult to compare and interpret. It would be relatively straightforward to create a series of maps of a particular variable of interest, for example the reported rate of lung cancer by health district where each map showed the rate for areas where the proportion of smokers was high, medium or low. This would be a form of control or "conditioning" on the information shown. This simple approach can be implemented within any GIS. The approach

can be extended further to two (or more) variables by crosstabulating the source data. If the crosstabulation is carried out on categorical data (e.g. sex, racial grouping) then once again a series of maps may be generated for each cell in the crosstabulation. However, with unclassified continuously varying data it is useful to examine the effect of specific levels of such data on the spatial distribution. Specialised visualisation tools have been developed recently to support operations of this type, including CCMaps and the most recent extensions to GeoDa (which have been derived from the ideas developed in the CCMaps project). They are known as interactive *Conditional Choropleth* mapping tools and may be dynamically linked to other visualisations such as box plots, histograms and scatterplots. The aim of such software, in the words of the original authors, is to:

• Stimulate analytical reasoning

• Detect the unexpected

- Discover the unexpected, and

- Stimulate hypothesis generation

Figure 5-8 illustrates this procedure using data on lung cancer mortality rates, by county, for the USA — see Carr *et al.* (2000, 2002) for a brief description of the method and this particular dataset. There are 9 maps in total. The coloured bar at the top shows how the counties have been classified, with for example 34% in the blue/dark grey (low) category, corresponding to 63.7-375 deaths per 100,000. The breakpoints on this scale may be dragged to provide alternative classification levels, the effects of which are dynamically updated in the map windows. Each map row represents one level of the percentage of the population below the USA designated poverty level (right hand slider scale) and each map column represents one level of the recorded annual precipitation level. The region in South-East of the USA (top right in the set of map windows) appears to have a high incidence of lung cancer mortality and a high score for both conditioning variables.

The figures in the top right of each mapped window show the weighted mean mortality rate, and the *R*-squared value in the lower right corner shows the percentage of the overall variability accounted for by these weighted means. By adjusting the two conditioning variable sliders (which are actually a form of box plot) or by using a built-in search facility (described as *cognostics*) a combination of slider values that maximise R^2 can be obtained — in this example to a value of just below 43%. For more details regarding the application of CCmapping and associated linked visualisations (e.g. conditional box-plots and conditional scatterplots) please refer to the CCMaps and/or GeoDa documentation and referenced articles.

Figure 5-8 Conditional Choropleth mapping

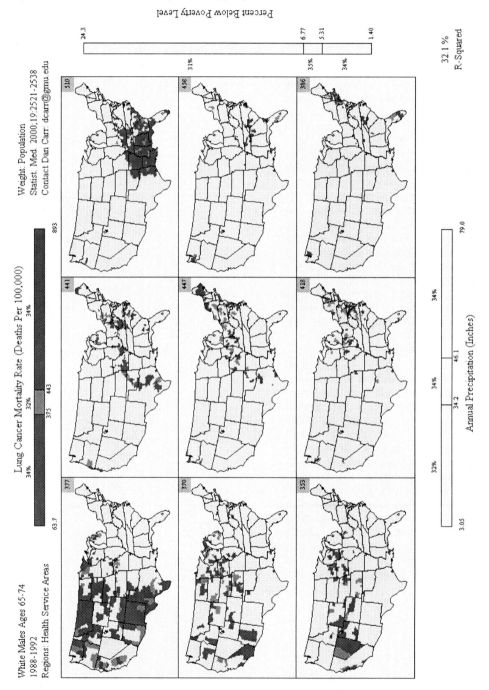

5.2.4 ESDA and mapped point data

Figure 5-9 illustrates a number of alternative methods for exploring variations in mapped point-based datasets with associated weights or continuously variable attributes. As in Figure 5-6 the (Sample 1) dataset consists of location information for 200 radioactivity monitoring stations in Germany, together with the levels of radiation recorded on a single day in 2004. A simple ESDA approach is to map the data and apply symbology to reflect the data values recorded. In Figure 5-9A and B this is illustrated by the use of variable symbol size and/or colour to reflect underlying values. In

Figure 5-9C a more sophisticated analysis has been conducted. In this case a semivariogram scatterplot of the dataset was generated (using ArcGIS Geostatistical Analyst) and the set of points on the scatterplot with the highest semivariance values were selected by brushing. Since these points in the scatterplot represent data pairs the linked map highlights the pairings. As can be seen the most extreme semivariance values are all related to just two of the original 200 source points in the lower right of the map. It may be that these latter points are of special interest (have unusually high or low radioactivity measurement) or that the data are incorrect and require adjustment or removal of selected values.

Figure 5-9 Exploratory analysis of radioactivity data

A. Variable point size	B. Variable colour	C. Semivariogram pairs	D. Voronoi analysis, cluster

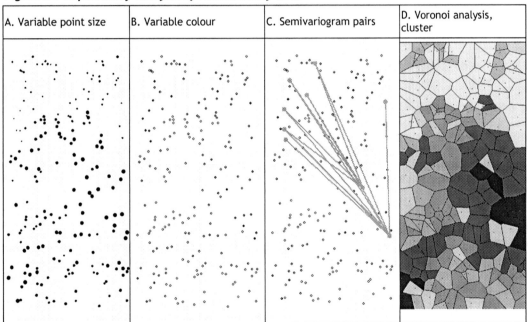

The last of the four images shown, Figure 5-9D, shows a rather different form of ESDA. Voronoi polygons (see Section 4.2.14.3) have been generated for the MBR of the entire point set and the radioactivity data mapped based on cell adjacencies. A set of 5 classes has been defined and cells that fall into a different class interval from all of their immediate neighbours are coloured grey, otherwise they are coloured by class interval. Note that these outliers are determined spatially, and will not in general correspond to extreme data values — they are local spatial outliers. Other Voronoi region statistics available as an alternative to this cluster approach include: mapping actual radioactivity values (described as "simple"

mapping); using a local neighbourhood mean or median for all cells; allocation based on a local neighbourhood entropy statistic (see Table 1.4); and spread-based statistics, such as standard deviation and range-based mapping.

Table 5.2 Voronoi-based ESDA

Local Smoothing	Mean, Mode, Median
Local Variation	Standard deviation
	Inter-quartile range
	Entropy
Local Outliers	Cluster
Local Influence	Simple

A summary of the interpretation of each of these choices is provided in Table 5.2 (these may be compared with some of the grid analysis methods described in Section 4.6).

5.2.5 Trend analysis of continuous data

A somewhat different approach to ESDA for continuous data represented as a point set with z-values is to examine whether any simple trends are present. For the radioactivity point set illustrated in Figure 5-9 a simple 3D trend analysis is shown in Figure 5-10.

Figure 5-10 Trend analysis of radioactivity dataset

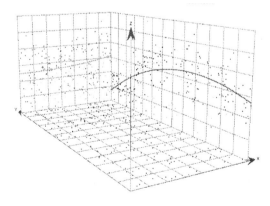

This diagram shows the point set mapped in the XY-plane, with the radioactivity levels (z-values) projected onto the XZ- and YZ-planes. Simple low-order (2nd degree) polynomial trend curves have then been fitted to these projected datasets, providing a useful pre-analysis visualisation of possible trends. Again, as with the forms of ESDA described earlier, points in this visualisation may be brushed and associated representations highlighted through linking. Trend analysis of the kind illustrated here uses the basic methods of curve fitting applied in regression modelling (see further, Section 5.6).

5.2.6 Cluster hunting

Cluster hunting is the term used to apply to a family of techniques that involve computationally intensive search procedures for point- and zone-based cluster identification. They aim to identify clusters based on the spatial arrangements of incidents combined with basic information on the background population. They then search for clusters (areas of unexpectedly high incidence) by exhaustively examining all possible locations on a fine grid covering the study area. The most well-known of these techniques, GAM, is best described in the authors' own words (from the GAM web site):

"The Geographical Analysis Machine (GAM) is an attempt at automated exploratory spatial data analysis of point or small area data that is easy to understand. The purpose is to answer a simple practical question; namely given some spatial data of something interesting where might there be evidence of localised geographic clustering if you do not know in advance where to look. [This may be] due to lack of knowledge of possible causal mechanisms, or if prior knowledge of the data precludes testing more hypotheses on the same database. Or more simply put, you send GAM a geographically referenced point or small area referenced database and it will indicate where there is evidence of localised clustering and how strong it is"

A Java version of GAM can be downloaded from the University of Leeds CCG web site (details are provided in the Appendix). This is a working version of the software but is, unfortunately, not the subject of ongoing support or development at present. The Java version of GAM takes as an input a text file of the form:

ID,easting,northing,incidence_count,population_count

The incidence_count value contains the variable of interest, e.g. disease incidence, and the population_count variable contains the population variable, e.g. the population at risk. Frequently the location data will be taken as the centres of small area statistics zones, and the population variable will be the total applicable value for that zone (e.g. census output area or tract). The study area is then divided into a grid, and circles are placed at every grid intersection. The number of incidents falling inside these circles are counted and checked to see if they are, by some measure, excessive. The notion of excessive could be user-specified (e.g. more than a certain number/rate of incidents, or pre-computed variable levels for each location) or can be computed by the program based on an expected level defined as the population at risk times the mean incidence rate.

The steps in the analytical procedure are as follows:

- Step 1. Read in X (easting), Y (northing), a variable of interest, and data for the population at risk

- Step 2 Identify the MBR containing the data, identify starting circle radius, and degree of overlap. If R is the radius of a circle of area equal to the MBR, then starting radius might be $r=R/100$ with increment $dr=R/100$

- Step 3 Generate a grid covering this rectangle so that circles of current radius

overlap by the desired amount – a range of radius values r will be used, varying over a user-specified range of distances

- Step 4 For each grid-intersection generate a circle of radius r

- Step 5 Retrieve two counts for the population at risk and the variable of interest

- Step 6 Apply some "significance" test procedure – a variety of alternative procedures, including Monte Carlo simulation and Bootstrap methods are supported (the latter is a form a sampling with replacement, similar to jack-knifing in some respects – see further, Section 6.7.2.2)

- Step 7 Keep the result if significant

- Step 8 Repeat Steps 5 to 7 until all circles have been processed

- Step 9 Increase circle radius by dr and return to Step 3 else go to Step 10

- Step 10 Create a smoothed density surface of excess incidence for the significant circles using a kernel smoothing procedure and aggregating the results for all circles (this step uses the Epanechnikov kernel function, Table 4.8)

- Step 11 Map this surface and inspect the results

GAM with kernel smoothing is generally referred to as GAM/K, and it is in this form that is currently used. It is by no means the only method of cluster hunting, but has been found in tests on synthetic data to be one of the most effective at locating genuine clusters in large test datasets, and also one of the least subject to finding false clusters. Further discussion relating to this topic is provided in Section 5.4.3

5.3 Grid-based Statistics

5.3.1 Overview of grid-based statistics

In the earlier discussion of local, focal, zonal and global analysis of grids (Section 4.6.1) we described how basic statistical measures may be obtained from grid datasets. These measures are similar to those applied to univariate attribute data, but may operate at various levels of spatial grouping of a single grid.

Table 5.3 Sample statistical tools for grid data — Idrisi

Facility	Description
Histo	Single grid — create histogram on n classes from cell values. Classes may cover a spatial subset of the grid (e.g. by masking), have excluded values (e.g. 0, -999), have user defined boundaries etc.
Pattern	Single grid — generate selected Landscape metrics (e.g. richness, dominance, diversity, fragmentation — as discussed in Section 5.3.4).
Centre	Computes the weighted mean centre of an input image (grid) where cells are regarded as point counts or weights (see also, Section 4.2.5.2). This function also generates the standard distance radius (see Section 4.5.3) for the point set, and the *Coefficient of Relative Dispersion* (CRD) which is the ratio of the standard distance measure to the radius of a circle having the same area as the study region, expressed as a percentage.
Crosstab	The standard form of Crosstab compares two classified images, typically matching in spatial extent (e.g. time slices of the same region, images of adjacent zones of the same size and resolution), each with up to 127 classes. The principal output is a crosstabulation table, showing each class in image 1 as the rows, each class of image 2 as the columns, and counts of values in the table. Hence if image 1 contains 234 cells classified as A and image 2 contains 370 cells classified as A, then *at most* 234 cells will appear in row A column A of the output table. For common classification schemes (with square crosstabs) the diagonal of this crosstab matrix shows matching (e.g. unchanged) grid values. Potentially up to MxN combinations of classes are possible (taking AB as different from BA). Class combinations may be mapped providing a spatial representation of the crosstabulation that is a form of spatial overlay (AND) operation. Two measures of association (correlation) between the images are provided: Cramer's V and the Kappa index (described in more detail in Section 5.3.2). The latter requires that the classification scheme for both images match.
Quadrat	Quadrat (grid based) point pattern analysis. In this case each grid cell is regarded as a separate quadrat with cell values corresponding to point counts. Masking may be applied, and may well be desirable. Standard statistical measures are provided, with the variance/mean (V/M) ratio being used as an indicator of spatial clustering. Values of $V/M=1$ may indicate a random pattern, assuming an underlying independent, homogeneous random process, whilst $V/M<1$ suggests more uniform patterns under this model. These interpretations are not valid if the process is not stationary. Note that the results are significantly affected by quadrat (grid cell) size and overall sampling area. If D=study area/number of points, ideal quadrat sizes are suggested as being in the range $[D,2D]$.
Regress Multireg Logisticreg	Provides regression modelling between multiple image files or attribute files, with or without image masking. Interpretation of image (grid) file regression may be difficult, since spatial autocorrelation within images is almost always strong (and may even have been generated by resampling or other interpolation methods), resulting in misleading values for degrees of freedom and associated statistical measures

Within many GIS packages additional statistical tools that apply to grid datasets are grouped together with other, non-statistical facilities, or embedded within application-specific functionality, such as image classification. For example, in ArcGIS the majority of these tools are located in the Spatial Analyst (Neighbourhood analysis and Multivariate analysis subsections); in ENVI statistics facilities are located under the Basic Tools|Statistics menu and the Basic Tools|Spatial Statistics menu, the latter providing global and local spatial autocorrelation measures of the types described in Section 5.5.3. Idrisi's statistics facilities, which are a rather mixed collection provided under a single menu, are listed in Table 5.3. Some, such as Histo and Centre are fairly straightforward descriptive statistical tools. Others, notably the Quadrat, Crosstab and Pattern tools, provide more advanced forms of analysis and warrant closer scrutiny. Regression methods are discussed in Section 5.6.

5.3.2 Crosstabulated grid data

The creation and analysis of crosstabulated data is a familiar process in many fields and often the subject of simple statistical analysis, principally through the use of Chi-square tests. These are covered in all basic statistics texts, but for completeness we will include a brief description here, before showing how such methods may be applied to specifically spatial problems.

A common crosstab arrangement is a table of rows, representing distinct treatments (e.g. A=no soil improvement, B=organic manure treatment, and C=non-organic treatment) tabulated against responses (R) or outcomes (e.g. classified levels of crop production, or presence/absence of a given pest or disease; Table 5.4). Tabulated cell values are counts of events, e.g. the number of treated plots falling into each response category. This kind of 2-way presentation of data is often analysed to identify whether the observed frequencies in each cell are significantly different from those that might be expected under the assumption that the outcomes are independent of the treatments — i.e. the row and column classifications are independent.

Let X be an N-row by M-column crosstabulation of frequency counts, x_{ij}, with overall total

$\Sigma x_{ij}=x..=T$ (the subscript dots indicate that we have summed across all subscript values — see Table 5.4, sometimes summation subscripts use a + symbol rather than a dot, as in x_{i+}). Now let $p=p_i$ be the proportion of the overall total T found in row i, and $q=p._j$ be the proportion of T found in column j, then the expected frequency in cell (i,j) under the assumption of independence, i.e. the treatments do not affect the responses, is simply Tpq.

Table 5.4 Simple 2-way contingency table

	R1	R2	R3	Totals
A	x_{11}	x_{12}	x_{13}	$x_1.$
B	x_{21}	x_{22}	x_{23}	$x_2.$
C	x_{31}	x_{32}	x_{33}	$x_3.$
Totals	$x._1$	$x._2$	$x._3$	$x..=T$

The difference between the observed and expected frequencies provides a measure of how close the observations are to a pattern that might arise assuming the rows and columns are independent. To remove the sign of these differences the values are squared, and then are standardised by dividing through by the expected frequencies, and finally summed. This provides an overall measure of differences that under a broad range of conditions follows the Chi-square (χ^2) distribution with $(N-1)(M-1)$ degrees of freedom. The computation of the statistic is often shown in the form:

$$\chi^2 = \sum \frac{(O-E)^2}{E} \text{, or}$$

$$\chi^2 = \sum_{i=1}^{N} \sum_{j=1}^{M} \frac{\left(x_{ij} - x_{i.}x_{.j} / N\right)^2}{x_{i.}x_{.j} / N}$$

If the computed value is large then it is less likely that the rows and columns are independent than if it is small. The probability that a particular computed value might have arisen by chance, given the size of the table (as represented by the degrees of freedom) can be obtained from tables of the Chi-square distribution or computed using built-in functions in many software packages, including Excel. Results in this analysis are also adversely affected by small counts and as a rule of thumb no more than 20% of cell values should be less than 5. In such cases it may be desirable to aggregate columns or rows (if such aggregation remains meaningful), or to adopt an alternative test procedure (e.g. simulation/permutation tests).

Within spatial analysis this approach to studying crosstabulated datasets has been utilised to provide an insight into classified spatial datasets — either matching pairs of images that are timeslices, or remote sensing imagery and ground truth datasets. The proportion, p, of cells in image 1 that match those in image 2, is the principal measure of interest. If p is close to 1 then the two images are likely to be very similar. The detailed pattern of differences can be interpreted from the crosstabulation of the classes and/or by generating a new image in which either every classification combination is presented, or binary coding is used to indicate the location of matching/unchanged cells (0) and non-matching/changed cells (1).

The crosstabulation (or cross-classification procedure) described in Table 5.3, results in a potentially large table comprising rows representing the classification of the cells in image 1 and columns showing corresponding classifications in image 2. Cell entries are then counts of observed combinations for co-located cells. In the context of automated classification using training data this crosstab is sometimes referred to as a *confusion matrix* and the diagonal elements as a proportion of

the row totals provide an indicator of the amount of confusion (100% would imply no confusion).

The cross-classification table may also be used to generate an index that describes the overall (global) similarity between the two images (a form of correlation measure). The two index values that are commonly computed (e.g. in Idrisi, ENVI and similar software) are the Kappa Index of Agreement and Cramer's V index. Both have a range of values that typically range from 0 to 1, with 1 indicating perfect agreement and 0 indicating a pattern arising by chance. Negative values are possible and occur where the proportion of matching cells is low. Both index measures are calculated using procedures developed from the Chi-square analysis of standard contingency tables.

The Kappa Index of Agreement (also known as KIA), is of the form:

$$\kappa = \frac{O - E}{1 - E}$$

where O is the observed accuracy or *proportion* of matching values (the matrix diagonal) and E is the expected proportion of matches in this diagonal assuming a model of classification independence derived from the observed row and column totals. Hence O is simply the sum of the diagonal elements divided by the overall total, T, and E is computed in a similar manner to the expected values for a Chi-square calculation, with each element being summed. Thus E is the sum for each diagonal cell of the row total times the column total divided by the overall total squared. If x_{ij} represents the observed entry in row i, column j, then row totals are given by $x_{i.}$, column totals are given by $x_{.j}$ and the overall total is given by $T = x_{..}$. Thus the expected values are obtained from the row and column proportions, $p_{i.} = x_{i.}/T$ and $p_{.j} = x_{.j}/T$, giving $e_{ij} = p_{i.}{}^*p_{.j}$

The Kappa index may be disaggregated into class or "per category" components by examining the row-wise expectations:

$$\kappa_i = \frac{O_{ii} - E_{ii}}{O_{i.} - E_{ii}}$$

Cramer's V statistic is similar to the Kappa index and is derived directly from the Chi-square statistic computed for a given crosstabulation (or contingency table). As with the Kappa statistic its value is reported in the range [0,1], with the same interpretation, but in this instance the source table can be an *MxN* array, thus the classification schemes do not have to be identical. The statistic is computed as:

$$V = \sqrt{\frac{1}{T} \frac{\chi^2}{\min\{(M-1),(N-1)\}}}$$

5.3.3   Quadrat analysis of grid datasets

As noted earlier, quadrat sampling is the term used to describe a procedure of sampling and recording point-based data within regularly shaped regions (typically a grid of square cells). Since grid data are already in this form, it is possible to analyse grids where the contents of grid cells (or blocks of cells) are regarded as counts of point objects (Figure 5-11). In the example illustrated a 5x5 grid has been used to collect data on point events shown by the symbol x (the larger * symbol in this diagram shows the position of the MAT point under the L_2 metric). The event distribution has then been coded as counts in the grid below. Simple statistics may then be computed, such as the mean number of points per cell/cell block (4 in this example), and the variance of this measure (4.59 in this case).

If the distribution of points across the set of grid cells is random, it can be modelled using the Poisson distribution. The Poisson distribution is applicable where events (points in our case) are independent, there are a large number of events (typically 100+), and the probability of an individual event occurring

(e.g. a point falling in a particular location) is small. It is derived as an approximation to the Binomial distribution by applying these conditions.

Figure 5-11 Quadrat counts

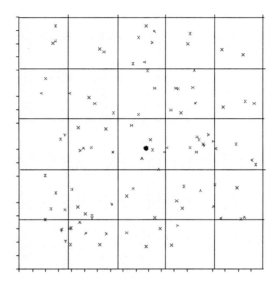

As noted in Table 1.4, the Poisson distribution has the form:

$$f(x) = \frac{m^x}{x!} e^{-m} ; \; x=0,1,2,...$$

where m is the mean and x is the count of events. In our example the (sample) mean is 4, so the individual terms of the distribution may be computed (i.e. for $x=0,1,2,3...$) and used as a set of "expected" values, under the null hypothesis that the observed frequency distribution is random. The set of n observed frequency values may then be compared to the set of expected values using a simple Chi-square test, to obtain an estimate of the probability that the data reflects a random distribution of events. For the example shown above the following table may be drawn up:

Freq	Obs, O	Exp, E	|O-E|	$|O-E|^2/E$
0	1	.5	.5	0.64
1	0	1.8	1.8	1.83
2	6	3.7	2.3	1.49
3	6	4.9	1.1	0.25
4	2	4.9	2.9	1.7
5	3	3.9	.9	0.21
6	4	2.6	1.4	0.75
7	2	1.5	.5	0.18
8	0	.7	.7	0.74
9	1	.3	.7	1.35
10	0	.1	.1	0.13
Sum	25			$\chi^2=9.3$

The *sum* row shows the total number of observations made (grid cells) and the value of the χ^2 statistic. The degrees of freedom in this case are 11-1-1=9, because there are 11 frequency classes, the total count is known (-1DF), and the mean (m) has been estimated from the sample (-1DF). The 5% probability level from tables or computed value of the Chi-square distribution is $\chi^2_{0.05,9}=16.9$, thus a value of 9.3 is well within the expectation for a random pattern and thus we cannot reject the null hypothesis on the basis of this information. Aggregating rows to ensure most table cell counts are greater than 5 — e.g. by grouping frequencies into four classes (0,1), (2,3), (4,5) and (6+) — gives a χ^2 statistic of 4.6 with 3 degrees of freedom, and $\chi^2_{0.05,2}=6$, confirming the previous result.

In principle the observed and expected frequency distributions could also be compared using the maximum absolute difference in their cumulative probability distributions, using the Kolmogorov-Smirnov test statistic. However, this procedure should really be reserved for expected distributions that are continuous (e.g. the Normal distribution and other, non-Normal continuous distributions), and there may be difficulties in estimation of the appropriate mean value for the expected distribution. Testing procedures of this type are rarely supported directly within GIS and related packages, but are widely supported in statistical packages such as SPSS and STATA and may be readily computed programmatically or by use of a generic tool such as Excel.

5.3.4 Landscape Metrics

There are many software packages that provide methods for analysing patterns observed in remote sensed image files and raster grids. Some of these focus on the process of classifying pixels or grid cells, in order to identify land use and vegetation cover. Once this classification process has been completed a second and often related procedure is that of describing and analysing the observed patterns. As we have seen in the preceding subsections, descriptive statistics may be simple summaries of the frequencies of particular classification values or categories, which are non-spatial attributes of the dataset, or they may provide spatial measures or *metrics*.

A substantial number of such metrics have been devised over the last 50 years, principally in the field of landscape ecology, and a number of software packages and toolkits developed to undertake such analysis. The most well-known package, Fragstats, provides much of the background for other packages, such as Fragstats for ArcView and Patch Analyst (which provides both vector and raster-based patch analysis). In this section, we provide basic details of both the non-spatial and spatial metrics supported by such software, based on an edited version of the Fragstats documentation (with their kind permission). Fragstats originated text is shown with a greyed background.

The common usage of the term "landscape metrics" refers exclusively to indices developed for categorical map patterns. Landscape metrics are algorithms that quantify specific spatial characteristics of patches, classes of patches, or entire landscape mosaics. A plethora of metrics has been developed to quantify categorical map patterns. These metrics fall into two general categories: those that quantify the composition of the map without reference to spatial attributes, and those that quantify the spatial configuration of the map, requiring spatial information for their calculation (McGarigal and Marks 1995, Gustafson 1998).

Although a large part of landscape pattern analysis deals with the identification of scale and intensity of pattern, landscape metrics focus on the characterisation of the geometric and spatial properties of categorical map patterns represented at a particular scale (grain and extent). Thus, while it is important to recognise the variety of types of landscape patterns and goals of landscape pattern analysis, Fragstats focuses on landscape metrics as they are applied in landscape ecology... Landscape is not necessarily defined by its size; rather, it is defined by an interacting mosaic of patches relevant to the phenomenon under consideration (at any scale).

P: **Patch-level metrics** are defined for individual patches, and characterise the spatial character and context of patches. In most applications, patch metrics serve primarily as the computational basis for several of the landscape metrics, for example by averaging patch attributes across all patches in the class or landscape; the computed values for each individual patch may have little interpretive value.

C: **Class-level metrics** are integrated over all the patches of a given type (class). These may be integrated by simple averaging, or through some sort of weighted-averaging scheme to bias the estimate to reflect the greater contribution of large patches to the overall index. There are additional aggregate properties at the class level that result from the unique configuration of patches across the landscape.

L: **Landscape-level metrics** are integrated over all patch types or classes over the full extent of the data (i.e., the entire landscape). Like class metrics, these may be integrated by a simple or weighted averaging, or may reflect aggregate properties of the patch mosaic. In many applications, the primary interest is in the pattern (i.e., composition and configuration) of the entire landscape mosaic.

5.3.4.1 Non-spatial metrics

Composition [of a grid map] is easily quantified and refers to features associated with the variety and abundance of patch types within the landscape, but without considering the spatial character, placement, or location of patches within the mosaic. Because composition requires integration over all patch types, composition metrics are only applicable at the landscape-level. There are many quantitative measures of landscape composition, including the proportion of the landscape in each patch type, patch richness, patch evenness, and patch diversity. Indeed, because of the many ways in which diversity can be measured, there are literally hundreds of possible ways to quantify landscape composition. Unfortunately, because diversity indices are derived from the indices used to summarise species diversity in community ecology, they suffer the same interpretative drawbacks. It is incumbent upon the investigator or manager to choose the formulation that best represents their concerns. The principal measures of composition are:

- **Proportional Abundance** of each Class: One of the simplest and perhaps most useful pieces of information that can be derived is the proportion of each class relative to the entire map

- **Richness:** Richness is simply the number of different patch types

- **Evenness:** Evenness is the relative abundance of different patch types, typically emphasizing either relative dominance or its complement, equitability. There are many possible evenness (or dominance) measures corresponding to the many diversity measures. Evenness is usually reported as a function of the maximum diversity possible for a given richness. That is, evenness is given as 1 when the patch mosaic is perfectly diverse given the observed patch richness, and approaches 0 as evenness decreases. Evenness is sometimes reported as its complement, dominance, by subtracting the observed diversity from the maximum for a given richness. In this case, dominance approaches 0 for maximum equitability and increases >0 for higher dominance

- **Diversity**: A composite measure of richness and evenness and can be computed in a variety of forms (e.g., Shannon and Weaver 1949, Simpson 1949), depending on the relative emphasis placed on these two components

5.3.4.2 Spatial metrics

Spatial configuration [of a grid map] is much more difficult to quantify and refers to the spatial character and arrangement, position, or orientation of patches within the class or landscape. Some aspects of configuration, such as patch isolation or patch contagion, are measures of the placement of patch types relative to other patches, other patch types,

or other features of interest. Other aspects of configuration, such as shape (see also, Section 4.2.8) and core area, are measures of the spatial character of the patches. There are many aspects of configuration and the literature is replete with methods and indices developed for representing them.

Configuration can be quantified in terms of the landscape unit itself (i.e., the patch). The spatial pattern being represented is the spatial character of the individual patches, even though the aggregation is across patches at the class or landscape level. The location of patches relative to each other is not explicitly represented. Metrics quantified in terms of the individual patches (e.g., mean patch size and shape) are spatially explicit at the level of the individual patch, not the class or landscape. Such metrics represent recognition that the ecological properties of a patch are influenced by the surrounding neighbourhood (e.g., edge effects) and that the magnitude of these influences is affected by patch size and shape. These metrics simply quantify, for the class or landscape as a whole, some attribute of the statistical distribution (e.g., mean, max, variance) of the corresponding patch characteristic (e.g., size, shape). Indeed, any patch-level metric can be summarized in this manner at the class and landscape levels. Configuration also can be quantified in terms of the spatial relationship of patches and patch types (e.g., nearest-neighbour, contagion). These metrics are spatially explicit at the class or landscape level because the relative location of individual patches within the patch mosaic is represented in some way. Such metrics represent recognition that ecological processes and organisms are affected by the overall configuration of patches and patch types within the broader patch mosaic.

A number of configuration metrics can be formulated either in terms of the individual patches or in terms of the whole class or landscape, depending on the emphasis sought. For example, perimeter-area fractal dimension is a measure of shape complexity (Mandelbrot 1982, Burrough 1986, Milne 1991) that can be computed for each patch and then averaged for the class or landscape, or it can be computed from the class or landscape as a whole by regressing the logarithm of patch perimeter on the logarithm of patch area. Similarly, core area can be computed for each patch and then represented as mean patch core area for the class or landscape, or it can be computed simply as total core area in the class or landscape. Obviously, one form can be derived from the other if the number of patches is known and so they are largely redundant; the choice of formulations is dependent upon user preference or the emphasis (patch or class/landscape) sought. The same is true for a number of other common landscape metrics. Typically, these metrics are spatially explicit at the patch level, not at the class or landscape level.

The principal aspects of configuration and a sample of representative metrics are:

- **Patch size distribution and density:** The simplest measure of configuration is patch size, which represents a fundamental attribute of the spatial character of a patch. Most landscape metrics either directly incorporate patch size information or are affected by patch size. Patch size distribution can be summarized at the class and landscape levels in a variety of ways (e.g., mean, median, max, variance, etc.), or, alternatively, represented as patch density, which is simply the number of patches per unit area.

- **Patch shape complexity:** Shape complexity relates to the geometry of patches--whether they tend to be simple and compact, or irregular and convoluted. Shape is an extremely difficult spatial attribute to capture in a metric because of the infinite number of possible patch shapes. Hence, shape metrics generally index overall shape complexity rather than attempt to assign a value to each unique shape. The most common measures of shape complexity are based on the relative amount of perimeter per unit area, usually

indexed in terms of a perimeter-to-area ratio, or as a fractal dimension, and often standardized to a simple Euclidean shape (e.g., circle or square). The interpretation varies among the various shape metrics, but in general, higher values mean greater shape complexity or greater departure from simple Euclidean geometry. Other methods have been proposed: radius of gyration (Pickover 1990); contiguity (LaGro 1991); linearity index (Gustafson and Parker 1992); and elongation and deformity indices (Baskent and Jordan 1995), but these have not yet become widely used (Gustafson 1998).

- **Core Area:** Core area represents the interior area of patches after a user-specified edge buffer is eliminated. The core area is the area unaffected by the edges of the patch. This "edge effect" distance is defined by the user to be relevant to the phenomenon under consideration and can either be treated as fixed or adjusted for each unique edge type. Core area integrates patch size, shape, and edge effect distance into a single measure. All other things equal, smaller patches with greater shape complexity have less core area. Most of the metrics associated with size distribution (e.g., mean patch size and variability) can be formulated in terms of core area.

- **Isolation/Proximity:** Isolation or proximity refers to the tendency for patches to be relatively isolated in space (i.e., distant) from other patches of the same or similar (ecologically friendly) class. Because the notion of "isolation" is vague, there are many possible measures depending on how distance is defined and how patches of the same class and those of other classes are treated. If d_{ij} is the nearest-neighbour distance from patch i to another patch j of the same type, then the average isolation of patches can be summarized simply as the mean nearest-neighbour distance over all patches. Alternatively, isolation can be formulated in terms of both the size and proximity of neighbouring patches within a local neighbourhood around each patch using the isolation index of Whitcomb *et al.*. (1981) or proximity index of Gustafson and Parker (1992), where the neighbourhood size is specified by the user and presumably scaled to the ecological process under consideration. The original proximity index was formulated to consider only patches of the same class within the specified neighbourhood. This binary representation of the landscape reflects an island biogeographic perspective on landscape pattern. Alternatively, this metric can be formulated to consider the contributions of all patch types to the isolation of the focal patch, reflecting a landscape mosaic perspective on landscape patterns.

- **Contrast:** Contrast refers to the relative difference among patch types. For example, mature forest next to younger forest might have a lower-contrast edge than mature forest adjacent to open field, depending on how the notion of contrast is defined. This can be computed as a contrast-weighted edge density, where each type of edge (i.e., between each pair of patch types) is assigned a contrast weight. Alternatively, this can be computed as a neighbourhood contrast index, where the mean contrast between the focal patch and all patches within a user-specified neighbourhood is computed based on assigned contrast weights. Relative to the focal patch, if patch types with high contrast lead to greater isolation of the focal patch, as is often the case, then contrast will be inversely related to isolation (at least for those isolation measures that consider all patch types).

- **Dispersion:** Dispersion refers to the tendency for patches to be regularly or contagiously distributed (i.e., clumped) with respect to each other. There are many dispersion indices developed for the assessment of spatial point patterns, some

of which have been applied to categorical maps. A common approach is based on nearest-neighbour distances between patches of the same type. Often this is computed in terms of the relative variability in nearest-neighbour distances among patches; for example, based on the ratio of the variance to mean nearest-neighbour distance. Here, if the variance is greater than the mean, then the patches are more clumped in distribution than random, and if the variance is less than the mean, then the patches are more uniformly distributed. This index can be averaged over all patch types to yield an average index of dispersion for the landscape. [*Guide editors note: this class of measure is not recommended as it is dimensionally inconsistent — V/M ratios are appropriate for count-based or quadrat data but not for distance-based measures*]. Alternative indices of dispersion based on nearest-neighbour distances can be computed, such as the familiar Clark and Evans (1954) index.

- **Contagion and Interspersion:** Contagion refers to the tendency of patch types to be spatially aggregated; that is, to occur in large, aggregated or "contagious" distributions. Contagion ignores patches per se and measures the extent to which cells of similar class are aggregated. Interspersion, on the other hand, refers to the intermixing of patches of different types and is based entirely on patch (as opposed to cell) adjacencies. There are several different approaches for measuring contagion and interspersion. One popular index that subsumes both dispersion and interspersion is the contagion index based on the probability of finding a cell of type i next to a cell of type j (Li and Reynolds 1993). This index increases in value as a landscape is dominated by a few large (i.e., contiguous) patches and decreases in value with increasing subdivision and interspersion of patch types. This index summarizes the aggregation of all classes and thereby provides a measure of overall clumpiness of the landscape. McGarigal and Marks (1995) suggest a complementary interspersion/juxtaposition index that increases in value as patches tend to be more evenly interspersed in a "salt and pepper" mixture. These and other metrics are generated from the matrix of pairwise adjacencies between all patch types, where the elements of the matrix are the proportions of edges in each pairwise type. There are alternative methods for calculating class-specific contagion using fractal geometry (Gardner and O'Neill 1991). Lacunarity is an especially promising method borrowed from fractal geometry by which contagion can be characterized across a range of spatial scales (Plotnick et al.. 1993 and 1996, Dale 2000). The technique involves using a moving window and is concerned with the frequency with which one encounters the focal class in a window of different sizes. A log-log plot of lacunarity against window size expresses the contagion of the map, or its tendency to aggregate into discrete patches, across a range of spatial scales.

- **Subdivision:** Subdivision refers to the degree to which a patch type is broken up (i.e., subdivided) into separate patches (i.e., fragments), not the size (per se), shape, relative location, or spatial arrangement of those patches. Because these latter attributes are usually affected by subdivision, it is difficult to isolate subdivision as an independent component. Subdivision can be evaluated using a variety of metrics already discussed; for example, the number, density, and average size of patches and the degree of contagion all indirectly evaluate subdivision. However, a suite of metrics derived from the cumulative distribution of patch sizes provide alternative and more explicit measures of subdivision (Jaeger 2000). When applied at the class level, these metrics can be used to measure the degree of fragmentation of the focal patch type. Applied at the landscape level, these metrics connote the graininess of the landscape; i.e., the tendency of the landscape to exhibit a fine- versus coarse-grain texture. A fine-grain landscape is characterized by many small patches (highly subdivided); whereas, a coarse-grain landscape is characterized by fewer large patches.

- **Connectivity:** Connectivity generally refers to the functional connections among patches. What constitutes a "functional connection" between patches clearly depends on the application or process of interest; patches that are connected for bird dispersal might not be connected for salamanders, seed dispersal, fire spread, or hydrologic flow. Connections might be based on strict adjacency (touching), some threshold distance, some decreasing function of distance that reflects the probability of connection at a given distance, or a resistance-weighted distance function. Then various indices of overall connectedness can be derived based on the pairwise connections between patches. For example, one such index, connectance, can be defined on the number of functional joinings, where each pair of patches is either connected or not. Alternatively, from percolation theory, connectedness can be inferred from patch density or be given as a binary response, indicating whether or not a spanning cluster or percolating cluster exists; i.e., a connection of patches of the same class that spans across the entire landscape (Gardner et al., 1987). Connectedness can also be defined in terms of correlation length for a raster map comprised of patches defined as clusters of connected cells. Correlation length is based on the average extensiveness of connected cells. A map's correlation length is interpreted as the average distance one might traverse the map, on average, from a random starting point and moving in a random direction, i.e., it is the expected traversibility of the map (Keitt et al., 1997).

5.3.4.3 Landscape metrics – table of metrics

Table 5.5 provides summary details of the 100+ metrics supported by Fragstats. Many of these measures are very similar, and can be considered as highly correlated. Full details of each can be found in the Fragstats documentation (web site links provided at the end of this Guide).

Table 5.5 Landscape metrics, Fragstats

Area/density/edge metrics
Patch Area
Patch Perimeter
Radius of Gyration
Total (Class) Area
Percentage of Landscape
Number of Patches
Patch Density
Total Edge
Edge Density
Landscape Shape Index
Normalized Landscape Shape Index
Largest Patch Index
Patch Area Distribution
Radius of Gyration Distribution
Total Area
Number of Patches
Patch Density
Total Edge
Edge Density
Landscape Shape Index
Largest Patch Index
Patch Area Distribution
Radius of Gyration Distribution
Shape metrics

Perimeter-Area Ratio
Shape Index
Fractal Dimension Index
Linearity Index
Related Circumscribing Circle
Contiguity Index
Perimeter-Area Fractal
Perimeter-Area Ratio Distribution
Shape Index Distribution
Fractal Index Distribution
Linearity Index Distribution
Related Circumscribing Circle Distribution
Contiguity Index Distribution
Perimeter-Area Fractal Dimension
Perimeter-Area Ratio Distribution
Shape Index Distribution
Fractal Index Distribution
Linearity Index Distribution
Related Circumscribing Circle Distribution
Contiguity Index Distribution
Core area metrics
Core Area
Number of Core Areas
Core Area Index
Average Depth Index
Maximum Depth Index
Total Core Area
Core Area Percentage of Landscape
Number of Disjunct Core Areas
Disjunct Core Area Density
Core Area Distribution
Disjunct Core Area Distribution
Core Area Index Distribution

Total Core Area	Landscape Division Index
Number of Disjunct Core Areas	Splitting Index
Disjunct Core Area Density	Effective Mesh Size
Core Area Distribution	Percentage of Like Adjacencies
Disjunct Core Area Distribution	Contagion
Core Area Index Distribution	Aggregation Index
Isolation/proximity metrics	Interspersion & Juxtaposition Index
Proximity Index	Landscape Division Index
Similarity Index	Splitting Index
Euclidean Nearest-neighbour Distance	Effective Mesh Size
Functional Nearest-neighbour Distance	**Connectivity metrics**
Proximity Index Distribution	Patch Cohesion Index
Similarity Index Distribution	Connectance Index
Euclidean Nearest-neighbour Distance Distribution	Traversability Index
Functional Nearest-neighbour Distance Distribution	Patch Cohesion Index
Proximity Index Distribution	Connectance Index
Similarity Index Distribution	Traversability Index
Euclidean Nearest-neighbour Distance Distribution	**Diversity metrics**
Functional Nearest-neighbour Distance Distribution	Patch Richness
Contrast metrics	Patch Richness Density
Edge Contrast Index	Relative Patch Richness
Contrast-Weighted Edge Density	Shannon's Diversity Index
Total Edge Contrast Index	Simpson's Diversity Index
Edge Contrast Index Distribution	Modified Simpson's Diversity Index
Contrast-Weighted Edge Density	Shannon's Evenness Index
Total Edge Contrast Index	Simpson's Evenness Index
Edge Contrast Index Distribution	Modified Simpson's Evenness Index
Contagion/interspersion metrics	
Percentage of Like Adjacencies	
Clumpiness Index	
Aggregation Index	
Interspersion & Juxtaposition Index	
Mass Fractal Dimension	

5.4 Point Sets and Distance Statistics

The aggregation of point-based observations into quadrats results in a substantial loss of information regarding the underlying distribution of events. The procedure is also highly susceptible to issues relating to the shape, size, orientation and contiguity of the quadrats. By retaining the specific locations of the original points, assuming that this information is available, more detailed analysis can be undertaken. Typically this form of spatial analysis examines the distances between point pairs — either between events, or between events and random or control points. Distances are typically computed as Euclidean, but spherical distances may also be provided by some packages. Network-based analyses are rarely if ever provided (the SANET software being an exception), other than indirectly through the use of the L_1 metric or other L_p-type measure (e.g. such metrics are supported in Crimestat and LOLA).

For analysis of this type of data, events should be distinct objects (possibly weighted) and not the result of computations (e.g. zone or point set centroids). All events in a meaningful delimited study region are normally included in an analysis — analysis of subsets of the data (samples) should be undertaken with caution. For more complete recent discussions of basic point pattern analysis see O'Sullivan and Unwin (2003, Ch.4), Mitchell (2005, Ch.3) and the free and extensive Crimestat Manual (most chapters). More specialised books covering this subject area include Cressie (1991), Diggle (1983) and Bailey and Gatrell (1995).

5.4.1 Basic distance-derived statistics

With point data a range of core global statistics can be computed. Commonly produced measures of central tendency and spread include: identification of the location of the centre of the point set (typically arithmetic mean, but optionally harmonic and geometric mean centres); identification of the median centre (with or without weights and constraints); and standard distance, weighted standard distance and standard deviational ellipses. Each of these measures has been described earlier in Sections 4.2.5 and 4.5.3. A further measure, provided within some packages, is a table of all pairs of inter-point distances. This matrix is typically symmetric and completely specifies the arrangement of points — i.e. given this distance matrix the point set can be re-constructed, subject only to simple Euclidean transformations (translation, rotation and reflection).

5.4.2 Nearest neighbour methods

In many disciplines the distance between events (e.g. trees, cases of mumps, bird's nests) reflects an underlying process, for example competition for food or nutrients, birth and dispersal processes, or contagion. Particular attention is therefore focused on the nearest event to a particular other event of point of interest. This nearest event is known as the nearest neighbour (NN) or first-order nearest neighbour. The second nearest event is then the second-order nearest neighbour and so forth to k^{th}-order NN. A global (whole area) measure of a point pattern is the mean distance to the kth-order nearest neighbour, and more typically for k=1. The steps involved in computing this measure are as follows:

- Input coordinates of all points $\{x_i, y_i\}$

- Compute (symmetric) distances matrix **D** between every pair of points

- For each i, sort the distances to identify the 1st, 2nd,...k^{th} nearest values

- Compute the mean of the observed 1st, 2nd, ...k^{th} nearest values

- Compare this mean with the expected mean under Complete Spatial Randomness (CSR or Poisson) model

To undertake the last step in this sequence the expected mean under the CSR hypothesis is needed. The distribution of NN distances can be obtained from the terms of the Poisson distribution (see Section 1.5.2.6) using the information in Figure 5-12. In this we envisage the point under consideration as lying at the centre of a circle, radius r.

Figure 5-12 Nearest Neighbour distribution

Area = $2\pi rdr$

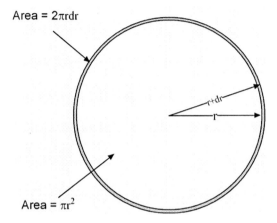

Area = πr^2

Assuming that the overall density of points in the study region is m, the expected number in a circle of radius r is simply $m\pi r^2$, which is where this term comes from in the expressions shown below. The model examines the probability that there are no other points within this circle, $p(0)$, but there is exactly one, $p(1)$, in the thin yellow/pale grey annular region shown, i.e. in the interval between r and $r+dr$ from the original point. We can now use this information to obtain the distribution of nearest neighbour distances under CSR, and hence the moments of this distribution, including the mean and variance.

The steps for deriving the first-order NN distribution are shown below, together with the first three central moments:

$$p(0) = (m\pi r^2)^0 e^{-(m\pi r^2)} / 0! = e^{-(m\pi r^2)}$$

$$p(1) = (m2\pi rdr)^1 e^{-(m2\pi rdr)} / 1! = m2\pi rdr$$

$$p(r) = p(0).p(1) = 2m\pi r e^{-(m\pi r^2)} dr$$

Let $x = m\pi r^2$ then $dx = 2m\pi rdr$, and $r = (x/m\pi)^{1/2}$, hence $p(r) = e^{-x} dr$, thus the mean, μ, may be obtained as the first moment, μ_1 or μ, by integration:

$$\mu = \int_0^\infty re^{-x} dx = \frac{1}{\sqrt{m\pi}} \int_0^\infty x^{1/2} e^{-x} dx$$

$$\mu = \frac{1}{\sqrt{m\pi}} \Gamma(3/2) = \frac{1}{2\sqrt{m}}$$

The variance, μ_2 or σ^2, can be obtained in a similar manner, giving the result:

$$\mu_2 = \frac{(4 - \pi)}{4\pi m}$$

The mean distance to NN under CSR is thus a simple function of the density, m. For a uniform (completely dispersed) distribution of points the expected mean distance to NN is simply double the CSR value. Assuming that the density can be estimated reliably (a significant issue in itself) then a simple index of global spatial randomness can be obtained by taking the ratio of the observed mean, distance to NN, $\bar{r}_o$, divided by the expected mean distance, μ or $\bar{r}_e$:

$$R = \bar{r}_o / \bar{r}_e$$

Values of $R<1$ suggest greater clustering (closer spacing) than would be expected under CSR, whilst values of $R>1$ suggest a more even distribution. The significance of the observed value can be tested, assuming we have a fairly large number of points, e.g. $n>100$. Using the z-transform (based on the standard error rather than the variance, as we are comparing mean values), we have:

$z = (\bar{r}_o - \bar{r}_e) / \sigma_e \sim N(0,1)$, where

$\sigma_e = \mu_2 / \sqrt{n} = 0.261358 / \sqrt{mn}$

Programs such as Crimestat provide index values and significance estimates, whilst ArcGIS Spatial Statistics toolbox simply provides index information. The general expression for the crude moments, μ_α, $\alpha = 1,2,3...$, of the kth-order nearest neighbour distribution in D-dimensions ($D=1,2,3..$) with population density, m, is:

$$\mu_\alpha(_D r_k) = \frac{1}{(k-1)! \, (m\Phi)^{\alpha/D}} \Gamma\left(\frac{kD + \alpha}{D}\right)$$

where Φ is the volume of a unit hypersphere:

$$\Phi = \frac{\pi^{D/2}}{\Gamma\left(\dfrac{D}{2} + 1\right)}, \text{ and}$$

$\Gamma(x)$ is the standard Gamma function.

In 2D this expression for the mean ($\alpha=1$, $D=2$) simplifies to:

$$\mu_k = \frac{\left(k - \dfrac{1}{2}\right)!}{(k-1)! \, \sqrt{m\pi}}$$

Hence for $k=1$ this simplifies further to:

$$\mu_1 = \frac{(1/2)!}{\sqrt{m\pi}} = \frac{1}{2\sqrt{m}}, \text{ as above}$$

A number of comments about this form of basic point-pattern analysis need to be made. The first is that the statistic is very dependent on the parameter, m. This is usually estimated by dividing the study area, A, by the number of points, N, i.e. $m=A/N$. However, deciding on the appropriate study area may alter the results substantially (Table 5.6). In this example three alternative region boundaries have been selected: SDC — standard distance circle; 2SDE — an ellipse of size 2 standard deviations; and the MBR of the point set —

Minimum Bounding Rectangle. These three regions result in substantially different ratio values and z-scores for the point set, primarily as a result of their different areal extents.

Table 5.6 NN Statistics and study area size

Region	SDC	2SDE	MBR
R-index	1.798	0.97	1.337
Z	7.163	0.267	3.026
Area, km^2	176	605	318

One solution to this problem is to be very careful about the boundary specification, ideally using a user-defined polygon that corresponds to a meaningful boundary (e.g. a physical habitat boundary), or possibly the convex hull of the point set plus a buffer zone (see below). Most available software does not support such boundary specification, although add-ins have been developed to facilitate this kind of region definition.

A related problem to that of region definition is that of edge-related errors (boundary effects). If the NN to a selected event is actually outside the study region the apparent NN found will be further away than the true value, leading to over-estimation of mean distances. This problem is most severe with low numbers of points and higher order NN analyses. It can be minimised by careful selection of the study boundary, the application of edge correction factors (e.g. as provided within Crimestat), by applying a boundary guard zone (e.g. a inner or outer buffer), by remapping the space onto a torus (e.g. so each of four MBR edges are treated as if they wrap onto the opposite pair) and lastly, by applying NN analysis only where the number of events is quite large and the NN order of analysis is not too high.

If there are very good reasons to assume that the event distribution is truly random over a particular study region, then the formula for the mean distance to NN can be inverted to obtain an estimate of the point density, m:

$$\mu = \frac{1}{2\sqrt{m}}, \text{ hence } m = \frac{1}{4\mu^2} \approx \frac{1}{4\bar{r}_o^2}$$

First-order NN analysis provides a simple global approach to point pattern analysis, but is of limited use for most real-world problems. Frequently real or apparent clustering is observed, and we may wish to consider questions such as:

- Is the observed clustering due to natural background variation in the population from which the events arise?

- Over what spatial scales does clustering occur?

- Are clusters a reflection of regional variations in underlying variables?

- Are clusters associated with some feature of interest, such as a refinery, waste disposal site or nuclear plant?

- Are clusters simply spatial or are they spatio-temporal?

These questions demand tools that are more exploratory in nature rather than strictly descriptive. Many tools have been developed to assist with answering such questions. The simplest, perhaps, is to plot the value of the kth order NN index against the order level, k. This option is supported with the Crimestat package, amongst others.

A rather better approach (making greater use of the underlying data, particularly with larger point sets) is to examine the observed frequency distribution of nearest neighbour distances. This can be achieved by dividing the observed nearest neighbour distances into evenly spaced distance bands, $0\text{-}d_1$, $d_1\text{-}d_2$, $d_2\text{-}...$ and then comparing these frequencies (expressed as proportions of the total) to the expected distribution under CSR. Comparison is usually carried out on the cumulative probability distribution, $G(r)$, which may be obtained by integration of the expression for $p(r)$ above:

$$G(r) = \int_0^r m2\pi r e^{-m\pi r^2} dr = 1 - e^{-m\pi r^2}$$

Typically both the observed and expected cumulative distributions are plotted and visually compared. A variation on this arrangement is to select random points in the study area and compute the distance from these to the nearest event. The observed cumulative distribution in this case is denoted $F(r)$, or commonly with a ^ (hat) symbol above the F to indicate that it is an estimate based on the observed data. It has an expected cumulative distribution as per $G(r)$. However, if the sample pattern is more clustered than random the observed $F(r)$ plot will differ from the observed $G(r)$ plot, rising far more slowly.

Support for $F(r)$ and $G(r)$ computation and associated Monte Carlo simulation (see below) is not provided within most GIS and related packages, but it is provided in the SPLANCS package (S-Plus and R-Plus implementations — FHAT and GHAT functions). It is also relatively straightforward to generate these functions programmatically.

In principle a measure of the difference between the observed and expected cumulative distributions can be computed (e.g. the Kolmogorov-Smirnov test statistic) and its significance determined. However, the set of nearest neighbour distance does not represent a set of independent samples — almost by definition NN distances are non-independent and frequently are reflexive — i.e. the NN of event A is event B and the NN of event B is event A. Even under CSR such reflexivity occurs for over 62% of first order neighbours. For this reason significance testing requires Monte Carlo simulation. For example, with N observed events in a study area A, one can simulate a set of N purely random events and compute the cumulative distribution for this set. The maximum absolute difference between this simulated cumulative distribution and the theoretical cumulative distribution can then be calculated and the value recorded. The process in then repeated a large number of times (e.g. $T=999$ times). Let X be the number of times the recorded difference was larger than that between the observed pattern and the expected cumulative distribution. Then $p=(1+X)/(1+T)$ is the estimated probability of observing a difference of the magnitude found. Thus if $X=99$ and $T=999$ we would have $p=0.1$ or 10%.

A similar and in many ways more powerful approach, known as Ripley's K (or L) function, is supported within several packages, including Crimestat and SPLANCS. Ripley's procedure utilises all event-event distances, not just nearest neighbour distances. It operates as follows (Figure 5-13):

- Construct a circle, radius d, around each point (event), i

- Count the number of other events, labelled j, that fall inside this circle

- Repeat these first two stages for all points i, and then sum the results

These steps equate to computing the sum:

$$K(d) = \frac{A}{N} \sum_i \sum_j \frac{I(d_{ij})}{N}, i \neq j$$

where $I(d_{ij})=1$ if the distance, d_{ij}, from i to j is less then d otherwise $I(d_{ij})=0$

- Increment d by a small fixed amount (e.g. $R/100$, where R is the radius of a circle of area equal to the study area, A)

- Repeat the computation, giving values of $K(d)$ for a set of distances, d

The function $K(d)$ can then be plotted against distance, d, in a manner similar to that described in Section 6.7 for correlograms and variograms. The graph of the (transformed) $K(d)$ function provides an excellent insight into localised clustering, with characteristic graphs reflecting particular observed patterns. Computation of (the transformed) $K(d)$ function for a mapped dataset is supported in

both SPLANCS (KHAT and KENV functions) and Crimestat, and in both cases Monte Carlo simulation of confidence envelopes is provided.

Figure 5-13 Ripley's K function computation

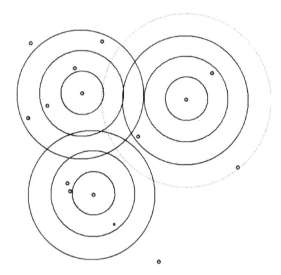

Under CSR the expected number of events within a circle of radius d is simply the event density, m, times the circle area. With m estimated as the overall number of events divided by the study area, N/A, we have

$$E(\# < d) = \frac{N}{A} \pi d^2$$

The observed $K(d)$ function minus the expected value above has an expected value of $L(d)=0$ for a given d. $L(d)$ measures the difference between the observed pattern and that expected under CSR. It is generally computed plotted with d subtracted from the expectation:

$$L(d) = \sqrt{\frac{K(d)}{\pi}} - d$$

If the component d is not subtracted (as per Ripley's original formulation of this method) then the $L(d)$ plot against d for a truly random distribution will be a line at 45 degrees (i.e. L(*d*)=*d*) whereas with *d* subtracted it will be approximately 0 for all *d*. This transformed expression is the form in which the *K*-statistic is analysed within Crimestat.

The sampling distribution of L(*d*) is unknown, but may be approximated by Monte Carlo simulation, as described earlier. Crimestat produces its simulations based on a rectangular region of area *A* and shape matching the MBR, whilst SPLANCS utilises a user-specified polygon, which is more satisfactory in most cases. Figure 5-14 shows the L(*d*) plot for the random point set shown in Figure 5-11. The envelope shown provides the maximum and minimum L(*d*) curves generated

from 300 simulation runs. The observed plot lies within this envelope, so it is reasonable to suppose that this pattern is broadly random rather than more clustered or even spaced than random.

As *d* is incremented the behaviour of the statistic becomes increasingly subject to border effects. Ideally the number of points utilised should be large (well in excess of 100 if possible) and where necessary edge correction procedures should be adopted. Simple edge correction can be implemented by weighting circular regions that lie partially outside the study boundary by the proportion lying within the study region.

Figure 5-14 Ripley K function, shown as transformed L function plot

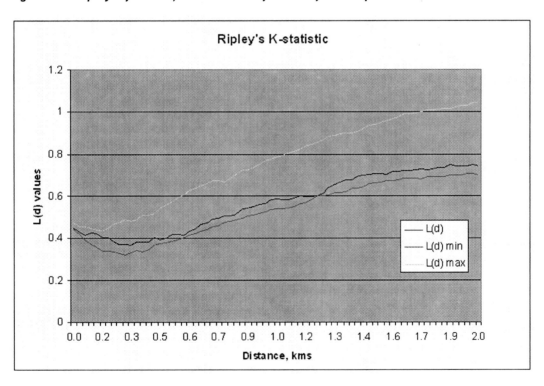

The ideas behind the production of the standard $K(d)$ statistic and associated Monte Carlo simulations can be used to provide additional measures that are often more useful. For example, instead of comparing the observed pattern to CSR, alternative models could be examined. SPLANCS, for example, supports comparison with a Poisson Cluster Process (PCP) rather than a simple Poisson (CSR) process. PCP is similar to a simple Poisson process in that it starts with a random point set. These are regarded as "parents". Each parent then produces a random number of offspring, and these are located around the parent according to a bivariate distribution function, $g(x,y)$, for example a circular Normal with variance (spread), σ^2:

$$g(x,y) = \frac{1}{k}e^{-(x^2+y^2)/k}, \text{where } k = 2\pi\sigma^2$$

The parents are then removed from the set and the PCP consists of the set of all offspring. SPLANCS implements this kind of model with three parameters:

RHO: intensity (density parameter, ρ) of the parent process

M: the average number of offspring per parent, and

S2: the variance of location of offspring relative to their parent (the σ^2 value in the expression above)

By generating many PCP simulations for a given study region, an alternative probability envelope can be computed and plotted against the observed K- or L-function.

A second important variant of the $K(d)$ function involves treatment of two or more point (event) patterns within the same region. For example, one set of points might represent disease cases and a second set might be matched cases (controls) selected at random from the background population. Similar procedures may be used for spatio-temporal analysis, for example comparing crime patterns or disease incidence over time. If both datasets exhibit similar clustering we would expect the plots for both to be similar, or the difference between the computed functions, $D=K_1(d)-K_2(d)$, to be approximately zero for all d. Where background data are only available at an aggregated level (e.g. census Output Areas), it may be acceptable to assign the second set of points to the OA centroids together with a weighting factor reflecting the zone population. Crimestat supports such weighting as an option in order to assist in studies with background variation.

Another variant of the $K(d)$ function, supported within SPLANCS, is the bivariate model (or Cross K function) in which the relationship between two separate point patterns is being examined. For example, in an ecological study this might involve examining collaborative or competitive behaviour between two species, or an analysis of infected trees and the "at risk" population. In this instance the $K(d)$ function measures the number of events in set B that lie within distance d of events A. The SPLANCS function K12Hat supports this model together with supporting simulation functions to provide an estimated probability envelope for the function.

Readers may have noticed the similarity between the procedures involved in generating Ripley's K, and the generation of kernel density maps (Section 4.3.4.1 and Figure 4-37). The latter involves assigning a proportion of each event or point to a circular neighbourhood, and then summing and normalising the weighted surface at grid intersection points. This provides a continuous density surface rather than a linear model, and can be used to highlight local clustering. However, with case/control data or case/background variation data, separate surfaces may be computed for each dataset and then these may be combined, e.g. by taking the difference or the ratio of the case surface/control surface (for densities greater than some minimum value) to provide a comparison surface. Values of 1 on the ratio surface represent matching densities, whilst values greater than 1 represent locations where the cases occur in higher densities than the controls or background data.

It is worth concluding this subsection with some general observations (cautionary notes) regarding point pattern analysis:

- The classical statistical model of CSR is inappropriate for many spatial datasets, but does provide a starting point for analysis and simulations in many instances

- A close examination of the source data is always advisable, checking questions such as: how the data were collected; how locations and attributes were recorded; when data were recorded; what data (e.g. true or surrogate) information is represented; what overall measurement error is associated with the data; etc. Examining the underlying dataset for duplicates or near-duplicates is often an importance exercise

- Multiple analytical approaches, including visualisation techniques (1D, 2D, 3D) are advisable since initial impressions and single approaches may be misleading

- The variables selectable in analysing point patterns (e.g. sample region, weights, order neighbour, maximum distance, clustering model, kernel function and bandwidth etc.) can yield very different results depending on the values and models chosen

- Borders, area definitions, metrics, background variation, temporal variation and non-spatial data issues are of major significance in describing and modelling point patterns

- Analysis of rare events presents particular problems — the small numbers involved may result in very low densities of events in large parts of a study region, if only by chance, and calculation based on zone counts are particularly subject to extremes and problems computing ratios

- Analysis should be seen as exploratory, and a part of an overall in-depth analysis process

- Analyses of this kind will not provide meaningful cause-effect or process-realisation models — in general pattern analysis tells us very little about process, although it may enable us to exclude certain processes from consideration. Many processes can result in the same or similar observed patterns, and different patterns can generate the same or very similar statistical measures, such as K(d) functions

5.4.3   Hot spot and cluster analysis

Identifying that clustering exists in spatial and spatio-temporal datasets does not provide a detailed picture of the nature and pattern of clustering. It is frequently helpful to apply simple hot-spot (and cold spot) identification techniques to such datasets. For example, Figure 5-15 shows the location of almost 1000 reported lung cancer cases in part of Lancashire, UK, over the period 1974-83.

Figure 5-15 Lung cancer incidence data

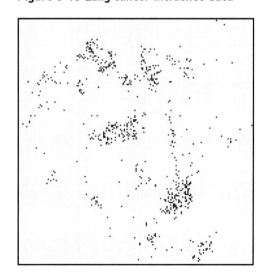

Visual inspection suggests that several clusters of different types and sizes exist, but the initial inspection of the mapped data is somewhat misleading. Examining the source dataset shows that roughly 50% of the dataset consists of duplicate locations, i.e. points are geocoded to the same coordinates — in several instances coincident locations with 5+ events. This is a very common feature of some types of event dataset and may occur for several reasons: (i) rounding to the nearest whole grid reference; (ii) measurement error and/or data resolution issues; (iii) genuinely co-located events — for example, in the infamous Dr Harold Shipman murder cases, 8 of the murders took place at the same nursing home; (iv) allocation of events to an agreed or surrogate location — for example, to the nearest road intersection or to the location at which an incident was reported rather than took place (e.g. a police station or a medical practice) or to a nominal address (e.g. place of birth); (v) deliberate data modification, e.g. for privacy or security reasons... the list is extensive and mapping the data as uniform sized points does not reflect this.

In this particular dataset, Figure 5-15, the observed clustering is essentially a reflection of population density, or more specifically the population at risk from another disease (cancer of the larynx), and was actually selected for use in this connection.

Very closely located events may also be difficult to detect, depending on their separation and the symbology used. ArcGIS incorporates a "collect events" tool that combines such duplicate data and creates a new "count" field containing the frequency of occurrences. The attribute table applies to a new feature with fewer elements in the new table. The count field may then be used as a weight and rendered as variable point sizes — although this may exacerbate overlap problems. The MapInfo add-in, HotSpot Detective, provides similar functionality.

Assuming the point set is unweighted, and exhibits marked clustering, it is then useful to identify factors such as: (i) where are the main

(most intensive) clusters located? (ii) are clusters distinct or do they merge into one another? (iii) are clusters associated with some known background variable, such as the presence of a suspected environmental hazard (e.g. a power station, a smoke plume from a chemical facility) or reflecting variations in land use (farmland, urban areas, water etc.), or variations in the background population or other regional variable? (iv) is there a common size to clusters or are they variable in size? (v) do clusters themselves cluster into higher order groupings? (vi) if comparable data are mapped over time, do the clusters remain stable or do they move and/or disappear? Again, there are many questions and many more approaches to addressing such questions.

Crimestat, Clusterseer and several other packages provide a very useful range of facilities to assist in answering some of these questions. Here we use Crimestat to illustrate a number of these. The first is identification and highlighting of the top N duplicate locations. In order to deal with possible uncertainty in georeferencing, a fuzzy variant of this facility is provided in Crimestat, enabling common locations to be regarded as those falling within a specified range of each other (e.g. within 10 metres). Having conducted this initial analysis Crimestat then provides a range of spatial clustering and hot-spot identification methods, as described in subsections 5.4.3.1 to 5.4.3.3.

5.4.3.1 Hierarchical nearest neighbour clustering

Crimestat provides a general purpose form of clustering based on nearest neighbour (NN) distances. This form of clustering can be single-level or multi-level hierarchical (NNh) and is of particular applicability if nearest neighbour distance is believed to be of relevance to the problem being considered. Events are considered to be a member of a level 1 cluster if they lie within the expected mean distance under CSR plus or minus a confidence interval value obtained from the standard error plus a user-definable tolerance. These parameters effectively define a search radius within which point pairs are combined into clusters. A further constraint can be applied, specifying the minimum number of events required to constitute a cluster. The mean centre and standard deviational ellipses for these clusters are calculated and may be saved in various GIS file formats. These mean centres are then regarded as a new point set, and are subjected to the same type of clustering in order to identify and generate second order and ultimately higher orders of clustering. Clearly the number of points in the initial sample, and the degree of clustering, have a major bearing on the way in which such clusters are identified. Figure 5-16 illustrates the results of applying this process for the lung cancer data shown earlier.

Figure 5-16 Lung cancer NNh clusters

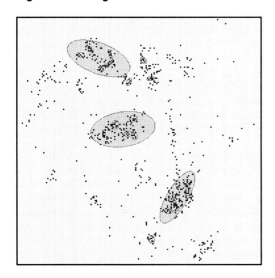

Crimestat also provides a variation on this clustering procedure to account for background or "baseline" variation. It describes the procedure as a risk adjusted NNh method, or RNNh. The background data is represented as a fine grid using kernel density estimation, and this is used to adjust the threshold distance for clustering the original point set, on a cell-by-cell basis (see also, Section 5.4.3.3). Note that with both NNh and RNNh not all events are assigned to clusters,

and each point is assigned to either one cluster at a given hierarchical level or none at all.

5.4.3.2 K-means clustering

A conceptually different clustering approach is point-set partitioning. Essentially this procedure is a form of K-means clustering, as described in Section 4.2.12 on classification. The user specifies the value K, and a set of K random points are placed in the study region as *seed points*. Each point in the dataset is then allocated to the nearest seed point. The set of points assigned to each seed point is then used to create a new set of seed points comprised from the centres of these initial groupings. The procedure continues until the sum of distances (or squared distances) from each point to its cluster centre seed cannot be reduced significantly by further iterations. The procedure only provides local optimisation and in the case of Crimestat is not repeated m-times with different random seeds as with normal K-means clustering. Instead it attempts to identify very good starting points for the initial seeds by a form of simple density analysis (placing a grid over the point set and identifying distinct areas of point concentrations). Unlike NNh, the K-means procedure assigns all events to a unique cluster and clusters do not form hierarchical groupings. Its dependence on the user selection of the K-value and the underlying sub-optimal algorithm used in this implementation are distinct weaknesses, but is nevertheless a useful exploratory tool.

5.4.3.3 Kernel density clustering

Kernel density estimation (KDE) may also provide an informative (exploratory) tool for hot-spot and cool-spot identification and analysis. Although not strictly a form of clustering, assignment of points to cells that have greater than a pre-determined density value provides a form of clustering in this case. Figure 5-17A illustrates the use of KDE for the same point dataset as before (lung cancer cases) using a quartic (finite extent) KDE model (see also, Figure 4-37, where a

Normal or Gaussian kernel has been applied to this dataset).

Figure 5-17 KDE cancer incidence mapping

A. Lung cancer incidence (controls)

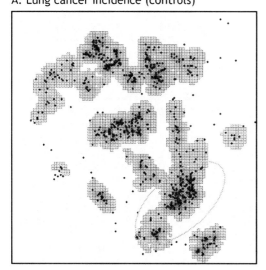

B. Larynx cancer incidence (cases)

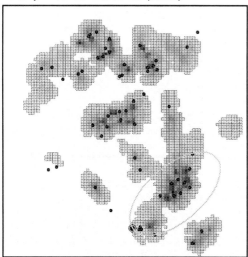

The pattern shown is largely a reflection of the distribution of population and associated infrastructure (roads etc.) in the region. The density grid illustrated has been overlain with the point set and an ellipse showing the possible relationship between the high density area in the south of the study area with an old incinerator plant (dot in lower left of ellipse) and hypothetical smoke plume. The principal interest in this particular dataset was in the relationship between the incinerator location and another, far rarer, form of cancer, affecting the larynx (Figure 5-17B). A small apparent cluster (4 events identified located very close to the incinerator site in Figure 5-17B) had been observed and the research sought to establish whether this was a real or apparent relationship. The incidence of lung cancer in this instance was being used as a form of control dataset, on the hypothesis that these data represented an estimate of the distribution of the underlying population at risk, and assuming that there was no relationship between lung cancer incidence and incinerator location (a working assumption only) — see Diggle (1990) and related papers for more details. As can been seen from Figure 5-17, the overall pattern of the larynx cancer cases seems to follow the pattern exhibited by the lung cancer cases, i.e. to be largely a reflection of underlying variation in the at-risk population. Whether there is a real and unexpected cluster in the neighbourhood of the old incinerator is difficult to determine.

Diggle's model and tests suggest that the cluster does appear to be significant. But he also notes the sensitivity of the model to the low number of cases — with deletion of just one of these cases there is a reasonable chance the result could have arisen by chance. He also notes the problem of formulating hypotheses based on examining specific apparent clusters. A wide range of comparable regions should be studied, without pre-conceptions, since clusters may well be observed which may or may not be associated with particular facilities (see also the earlier discussion of exploratory *cluster hunting* in Section 5.2.6).

5.5 Spatial Autocorrelation

5.5.1 Autocorrelation, time series and spatial analysis

The term spatial autocorrelation owes its origins to work in a related field, time series analysis (TSA), and in turn to the notion of correlation in univariate statistics. Since spatial autocorrelation follows these concepts so closely we will introduce the subject using these more familiar origins.

As we saw in Table 1.4, if we have a sample set $\{x_i,y_i\}$ of n pairs of data values the correlation between them is given by the ratio of the covariance (the way they vary jointly) to the square root of the variance of each variable. This is effectively a way of standardising the covariance by the average spread of each variable, to ensure that the correlation coefficient, r, falls in the range [-1,1]. The formula used for this ratio is:

$$r = \frac{\sum_{i=1}^{n}(x_i - \overline{x})(y_i - \overline{y})}{\sqrt{\sum_{i=1}^{n}(x_i - \overline{x})^2}\sqrt{\sum_{i=1}^{n}(y_i - \overline{y})^2}}$$

Now suppose that instead of a set of data pairs $\{x_i,y_i\}$ we have a set of n values, $\{x_t\}$, which represent measurements taken at different time periods, $t=1,2,3,4,...n$, for example daily levels of rainfall at a particular location, or the closing daily price of crude oil per barrel on the London market. The pattern of values recorded and graphed might show that rainfall, or oil prices, exhibits some regularity over time. For example, it might show that days of high rainfall are commonly followed by another day of high rainfall, and days of low rainfall are also often followed by days of low rainfall. In this case there would be a strong positive correlation between the rainfall on successive days, i.e. on days that are one step or *lag* apart. We could regard the set of "day 1" values as one series, $\{x_{t,1}\}$ $t=1,2,3...n-1$, and set of "day 2" values as a second series $\{x_{t,2}\}$

$t=2,3...n$, and compute the correlation coefficient for these two series in the same manner as for the r expression above. Each series has a mean value, which is simply:

$$\overline{x}_{.1} = \frac{1}{n-1}\sum_{t=1}^{n-1}x_t \text{ and } \overline{x}_{.2} = \frac{1}{n-1}\sum_{t=2}^{n}x_t$$

Using these two mean values we can then construct a correlation coefficient between our two series. This is essentially the same formula as for r:

$$r_{.1} = \frac{\sum_{t=1}^{n-1}(x_t - \overline{x}_{.1})(x_{t+1} - \overline{x}_{.2})}{\sqrt{\sum_{t=1}^{n-1}(x_t - \overline{x}_{.1})^2}\sqrt{\sum_{t=1}^{n-1}(x_{t+1} - \overline{x}_{.2})^2}}$$

If n is reasonably large then the value $1/(n-1)$ will be very close to $1/n$, and the values of the two means and standard deviations will be almost the same, so the above expression can be simplified under these circumstances to:

$$r_{.1} = \frac{\sum_{t=1}^{n-1}(x_t - \overline{x})(x_{t+1} - \overline{x})}{\sum_{t=1}^{n}(x_t - \overline{x})^2}$$

This expression is known as the *serial correlation coefficient* for a lag of 1 time period.

It may be generalised for lags of 2, 3, ...,k steps as follows:

$$r_{.k} = \frac{\sum_{t=1}^{n-k}(x_t - \overline{x})(x_{t+k} - \overline{x})}{\sum_{t=1}^{n}(x_t - \overline{x})^2}$$

The term *autocorrelation* coefficient has been used since the 1950s to describe this expression, rather than serial correlation coefficient. The top part of this expression is like the covariance, but at a lag of k, and the bottom is like the covariance at lag of 0. These two components are sometimes known as the *autocovariance* at k and 0 lags. In time series analysis it is usual for the time spacing, or distance, to be in equal steps. The set of values $\{r_{.k}\}$ can then be plotted against the lag, k, to see how the pattern of correlation varies with lag. This plot is known as a *correlogram*, and provides a valuable insight into the behaviour of the time series at different lags or "distances". For a random series the values for the $r_{.k}$ will all be approximately 0 — in fact they are distributed as $N(0,1/n)$. If there is short term correlation, as in our rainfall example, the $r_{.k}$ will start high (close to +1) and decrease to roughly 0 when the number of lags exceeds the length (or range) of this correlation. It is possible of course, that the overall pattern of rainfall shows a steady increase over time, in which case the correlograms will not tend to zero in the manner expected. In this case the series is described as *non-stationary* (see further, Section 6.7.1.8) and before carrying out such analysis an attempt to remove the trend component should be undertaken. Typically this involves fitting a trend curve (e.g. a best fit straight line) to the original data points and subtracting values for this curve from the original dataset at lags 1,2,3... before carrying out the analysis. The original data may also contain outliers, which if left in may distort the analysis, so inspection of the source data and outlier adjustment or removal (e.g. of data errors) may be advisable.

Having identified these factors, adjusted the data if necessary, and computed the correlograms, the next step is to examine the results and attempt to interpret the observed patterns. Unfortunately, as is the case with many patterns observed in time or space, more than one process can generate an identical pattern. However, modelling an observed pattern may provide an effective means of estimating missing or sparse data, or predicting values beyond the observed range, despite the fact that the process generating the model may not be unique.

These comments apply to series which follow a clear sequence of steps in a single dimension, time. At first sight such methods do not translate easily to spatial problems, since there is no obvious single direction to follow. Of course one could select a single transect and take measurements at fixed intervals to produce a well-ordered series, which could then be analysed in exactly the manner described. But more general procedures are needed if a wide range of practical spatial problems are to be subjected to such analysis. These procedures need to model space in a manner that results in well-ordered data series, ideally in evenly spaced steps, using the general notion of proximity bands (see further, Sections 5.5.1 and 5.5.2).

5.5.2 Global spatial autocorrelation

The procedures adopted for analysing patterns of *spatial* autocorrelation depend on the type of data available. There is considerable difference between: (a) a set of 100 values obtained for a 10x10 grid of (100mx100m) squares which covers a 1000mx1000m region; (b) a set of values obtained for 100 contiguous but arbitrarily shaped polygons which again cover the same region; and (c) a set of data values obtained from 100 arbitrarily distributed sample points in our study region. Each case warrants a slightly different approach, but each utilises the notion of proximity bands as a means of imposing some form of serial behaviour to the data. The idea is then to examine the correlation between areas or points at given levels of separation, to obtain a similar measure to that used within time series analysis.

5.5.2.1 Join counts

Consider a regular grid of cells that completely covers a sampled region with a value associated with each grid square. The set of values in this case could be binary, e.g. presence/absence; they could be classified into one of k classes; or they could be integer or real-valued data. If we assume the data are binary, perhaps representing the presence or absence of a particular species of insect or plant variety in sample quadrats, a range of possible patterns might be observed (Figure 5-18A-D). In Figure 5-18A all the observed values for presence are in one half of the 6x6 grid (strong positive autocorrelation), whilst in B they are perfectly evenly distributed (strong negative autocorrelation). Figure 5-18C gives a particular case of a random pattern (of which there are many). In each case 50% of the cells show presence, but this value could easily be 10% or any other value depending on the data being studied. Figure 5-18D is an example of a real-world dataset showing the presence or absence of desert holly in a 16x16 cell sample region.

Figure 5-18 Join count patterns

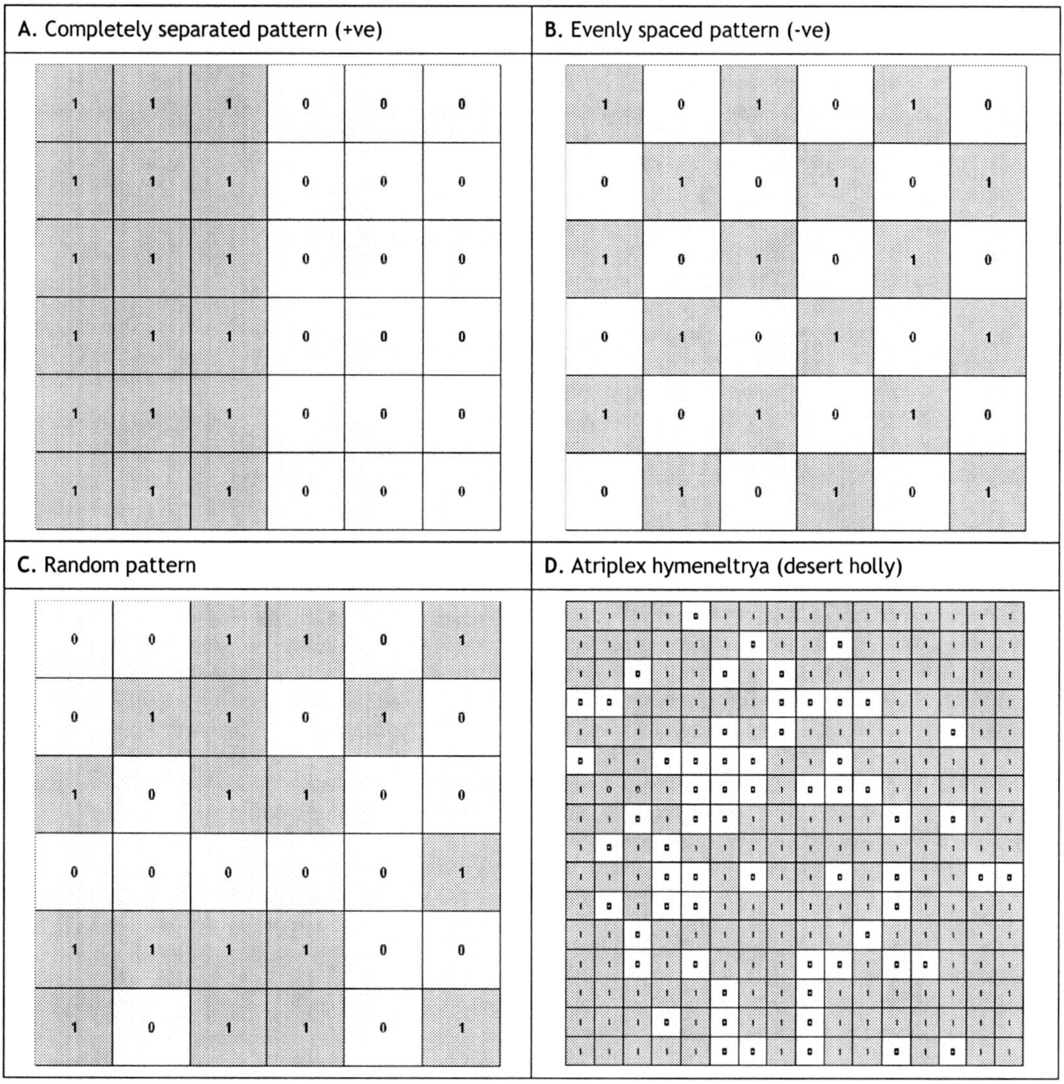

A. Completely separated pattern (+ve)

B. Evenly spaced pattern (-ve)

C. Random pattern

D. Atriplex hymeneltrya (desert holly)

One way of analysing these patterns is to ask "what is the chance that a particular pattern could occur at random?" In each of the three sample patterns shown we can look at the spatial equivalent of one time step or lag, the patterns observed at one cell step, i.e. adjacent cells. If steps are restricted to rook's moves we can count the number of instances of 1-1, 0-0, 1-0 and 0-1 occurring, and compare these to the number we might expect if the pattern was random.

For smaller regions edge effects will be significant, so calculations need to be adjusted to reflect the fact that along the borders not all of the four directions are possible. For example, only two adjacent cells exist for the four corner positions, and only three adjacencies for other border cells. Row and column totals for the adjacencies, or joins, are shown in Figure 5-19 with the overall total being 120/2=60 joins. For our patterns, with 50% occupancy, we might expect 15 of these to be 1-1 joins, 15 to be 0-0 joins and the remaining 30 to be 0-1 or 1-0 joins. We can count up the number of each type of join and compare this to our expected values to judge how special (significant) or not our patterns are. In Figure 5-18A there are 27 1-1 joins, 27 0-0 joins and only 6 0-1 or 1-0 joins – this seems very unlikely, and is indeed most unlikely. Similar calculations can be undertaken for cases B and C, and as expected for case B all 60 joins are of type 1-0 or 0-1, which is again extremely unlikely to occur by chance. Case C has 35 1-0/0-1 joins compared with perhaps 30 expected, with 1-1 joins being 13 and 0-0 being 12, as against perhaps 15 in each case. A test for the significance of the results we have observed can be produced using a Normal or z-transform of the data. In practice three separate z-transforms are needed, one for the 1-1 case, one for 0-0, and one for 0-1 and 1-0. These transforms evaluate expressions of the form $z=(O-E)/SD$ where O is the observed number of joins of a given type, E is the expected number based on a random model, and SD is the expected standard deviation. Details of the formulas required for computing E and SD for regular lattices of the type described above, and for non-regular

lattices and k-classes, are summarised in Dacey (1968).

Figure 5-19 Join count computation

The procedure and formulas can be implemented in software scripts for use within mainstream GIS packages or data may be externally analysed first and then the results mapped within a GIS – for example using the Rookcase Excel add-in produced by Sawada (1999) and obtainable from the University of Ottawa, Laboratory for Paleoclimatology and Climatology. Results for the four examples given in Figure 5-18A-D generated using the Rookcase add-in are shown in Table 5.7 ("B" and "W" here refer to Black and White, rather than 1 or 0; "Rand." Refers to the randomisation model, which is discussed further below; and # refers to "number").

Two features of these results should be noted: (i) the expected number of joins is not 30,15,15 as we suggested earlier, but 30.86, 14.57,14.57 these being adjusted values to take account of what is known as non-free sampling (i.e. sampling without replacement); (ii) the z-statistic in case B shows a large positive value for BW joins, and large negative values for BB and WW joins (absolute values of >1.96 would be significant at the 5% probability level). This mixed pattern can be confusing, especially if one or two of the z-scores shows a significant value whilst others do not, as in Table 5.7D.

Table 5.7 Join count mean and variance results

A. Positive autocorrelation

	BW Joins	BB Joins	WW Joins
# of Joins	6	27	27
z-Rand. statistic	-6.58	5.80	5.80
Variance Rand.	14.26	4.59	4.59
Expected # of Joins	30.86	14.57	14.57
Number of Observations	# of B's	# of W's	Total Joins
36	18	18	60

B. Negative autocorrelation

	BW Joins	BB Joins	WW Joins
# of Joins	60	0	0
z-Rand. statistic	7.72	-6.80	-6.80
Variance Rand.	14.26	4.59	4.59
Expected # of Joins	30.86	14.57	14.57
Number of Observations	# of B's	# of W's	Total Joins
36	18	18	60

C. Random model — no discernable autocorrelation

	BW Joins	BB Joins	WW Joins
# of Joins	35	13	12
z-Rand. statistic	1.10	-0.73	-1.20
Variance Rand.	14.26	4.59	4.59
Expected # of Joins	30.86	14.57	14.57
Number of Observations	# of B's	# of W's	Total Joins
36	18	18	60

D. Atriplex hymeneltrya — positive BB autocorrelation

	BW Joins	BB Joins	WW Joins
# of Joins	173	39	268
z-Rand. statistic	-1.14	2.00	0.24
Variance Rand.	70.67	17.69	22.97
Expected # of Joins	182.57	30.59	266.84
Number of Observations	# of B's	# of W's	Total Joins
256	65	191	480

The joints count method has been utilised in a variety of application areas including: ecological data analysis; to analyse patterns of voting (for example voting for one of two parties or individuals); for analysis of land use (e.g. developed or undeveloped land); and for examination of the distribution of rural settlements. However, the complexity of computing the theoretical means and variances in more realistic spatial models, the difficulty of interpreting the multiple z-scores,

coupled with the availability of alternative approaches to analysis, have meant that join count procedures are not widely used and are rarely implemented in mainstream GIS packages. Metrics based on these concepts are provided in Fragstats and similar packages (see further, Section 5.3.4), for example in the computation of Connection and Contagion indices.

Additional questions might reasonably be asked about this analytical procedure that point the way to more realistic and useful extensions and developments of these ideas. Key questions include:

- why choose rook's moves — why not permit queen's moves, which would extend the idea of contiguity but increase the complexity of computation?

- why restrict the notion of contiguity to directly adjacent cells (a lag of 1) — why not examine longer range effects, i.e. higher order spatial lags?

- why should every cell have the same level of importance or weight — why not permit cells to have differing weights, depending on the kind of model of spatial association we are considering?

- should the analysis be restricted to regular grids, or could it be extended to irregular grids, polygonal regions and even pure point data?

- is a single statistic for a whole area meaningful, or should separate statistics be computed for sub-regions and then plotted?

- how sensitive is this technique (and of course other techniques) to the particular size of grid cell used and the number of cells?

The first point raised in the last of these questions can be addressed directly by examining our example in Figure 5-18 again. If the grid had been sampled at half the frequency in each direction there would be 8x8=64 cells rather than 16x16=256 cells, and in this instance only one new larger cell contains no data, i.e. a 0, all others showing a desert holly plant present (Figure 5-20A, empty 2x2 cell shown in white).

Figure 5-20 Grouping and size effects

A. 2x2 grouping of Atriplex hymeneltrya grid

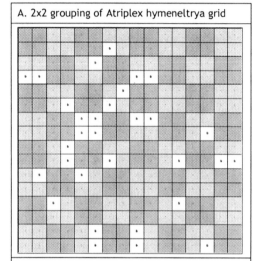

B. 128x128 grid of Calluna vulgaris presence

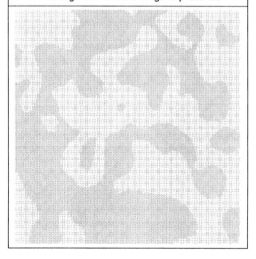

With large grids, e.g. 128x128 cells, the z-scores may become huge, raising doubts over

the sensitivity and interpretation of the technique as the resolution of sampling or the size of the area covered is increased (Figure 5-20B, 16384 cells, all three z-scores are over 160). A critical question here is whether the data in such cases reflects the result of direct observation, or whether they represent results from re-sampling or similar computational procedures. In the latter instance test scores may not be valid.

5.5.2.2 Moran I and Geary C

Many of the remaining questions from the preceding subsection relating to join count procedures can be addressed by considering a more general case than presence/absence data, with a region that is not a regular lattice. Figure 5-21A shows a set of 10 cells with real-valued entries. Instead of displaying this as an irregular lattice or grid, it can be regarded as a table of (x,y,z) triples, where rows are numbered 1,2... from the top and columns are likewise numbered 1,2... from the left, as shown below.

x	y	z
1	2	4.55
1	3	5.54
2	1	2.24
2	2	-5.15
2	3	9.02
3	1	3.1
3	2	-4.39
3	3	-2.09
4	2	0.46
4	3	-3.06

This coordinate based description of the data immediately suggests a generalisation that may be applied to point datasets, with the (x,y) pairs providing the coordinate information. These (x,y) pairs could be well-defined points, or grid cells, or even polygon centroids. The tabular form loses information, in that cell adjacencies are not explicitly captured. Figure 5-21B provides this adjacency information for the sample lattice as a binary matrix using rook's move definitions.

Figure 5-21 Irregular lattice dataset

A. Source data

	+4.55	+5.54
+2.24	-5.15	+9.02
+3.10	-4.39	-2.09
	+0.46	-3.06

B. Adjacency matrix, W

	1	2	3	4	5	6	7	8	9	10
1	0	1	0	1	0	0	0	0	0	0
2	1	0	0	0	1	0	0	0	0	0
3	0	0	0	1	0	0	1	0	0	0
4	1	0	1	0	1	0	0	1	0	0
5	0	1	0	1	0	1	0	0	1	0
6	0	0	0	0	1	0	0	0	0	1
7	0	0	1	0	0	0	0	1	0	0
8	0	0	0	1	0	0	1	0	1	0
9	0	0	0	0	1	0	0	1	0	1
10	0	0	0	0	0	1	0	0	1	0

C. Revised source data

	+4.55	-5.15
+2.24	+5.54	-4.39
+3.10	+9.02	-2.09
	+0.46	-3.06

In this example cells in the original dataset are numbered from 1 to 10, starting with the first cell being in row 2 column 1 of the source data and continuing down the rows and across the columns. This matrix can be thought of as a specific example of a spatial weights matrix, $W=\{w_{ij}\}$, indicating elements of computations that are to be included or excluded. In this example $w_{ij}=1$ if two cells are adjacent (joined) and $w_{ij}=0$ otherwise. This matrix is symmetric and binary, but these features are not pre-requisites. For example, one might choose to create a matrix W in which rook's move adjacencies were given a weight of 0.9 and bishop's moves (the diagonal connections) a weight of 0.1. More generally, if the source dataset was polygonal, e.g. well-defined agricultural tracts, then adjacency weights might be chosen to reflect the relative lengths of shared boundaries.

We now have a set of values $\{z_i\}$ and a set of weights $\{w_{ij}\}$. We are looking for some way of combining this information using a function, f(), which satisfies the following criteria:

- If the pair (z_i,z_j) are both +ve or both -ve $f(z_i,z_j)>0$

- If the pair (z_i,z_j) are a mix of +ve and -ve $f(z_i,z_j)<0$

- If the pair (z_i,z_j) are both large values $f(z_i,z_j)$ is large

- Patterns of adjacency reflected in the matrix W must be accounted for in the computation

One of the simplest ways to fulfil these criteria is to multiply the zone values together, optionally adjusting for the overall mean value for all zones and including the adjacency information, which gives:

$$\sum\sum z_i z_j$$

or with mean adjustment and weights included:

$$\sum\sum w_{ij}(z_i - \bar{z})(z_j - \bar{z})$$

As noted above, assuming the weights w_{ij} are binary, they simply identify which elements of the computation are to be included or excluded in the calculation. This expression looks quite like the covariance of the selected data. To adjust for the number of items included in this sum, and produce a covariance value, we need to divide through by the sum of the weights used, i.e.

$$\frac{\sum_i \sum_i w_{ij}(z_i - \bar{z})(z_j - \bar{z})}{\sum_i \sum_i w_{ij}}$$

To standardise this covariance expression we divide it by the variance of the data, which is simply:

$$\frac{\sum_i (z_i - \bar{z})^2}{n}$$

The ratio of these two expressions gives us an index, known as Moran's I, and has values that typically range from approximately +1, representing complete positive spatial autocorrelation, to approximately -1, representing complete negative spatial autocorrelation:

$$I = \frac{1}{p} \frac{\sum_i \sum_j w_{ij}(z_i - \bar{z})(z_j - \bar{z})}{\sum_i (z_i - \bar{z})^2}, \text{ where}$$

$$p = \sum_i \sum_j w_{ij} / n$$

The upper and lower limits of this index may vary depending on the configuration of the weights used. It is the spatial equivalent of the correlation coefficient, r, and very similar to the (unweighted) time series autocorrelation coefficient, $r_{.1}$ derived in Section 5.5.1:

$$r_{.1} = \frac{\sum_t (x_t - \bar{x})(x_{t+1} - \bar{x})}{\sum_t (x_t - \bar{x})^2}$$

If the observations, z_i, are spatially independent Normal random variables, the expected value of I is:

$$E(I) = -\frac{1}{n-1}$$

hence for large n the expected value is approximately 0. The variance of I under these conditions is quite complicated:

$$Var(I) = \frac{n^2(n-1)A - n(n-1)B - 2C^2}{(n+1)(n-1)C^2}, \text{ where}$$

$$A = \frac{1}{2}\sum_i \sum_j \left(w_{ij} + w_{ji}\right)^2, i \neq j$$

$$B = \sum_k \left(\sum_j w_{jk} + \sum_i w_{ik}\right)^2, \text{ and}$$

$$C = \sum_i \sum_j w_{ij}, i \neq j$$

The variance of I under the assumption of randomisation (i.e. randomly re-distributed over the support, without replacement) is yet more complicated, but is provided in a number of software packages, such as Clusterseer, Rookcase and Geomed (from BioMedware). In this case the relevant formulas, using the expressions for A, B and C above are:

$$Var(I) = \frac{nD - 6EC^2}{(n+1)(n-1)C^2}, \text{ where}$$

$$D = (n^2 - 3n + 3)A - nB + 3C^2, \text{ and}$$

$$E = \frac{\sum_i (z_i - \bar{z})^4 / n}{\left(\sum_i (z_i - \bar{z})^2 / n\right)^2}$$

From the expected values and variances z-scores and significance levels for I can then be computed under the alternative distributional assumptions.

The computation of Moran's I for the sample dataset provided in Figure 5-21A illustrates the procedure. First we compute the matrix of unadjusted variance- and covariance-like quantities (Figure 5-22). There are 10 cells, and values for each cell are shown together with their deviations from the overall mean value, 1.022, in the column and row bounding the 10x10 matrix of computed values, C. The diagonal of the 10x10 matrix gives the variance-like quantities whilst the off-diagonal components give the covariance-like quantities.

The weighting adjusted matrix includes only the off-diagonal elements, and only those for which adjacencies have been included. The total of these values is 16.19, compared with the sum of the diagonal values which is 196.68. The computation of I requires one final element, adjustment by $1/p$, which in this

case is the number of cells (10 in our case) divided by the sum of the elements in W, which is 26, giving:

$I=10*16.19/(26*196.68)=0.0317$

If I is close to 0 there is very little spatial autocorrelation, which is what we have found in this example. If we rearrange the cell values the index value will change accordingly. For example, if the negative values in our example are swapped with the positive values in column 3, so that column 2 is entirely positive and column 3 entirely negative (Figure 5-21C) the index value computed is $I=0.26$.

Figure 5-22 Moran's I computation

A. Computation of variance/covariance-like quantities, matrix C												
Var/Covar matrix			1	2	3	4	5	6	7	8	9	10	Diagonal values:
	z(i)	z(i)-mean	1.22	2.08	3.53	-6.17	-5.41	-0.56	4.52	8.00	-3.11	-4.08	(z(i)-mean)^2
1	2.24	1.22	1.48	2.53	4.30	-7.52	-6.59	-0.68	5.50	9.74	-3.79	-4.97	1.48
2	3.10	2.08	2.53	4.32	7.33	-12.83	-11.25	-1.17	9.39	16.62	-6.47	-8.48	4.32
3	4.55	3.53	4.30	7.33	12.45	-21.77	-19.09	-1.98	15.94	28.22	-10.98	-14.40	12.45
4	-5.15	-6.17	-7.52	-12.83	-21.77	38.09	33.40	3.47	-27.89	-49.36	19.21	25.19	38.09
5	-4.39	-5.41	-6.59	-11.25	-19.09	33.40	29.29	3.04	-24.45	-43.29	16.84	22.09	29.29
6	0.46	-0.56	-0.68	-1.17	-1.98	3.47	3.04	0.32	-2.54	-4.49	1.75	2.29	0.32
7	5.54	4.52	5.50	9.39	15.94	-27.89	-24.45	-2.54	20.41	36.13	-14.06	-18.44	20.41
8	9.02	8.00	9.74	16.62	28.22	-49.36	-43.29	-4.49	36.13	63.97	-24.89	-32.65	63.97
9	-2.09	-3.11	-3.79	-6.47	-10.98	19.21	16.84	1.75	-14.06	-24.89	9.68	12.70	9.68
10	-3.06	-4.08	-4.97	-8.48	-14.40	25.19	22.09	2.29	-18.44	-32.65	12.70	16.66	16.66
Mean	1.02											SSD	196.68
SD	4.67												4.67

B. C*W: Adjustment by multiplication of the weighting matrix, W											
1	0.00	2.53	0.00	-7.52	0.00	0.00	0.00	0.00	0.00	0.00	-4.99
2	2.53	0.00	0.00	0.00	-11.25	0.00	0.00	0.00	0.00	0.00	-8.72
3	0.00	0.00	0.00	-21.77	0.00	0.00	15.94	0.00	0.00	0.00	-5.84
4	-7.52	0.00	-21.77	0.00	33.40	0.00	0.00	-49.36	0.00	0.00	-45.25
5	0.00	-11.25	0.00	33.40	0.00	3.04	0.00	0.00	16.84	0.00	42.04
6	0.00	0.00	0.00	0.00	3.04	0.00	0.00	0.00	0.00	2.29	5.34
7	0.00	0.00	15.94	0.00	0.00	0.00	0.00	36.13	0.00	0.00	52.07
8	0.00	0.00	0.00	-49.36	0.00	0.00	36.13	0.00	-24.89	0.00	-38.12
9	0.00	0.00	0.00	0.00	16.84	0.00	0.00	-24.89	0.00	12.70	4.66
10	0.00	0.00	0.00	0.00	0.00	2.29	0.00	0.00	12.70	0.00	15.00
										Covar SS	16.19

Moran's I is one of the most frequently implemented statistics within GIS packages, including ArcGIS Spatial Statistics Toolbox, the AUTOCORR function in Idrisi, and GIS-related analytical tools such as Crimestat, Rookcase, Clusterseer, GS+ and GeoDa. This is in part due to its ease of computation across a range of datasets and weightings, and because of its relationship to well-understood measures used in univariate statistics and time series analysis. It is also readily extended in a number of ways that increase its application and usefulness. These extensions include: adjustment for differences in underlying population size; treatment of spatial point data utilising distance bands rather like time lags; and computation of local I values rather than a single whole map or global I value. We examine each of these in turn.

Oden (1995) proposed an adjustment to Moran's I for zonal data where population size data (e.g. population at risk) is available. This statistic is supported within Clusterseer, but is not generally provided in other packages. Under this adjusted version the null hypothesis is that measured rates (e.g. disease rates, crime rates) in adjacent areas are independent, with the spatial variation observed simply reflecting the spatial variation in the underlying population. The computation of this statistic is quite involved, but essentially utilises the difference between the proportion of cases or incidents in a given zone (i.e. as a proportion of the total cases in the study region) and the expected proportion in the population at large (again, computed as the population in zone i/total population at risk). The expected value for the adjusted statistic is similar to that for the standard statistic, being $E(I_{pop})=-1/(N-1)$ where N is now the total population at risk. This statistic is not independent of population size, but can be readily standardised by dividing the computed value by the overall incidence rate for the study region.

We now turn to treatment of spatial point data rather than zoned or lattice data. Suppose that instead of zones we had a set of points (x,y) with associated data values, z (which could be values assigned to zone centroids). We could then compute:

$$I(h) = N(h) \frac{\sum_i \sum_j z_i z_j}{\sum_i z_i^2}$$

where $N(h)$ is the number of points in a distance band of width d and mean distance of h; z_i is the (standardised) value at point i; and z_j is the (standardised) value at a point j distance h from i (in practice within a distance band from $h-0.5d$ to $h+0.5d$). The summations apply for all point pairs in the selected band. Hence the adjacency matrix $\mathbf{W}$ is replaced by distance, h, and by using a number of bands a set of I values can be computed and plotted against h. This form of graph or diagram showing correlation information, as noted earlier, is known as a *correlogram*. To be meaningful the width of distance bands, h, should be greater than the minimum inter-point distance in the dataset, and the initial band should commence at $0.5d$ in order to give a better representation to values near the origin.

For example, the GS+ demonstration dataset comprises 132 separate spatial measurements of lead and aluminium levels in soil samples. An extract of the first 10 records is shown in Figure 5-23A with missing values being blank entries (coded on input in this case as -99). Figure 5-23B shows a map of the point set, with separate symbols indicating the four quartiles of the overall dataset that each point lies within (the triangular shape is the upper quartile, the square is the lower quartile). There are well over 10,000 possible point pairings in this sample dataset, but when grouped into distance bands only a subset of pairings is considered (Figure 5-23C). For example, in the first band, lag class 1, there are 226 point pairs, the average distance between pairs is 5.77 units, the bandwidth, h, is fixed for all lags at 8.21 units (defaulted to the maximum pairwise distance/10), and the Moran I value is 0.4179. This value decreases for the first few lags and then becomes closer to 0 and somewhat random. These values may be plotted, as shown in Figure 5-23D. Spatial autocorrelation (positive) as measured by this function, appears to be present for separation distances up to around 20-25 metres.

Figure 5-23 Sample dataset and Moran I analysis

A. Dataset, first 10 records				B. Coordinates of data points

A. Dataset, first 10 records

X-Coor	Y-Coor	Z	
m east	m north	Pb	Al
4.50	11.90	0.42	0.42
2.70	29.40	0.60	0.45
1.60	32.60	0.60	0.08
4.10	44.50	0.43	
0.60	64.00	0.51	0.14
2.40	71.80	0.34	0.32
7.80	3.50	0.37	0.12
6.70	10.20	0.61	
6.70	16.30	0.46	0.49
6.50	27.20	0.44	0.07

B. Coordinates of data points

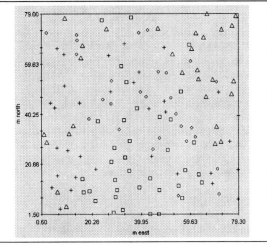

C. Moran *I* statistics, by lag

Lag Class	Average Distance	Moran's I	Pairs
1	5.77	0.4179	226
2	12.74	0.3691	675
3	20.74	0.1321	979
4	28.69	0.0616	1174
5	37.00	-0.0772	1214
6	45.15	-0.1483	1174
7	53.27	-0.1997	989
8	61.37	-0.1907	785
9	69.42	0.0037	553
10	77.39	-0.0237	241

D. Isotropic Correlogram

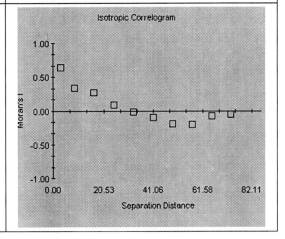

Each square on the graph in Figure 5-23D represents an index value generated from a large number of point pairs, and the individual contributions to the covariance can be plotted as a cloud of (in this case) 226 values (Figure 5-24). For example, the largest value in this cloud has a value of 0.4276 and is derived from points 100 and 115, whose separation distance is 6.44 units, and whose data values are 0.17 and 1.25 respectively. Now $(0.17-1.25)^2$ is not 0.4276, so we must look rather more carefully at what GS+ is doing with the data. The input dataset, like many such samples, is strongly left skewed — there are many small values recorded and only a few larger ones, with 1.25 being the maximum z-value. By default GS+ automatically transforms the data to make it more Normal, using the log transform $z^*=\ln(z+1)$. This gives the values

$z^*_{100}=0.157$ and

$z^*_{115}=0.811$ and thus $(0.157-0.811)^2=0.4276$.

The correlogram shown is isotropic, i.e. it has been calculated for all points, irrespective of direction. Spatial patterns may well exhibit directional bias (see also, Section 4.5.3) and hence it may be of interest to divide the circular lag regions into sectors and investigate the correlograms for each sector. Due to the symmetry of orientations, effectively only half the possible bearings need to be considered. Taking a sector window of, say, 45 degrees (i.e. X +/- 22.5 degrees) results in a set of 4 possible sectors to be analysed. The choice of initial bearing, X, can be selected by reference to the problem at hand, or by examining the dataset to locate the orientation of the principal axis of the standard deviational ellipse. Software packages often identify this for the user and then generate a set of anisotropic correlograms to enable the data to be examined sector-by-sector. If all are broadly similar it is likely that directional effects are small. Again, looking at this sample dataset, the principal axis is at 130 degrees, and the four anisotropic correlograms do not differ greatly.

Figure 5-24 Moran I (co)variance cloud, lag 1

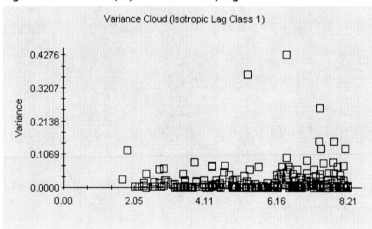

One alternative to the Moran I form of computation involves returning to the original set of criteria for a suitable function. Instead of computing the product sum

$$A = \sum\sum z_i z_j$$

We could equally well have computed the sum of squared differences

$$B = \sum\sum (z_i - z_j)^2$$

Or the absolute differences

$$C = \sum\sum |z_i - z_j|$$

since these have similar desirable properties.

These expressions once again can be standardised by dividing by the variance and adjusted for the ratio of zones or point pairs sampled. If the expression B above is used the resulting measure is known as the Contiguity ratio or Geary's ratio.

$$C = \frac{1}{p} \frac{\sum\sum w_{ij}(z_i - z_j)^2}{\sum(z_i - \bar{z})^2}$$

$$p = 2\frac{\sum\sum w_{ij}}{n-1}$$

This expression may be compared to the formula for Moran's I provided earlier. Unfortunately this ratio is not particularly easy to interpret: its values vary around 1, which indicates no spatial autocorrelation; to values as low as 0, which indicate positive autocorrelation; and values greater than 1 which indicate negative autocorrelation. Perhaps for this reason it is rarely implemented in GIS packages in this form, but expression B turns out to be extremely useful in what is known as semivariance analysis as we shall see in our discussion on geostatistics (see Section 6.7).

Due to the inconvenience of working with absolute values, expression C is rarely used,

but it is applied in semivariance analysis and is less sensitive to extreme values than computations based on B. In this case the correlograms are known as Madograms — for more details see Deutsch and Journel (1992) and Goovaerts (1997).

5.5.2.3 Weighting models and lags

Our discussion on correlograms highlights the way in which distance bands can be used to generate data series in a manner similar to the lags used in time series analysis and Ripley's K function. Weighting matrices that incorporate values other than 0 and 1 were introduced by Cliff and Ord (1973). One way in which distance-based weighting is applied is to use the actual distances, rather than distance bands, as the means by which data pairs are "ordered". The usual approach (for example that adopted within the Crimestat package) is to use inverse distance, so that pairs that are further apart contribute less to the overall index value than those that are close together. The distance weighted Moran's I is then given by:

$$I = \frac{1}{p} \frac{\sum\sum (z_i - \bar{z})(z_j - \bar{z})/(d_{ij} + \varepsilon)}{\sum(z_i - \bar{z})^2}, \text{ where}$$

$$p = \frac{1}{n}\sum\sum 1/(d_{ij} + \varepsilon)$$

The ε here represents an increment applied to the distance to ensure that when d_{ij} is small, results are not overly distorted or overflow. Other software packages implement a range of such distance weightings. For example, ArcGIS includes: inverse distance; inverse distance squared; and specified distance bands amongst its options.

With datasets that represents zones rather than points, as in our earlier examples, GIS tools may be used to map the data and compute spatial weights. Amongst the most extensive facilities available at present for conducting such calculations are provided by the GeoDa and ArcGIS packages, so we will use these to illustrate the range of facilities that have been implemented in readily available software.

Simple rook's move and queen's move contiguity weights are supported within GeoDa for polygonal regions. Typically these will be first order contiguity values, but 2nd or higher order weights can also be generated. Secondly, distance-based weights can be generated based on polygon centroid locations, or a defined (x,y) pair falling within the selected polygon. This requires a threshold value to be set, and all points (representing polygons) within this distance band are included within the weighting (as per ArcGIS distance bands). ArcGIS also supports a combination of fixed distance bands and distance decay, which it calls *zone of indifference weighting*. Distance computations in both GeoDa and ArcGIS are based on projected (Euclidean) distance, plus a second option: in GeoDa's case unprojected (approximate arc distance); and in ArcGIS' case city block calculations (L_1 metric).

A further method, based on the number of nearest-neighbours, is also supported in GeoDa. This requires the number of neighbours to be included as a parameter, and then varies the distances to reflect the value set (i.e. an adaptive method). The GeoDa code uses the Approximate Nearest Neighbour (ANN) software library developed at the University of Maryland to determine such neighbourhood relations.

The advantage of this nearest-neighbours approach is that the spatial weights for tightly packed small zones will be more similar to those of larger zones within the same study region, which tends not be the case with pure distance-based selection. Finally, a user-defined spatial weights matrix may be defined and read into either of these two packages — of course this requires very careful construction to ensure the weights matrix reflects the topology and weighting patterns intended. Neither of these packages explicitly supports weights based on shared boundary lengths. With initial "point set data" weights could be determined by construction of

Voronoi polygons and computing the lengths of shared boundaries, but facilities to perform such operations directly are not provided within existing GIS and related software. Information from the Delaunay triangulation and/or Voronoi polygon creation could be used to generate a spatial weights file, which ArcGIS and GeoDa, for example could read in. In this case, and by modifying or augmenting a program-generated standard weights file, weights with closer application-specific meaning may be created, for example based on measured interactions or flows between the sampled regions.

Idrisi supports simple autocorrelation computations for raster datasets using Moran's *I*. Its weights matrix options include rook's move and king's move (we would call this queen's move), with weights of 1 and 0.7071... for the direct and diagonal moves respectively. A form of support for higher order lags is indirectly supported in Idrisi by resampling the grid.

5.5.3 Local indicators of spatial association (LISA)

If we look more closely at the Moran *Index* we see that it can be disaggregated to provide a series of local indices. We start with the standard formulation we provided earlier:

$$I = \frac{1}{p} \frac{\sum_i \sum_j w_{ij}(z_i - \bar{z})(z_j - \bar{z})}{\sum_i (z_i - \bar{z})^2}, \text{ where}$$

$$p = \sum_i \sum_j w_{ij} / n$$

Examining the computation of this "global" index provided earlier in Figure 5-22B, we can see that each row in the computation contributes to the overall sum (Figure 5-25A), with the covariance component being 16.19 and overall Moran *I* value 0.03167. Each row item could be standardised by the overall sum of squared deviations and adjusted by the number of cells, *n*, as shown. The total of these local contributions divided by the total number of joins, $\Sigma\Sigma w_{ij}$, gives the overall or Global Moran I value. These individual components, or Local Indicators of Spatial Association (LISA), can be mapped and tested for significance to provide an indication of clustering patterns within the study region.

However, these calculations require some adjustment in order to provide Global Moran's I and LISA values in their commonly used form. The first adjustment is to the weights matrix. It is usual to standardise the row totals in the weights matrix to sum to 1. One advantage of row standardisation is that for each row (location) the set of weights corresponds to a form of weighted average of the neighbours that are included. Row standardisation may also provide a computationally more stable weights matrix. This procedure alters the Global Moran I value when rows have differing totals, which is the typical situation. For the example above the effect is to increase the computed Moran I to 0.0736. The second adjustment involves using (n-1) rather than n as the row multiplier, and then the sum is re-adjusted by this factor to produce the Global index value. These changes are illustrated in Figure 5-25B and Figure 5-26, and correspond to the computations performed within GeoDa, R and SpaceStat (Anselin, personal communication).

Figure 5-25 Local Moran's I computation

A. C*W: Adjustment by multiplication of the weighting matrix, W

											Row sum, S	R=S/SSD	I(i)=nR	
1	0.00	2.53	0.00	-7.52	0.00	0.00	0.00	0.00	0.00	0.00	-4.99	-0.03	-0.25	
2	2.53	0.00	0.00	0.00	-11.25	0.00	0.00	0.00	0.00	0.00	-8.72	-0.04	-0.44	
3	0.00	0.00	0.00	-21.77	0.00	0.00	15.94	0.00	0.00	0.00	-5.84	-0.03	-0.30	
4	-7.52	0.00	-21.77	0.00	33.40	0.00	0.00	-49.36	0.00	0.00	-45.25	-0.23	-2.30	
5	0.00	-11.25	0.00	33.40	0.00	3.04	0.00	0.00	16.84	0.00	42.04	0.21	2.14	
6	0.00	0.00	0.00	0.00	3.04	0.00	0.00	0.00	0.00	2.29	5.34	0.03	0.27	
7	0.00	0.00	15.94	0.00	0.00	0.00	0.00	36.13	0.00	0.00	52.07	0.26	2.65	
8	0.00	0.00	0.00	-49.36	0.00	0.00	36.13	0.00	-24.89	0.00	-38.12	-0.19	-1.94	
9	0.00	0.00	0.00	0.00	16.84	0.00	0.00	-24.89	0.00	12.70	4.66	0.02	0.24	
10	0.00	0.00	0.00	0.00	0.00	2.29	0.00	0.00	12.70	0.00	15.00	0.08	0.76	
										Covar S	16.19		0.82	0.03167

B. C*W1: Adjustment by multiplication of the row-adjusted weighting matrix, W1

											Row sum, S	R=S/SSD	I(i)=(n-1)R	
1	0.00	1.27	0.00	-3.76	0.00	0.00	0.00	0.00	0.00	0.00	-2.49	-0.0127	-0.1141	
2	1.27	0.00	0.00	0.00	-5.62	0.00	0.00	0.00	0.00	0.00	-4.36	-0.0222	-0.1994	
3	0.00	0.00	0.00	-10.89	0.00	0.00	7.97	0.00	0.00	0.00	-2.92	-0.0148	-0.1335	
4	-1.88	0.00	-5.44	0.00	8.35	0.00	0.00	-12.34	0.00	0.00	-11.31	-0.0575	-0.5177	
5	0.00	-2.81	0.00	8.35	0.00	0.76	0.00	0.00	4.21	0.00	10.51	0.0534	0.4810	
6	0.00	0.00	0.00	0.00	1.52	0.00	0.00	0.00	0.00	1.15	2.67	0.0136	0.1221	
7	0.00	0.00	7.97	0.00	0.00	0.00	0.00	18.07	0.00	0.00	26.04	0.1324	1.1915	
8	0.00	0.00	0.00	-16.45	0.00	0.00	12.04	0.00	-8.30	0.00	-12.71	-0.0646	-0.5814	
9	0.00	0.00	0.00	0.00	5.61	0.00	0.00	-8.30	0.00	4.23	1.55	0.0079	0.0710	
10	0.00	0.00	0.00	0.00	0.00	1.15	0.00	0.00	6.35	0.00	7.50	0.0381	0.3431	
										Covar S	14.48		0.6625	0.073612

ArcGIS offers the option to have no weights matrix standardisation, row standardisation (dividing rows by row totals) or total weights standardisation (dividing all cells by the overall cell total). The Geary C contiguity index may also be disaggregated to produce a LISA measure, in the same manner as for the Moran I.

Figure 5-26 LISA map, Moran's I

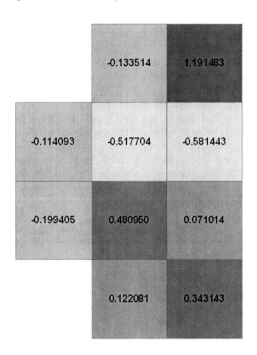

$$G = \frac{\sum_i \sum_j w_{ij} z_i z_j}{\sum_i \sum_j z_i z_j}, i \neq j$$

Whilst the local variant is

$$G_i = \frac{\sum_j w_{ij} z_j}{\sum_j z_j}, i \neq j$$

The weights matrix, **W**, is typically defined as a symmetric binary contiguity matrix, with contiguity determined by a (generalised) distance threshold, *d*. A variant of this latter statistic, which is regarded as of greater use in hot-spot analysis, permits *i=j* in the expression (so includes *i* in the summations, with $w_{ii}=1$ rather than 0 in the binary weights contiguity model) and is known as the G_i^* statistic rather than G_i. Note that these local statistics are not meaningful if there are no events or values in the set of cells or zones under examination.

Another statistic, "G", due to Getis and Ord (1992, 1995) may also be used in this way. The latter is supported within ArcGIS as both a global index and in disaggregated form for identifying local clustering patterns, which ArcGIS classifies as a form of hot-spot analysis. The ArcGIS implementation facilitates the use of simple fixed Euclidean distance bands (threshold distances) within binary weights (as described below), but also permits the use of the Manhattan metric and a variety of alternative models of contiguity such as: inverse distance; inverse distance squared; so-called Zones of Indifference (a combination of inverse distance and threshold distance); and user-defined spatial weights matrices.

Unlike Moran's *I* and Geary's *C* statistics, the *G* statistic identifies the degree to which high or low values cluster together. The standard global form of the statistic is:

The expected value of *G* under Complete Spatial Randomness (CSR) is:

$$E(G) = \frac{\sum_i \sum_j w_{ij}}{n(n-1)} = \frac{W}{n(n-1)}, i \neq j$$

where *n* is the number of points or zones in the study area and *W* is the sum of the weights, which for binary weights is simply the number of pairs within the distance threshold, *d*. The equivalent expected value for the local variant is:

$$E(G_i) = \frac{\sum_j w_{ij}}{(n-1)} = \frac{W_i}{(n-1)}$$

where *n* is the number of points or zones within the threshold distance, *d*, for point/region *i*. Closed formulas for the variances have been obtained but are very lengthy. They facilitate the construction of z-scores and hence significance testing of the global and local versions of the G statistics. For the local measure both computed values and z-scores may be useful to map.

The Rookcase add-in provides support for all three LISA measures with variable lag distance and number of lags. Further discussion of these methods is provided in several other sections of this Guide: in Section 5.2 covering exploratory spatial data analysis (ESDA); in Section 5.6.5 in the context of regression modelling; and in Sections 4.6 and 6 relating to grid files, since there are parallels between several of the concepts applied.

5.5.4 Significance tests for autocorrelation indices

The various global and local spatial autocorrelation coefficients discussed in the preceding subsection can be tested for statistical significance under two, rather different, model assumptions. The first is the classical statistical assumption of Normality, whereby it is assumed that the observed value of the coefficient is the result of the set $\{z_i\}$ of values being independent and identically distributed drawings from a Normal distribution, implying that variances are constant across the region. The second model is one of randomisation, whereby the observed pattern of the set $\{z_i\}$ of values is assumed to be just one realisation from all possible

random permutations of the observed values across all the zones. Both models have substantial weaknesses, for example as a result of underlying population size variation, but are widely implemented in software packages to provide estimates of the significance of observed results. In the case of the randomisation model most software packages generate a set of N random permutations of the input data (Monte Carlo tests), where N is user specified. For each simulation run index values are computed and the set of such values used to provide a pseudo-probability distribution for the given problem, against which the observed value can be compared. A z-transform of the coefficients under Normality or Randomisation assumptions is distributed approximately $N(0,1)$, hence this may be compared to percentage points of the Normal distribution to identify particularly high or low values.

The Rookcase add-in for Excel computes both values, for the global Moran I and Geary C, and the summary results in Figure 5-27 illustrate this for the test dataset shown in Figure 5-21C. Note that these results are calculated using overall weight adjustment rather than row-wise adjustment. The results suggest the observed index value of 0.26 could reasonably be expected to have occurred by chance. With the local variant of Moran's I the same procedures as for the Global statistic can be adopted.

Figure 5-27 Significance tests for revised sample dataset

Summary: Moran's I and Geary's C				
Moran's I =	0.26		Geary's C =	0.82
z-Normal I =	1.63		z-Normal C =	0.72
Var. Normal I =	0.05		Var. Normal C =	0.06
z-Random I =	1.57		z-Random C =	0.74
Var. Random I =	0.06		Var. Random C =	0.06
# Obs	Mean	SD	# Neighbours	
10	1.02	4.67	13	

Crimestat provides the option to compute z-scores for this statistic but notes that it may be very slow to calculate, depending on the number of zones. GeoDa uses random permutations to obtain an estimated mean and standard deviation, which may then be used to create an estimated z-score.

The local variants of these statistics utilise values in individual cells or zones over and over again. This process means that the values obtained for adjacent neighbours will not be independent, which breaches the requirements for significance testing, a problem exacerbated by the fact that such cells are also likely to exhibit spatial autocorrelation. There are no generally-agreed solutions to these problems, but one approach is to regard each test as one of a multiple set. In classical statistics multiple tests of *independent* sets (the members of which are not correlated) are made using a reduced or corrected significance level, α. The most basic of these adjustments (known as a Bonferroni correction, after its original protagonist) is to use a significance level defined by α/n, where n is the number of tests carried out, or in the present case, the number of zones or features being tested. If n is very large the use of such corrections results in extremely conservative confidence levels that may be inappropriate. Monte Carlo randomisation techniques that seek to estimate the sampling distribution may be more appropriate in such cases.

5.6 Regression Methods

5.6.1 Regression overview

Regression analysis is the term used to describe a family of methods that seek to model the relationship between one (or more) *dependent* or *response* variables and a number of *independent* or *predictor* variables. A typical (linear) relationship might be of the form:

$$y = \beta_0 + \beta_1 x_1 + \beta_2 x_2 + \beta_3 x_3 + \dots, \text{ or}$$
$$y = x\beta$$

where β is a column vector of p parameters to be determined and x is a row vector of independent variables with a 1 in the first column. It should be noted that the expression shown below is also described as linear since it remains linear in the coefficients, β, even if it is not linear in the predictor variables, x, e.g.:

$$y = \beta_0 + \beta_1 x_1 + \beta_2 x_1^2 + \beta_3 x_2 + \beta_4 x_1 x_2 + \dots$$

If $y = \{y_i\}$ is a set of n observations or measurements of the response variable, with corresponding recording of matching values for the set of independent variables, then a series of linear equations of the type shown above can be constructed. However, typically the number of observations will be greater than the number of coefficients (the system is said to be *over-determined*). One approach to this problem is to seek a best fit solution, where best fit is taken to mean a solution for the vector β that minimises the difference between the fitted model and observed values at these data points. In practice the procedure most widely applied seeks to minimise the sum of squared differences, hence the term *Least Squares*.

Each linear expression requires an additional "error" component to allow for the degree of fit of the expression to the particular data values. If we denote these error components by ε_i, the general form of the expression above now becomes:

$$y_i = \beta_0 + \beta_1 x_{1i} + \beta_2 x_{2i} + \beta_3 x_{3i} + \dots + \varepsilon_i, \text{ or}$$
$$y = X\beta + \varepsilon$$

where X is now a matrix with 1's in column 1, and ε is a vector of errors that in conceptual terms is assumed to represent the effects of unobserved variables and measurement errors in the observations.

This set of n equations in p unknowns ($n > p$) is typically solved for the vector β by seeking to minimise the sum of the squared error terms, $\varepsilon^T \varepsilon$ (*Ordinary Least Squares*, or *OLS*). The solution for the coefficients using this approach is obtained from the matrix expression:

$$\hat{\beta} = \left(XX^T \right)^{-1} X^T y, \text{ and } \operatorname{var}(\hat{\beta}) = \sigma^2 \left(XX^T \right)^{-1}$$

where the hat (^) symbol denotes concern with sample estimates of population parameters and the sample estimates of the population error (ε) terms are described as 'residuals'.

The variance for such models is usually estimated from the residuals of the fitted model, using the expression:

$$\hat{\sigma}^2 = \frac{\sum_{i=1}^{n} \left(y_i - \hat{y}_i \right)^2}{n - p}$$

Methods such as these may be used in the context of spatial description and mapping, as part of a process of data exploration, or for explanatory and/or predictive purposes.

The broad principles of regression analysis and modelling, as described in basic statistics texts, also apply to spatial datasets. In a spatial context one is seeking to model the variation in some spatially distributed dependent variable, for example house prices in a city, from a set of independent variables such as size of property, age of property, distance from transport facilities, proportion of green space in the region etc. (see further Table 5.8, *hedonic regression*). Spatial coordinates might be explicitly or implicitly incorporated into such models, for example including the coordinates of the centroid of a census tract or event location into the modelling process, or by taking account of the pattern of adjacency of neighbouring zones.

Similar methods are also widely supported in raster-based packages, with the dependent variable being taken as one grid from a set of matching layers or bands, and the independent variables being taken as the matching grids. Grid cell entries are treated as the sample values (responses) and the measured predictor values. In most packages this technique is used to aid grid combination, but others (e.g. MapCalc) suggest that it may be used for predictive purposes (e.g. predicting crop yield from a series of measured grids that provide details of soil structure and chemical makeup). This kind of modelling is generally unsafe due to the high degree of spatial autocorrelation between nearby cells in each grid (possibly generated by interpolation or resampling).

Ideally in regression analysis the form of the chosen model should be as simple as possible, both in terms of the expression employed and the number of independent variables included — these are sometimes referred to as the *simplicity* and *parsimony* objectives. In addition, the proportion of the variation in the dependent variable(s), **y**, explained by the independent variables, **x**, should be as high as possible, and the correlation between separate independent variables should be as low as possible. These criteria form part of a *strength of evidence* objective. If some, or all, of the **x** are highly correlated (typically having a correlation coefficient above 0.8) the model contains redundant information and may be described as being *over-specified*. If a strong relationship exists between selected independent variables and is also broadly linear, the variables are said to exhibit *multi-collinearity*. There are a number of standard techniques for reducing multi-collinearity, including: applying the so-called centering transform (deducting the mean of the relevant independent variable from each measured value); increasing the sample size (especially if

samples are quite small); removing the most inter-correlated variable; and combining inter-correlated variables into a new single composite variable (e.g. using principal components analysis or other data reduction techniques). The latter procedures can be viewed as forms of model re-specification.

With higher-order spatial trend analyses, multi-collinearity is to be expected since various combinations of the coordinates are themselves the independent variables. Technically, imperfect multi-collinearity does not result in the OLS assumptions being violated, but does lead to large standard errors for the coefficients involved, also referred to as *variance inflation* (VI). This limits their usefulness. As a rule of thumb, if the measure known as the Multi-collinearity Condition Number, MCN, is greater than 30 this problem is quite serious and should be addressed. The MCN is derived from the ratio of the largest to the smallest eigenvalue of the matrix XX^T. Generally the square root of this ratio is reported, but it is important to check this for any specific software package (if no square root is taken values greater than 1000 are taken as significant).

When fitting a model using OLS it is generally assumed that the errors (residuals for sample points) are identical and independently distributed (denoted *iid*). If the spread of errors is not constant, for example if in some parts of the study area the residuals are much more variable than in others these errors are said to exhibit *heteroskedasticity*. If the latter is the case (which may be detected using a range of standard tests) the estimated variance under OLS will be biased, either being too large or too small (it is not generally possible to know which direction the bias takes). As a result the specification of confidence intervals and the application of standard statistical significance tests (e.g. F tests) will not be reliable.

Many (most) spatial datasets exhibit patterns of data and/or residuals in which neighbouring areas have similar values (*positive spatial autocorrelation*) and hence violate the core assumptions of standard regression models. One result of this feature of spatial data is to greatly over-estimate the effective sample size and hence the degrees of freedom that may be indicated (for example in multiple grid regressions) will be stated as being far higher than they should be. One possible (partial) solution to this problem, in the case of grids, is to take (stratified or equalised) random samples from the response grid and apply regression to the subset of grid cells that correspond to these samples in the predictor grids. This approach is sometimes (rather confusingly) referred to as resampling. A common additional requirement for standard regression modelling is that the distribution of the errors is not only identical but also Normal, i.e.

$\varepsilon \sim N(0,\sigma^2 I)$

If the distribution of the errors is Normal and the regression relationship is linear, this equates to the response variable being Normally distributed, and vice versa, i.e.

$y \sim N(X\beta,\sigma^2 I)$, or setting the vector $\mu = X\beta$

$y \sim N(\mu,\sigma^2 I)$

The assumption of Normality permits the analytical derivation of confidence intervals for the model parameters and significance tests on their values (against the null hypothesis that they are zero). In order to satisfy the Normality conditions required in many models it is common to transform (continuous valued) response data using Log, Square Root or Box-Cox functions (Table 1.4) prior to analysis where this can be shown to improve the approximation to the Normal. It is also advisable to examine and, where appropriate, remove outliers even if these are not errors in the dataset but represent valid data that would otherwise distort the modelling process. Exclusion must be made explicit and separate investigation of any excluded data carried out. An alternative to exclusion is to weight observations according to some rule. Spatial declustering (see 5.1.2.2)

is an example of such weighting applied to spatial datasets.

If spatial autocorrelation has been detected a more reasonable assumption is:

$y \sim N(\mu,C)$

where C is a positive definite covariance matrix. Typically C would be defined in such a way as to ensure that locations that are close to each other are represented with a high value for the modelled covariance, whilst places further apart have lower covariance. This may be achieved using some form of distance-decay function or, for zoned data, using a contiguity-based spatial weights matrix. This kind of arrangement is similar to *Generalised Least Squares* (GLS), which allows for so-called second-order effects via the inclusion of a covariance matrix. The standard form of expression for GLS is:

$y = X\beta + u$

where u is a vector of random variables (errors) with mean 0 and variance-covariance matrix C. The GLS solution for β is then:

$$\hat{\beta} = \left(XC^{-1}X^T\right)^{-1}X^TC^{-1}y, \ \mathrm{var}(\hat{\beta}) = \left(X^TC^{-1}X^T\right)^{-1}$$

In spatial analysis this kind of GLS model is often utilised, with the matrix C modelled using some form of distance or interaction function. Note that C must be invertible, which places some restrictions on the possible forms it may take. Simple weighted least squares (WLS) is a variant of GLS in which the matrix C is diagonal and the diagonal values are not equal (hence the errors are heteroskedastic). In this case the weights are normally set to the inverse of the variance, thereby giving less weight to observations which have higher variance.

In addition to regression on dependent variables that take continuous values, similar methods have been developed for binary and count or rate data. In the case of binary data the data typically are transformed using the logit function (see Table 1.4). The central assumption of this model is that that the probability, p, of the dependent variable, y, taking the value 1 (rather than 0) follows the logistic curve, i.e.

$$p = prob\langle y = 1 | \mathbf{x} \rangle = \frac{e^{Bx}}{1 + e^{Bx}}$$

This model can be linearised using the logit transform, $q = \ln(p/(1-p))$. The regression model then becomes

$$q_i = \beta_0 + \beta_1 x_{1i} + \beta_2 x_{2i} + \beta_3 x_{3i} + \ldots + \varepsilon_i, \text{ or}$$
$$\mathbf{q} = \mathbf{X\beta} + \varepsilon$$

As with standard linear regression the errors in this model are assumed to be *iid* and the predictor variables are assumed not to be co-linear. The parameters in such models are determined using non-linear (iterative) optimisation methods, involving maximisation of the likelihood function.

With count or rate data Poisson regression may be applicable. The model is similar in form to the binary logit model, but in this case the response variable, y, is count (integer) data (hence non-continuous). The basic model in this case is of the form:

$$y = Ae^{Bx}$$

where A is an (optional) offset or population-based value (e.g. the underlying population at risk in each modelled zone). This model may be linearised, as per the logit model, but this time by simply applying a log transform: $q = \ln(y)$. The regression model then becomes:

$$q_i = (\ln(A) + \beta_0) + \beta_1 x_{1i} + \beta_2 x_{2i} + \beta_3 x_{3i} + \ldots + \varepsilon_i, \text{ or}$$
$$\mathbf{q} = \mathbf{X\beta} + \varepsilon$$

There are many variants of linear and non-linear regression analysis and a wealth of associated terms applied in this field — a number of these are summarised in Table 5.8.

In some instances the techniques are not always described as regression analysis *per se*, but share many underlying statistical assumptions and associated tests.

Choosing between alternative models presents a particular challenge. Each additional parameter will improve the fit of the model to the sample dataset (or data set), but at the cost of increasing model complexity. Model building should be guided by the principal of *parsimony* — that is the simplest model that is fit for the purpose for which it is built. Overly complicated models frequently involve a loss of information regarding the nature of key relationships. Drawing on this *information content* view of modelling, an information theoretic statistic is often used as a guide to model selection. The most common statistic now used is known as the Akaike Information Criterion (AIC) and is often computed in its small-sample corrected version (AICc) since this is asymptotic to the standard version. The measures are defined as:

$$AIC = -2\ln(L) + 2k$$
$$AICc = -2\ln(L) + 2k\left(\frac{n}{n-k-1}\right)$$

where n is the sample size, k is the number of parameters used in the model, and L is the likelihood function. It is recommended that the corrected version be used unless $n/k > 40$. If the errors are assumed to be Normally distributed then the standard expression can be re-written as:

$$AIC = n(\ln(2\pi\sigma^2) + 1) + 2k$$

The AIC measure is utilised in several of the software packages described in this Section. For example GeoDa uses the version above in OLS estimation, whilst Geographically Weighted Regression (see further Section 5.6.3) uses a variant of the version corrected for small sample sizes. Both packages use appropriate alternative expressions in cases where the likelihood functions are different from the Normal model.

Table 5.8 *Selected regression analysis terminology*

Form of model	Notes
Simple linear	A single approximately continuous response (dependent) variable and one or more predictor variables related by an expression that is linear in its coefficients (i.e. not necessary linear in the predictor variables). Note the similarity to weighted averages and geostatistical models
Multiple	This term applies when there are multiple predictor variables and all are quantitative
Multivariate	Regression involving more than one response variable. If, in addition, there are multiple predictor variables the composite term multivariate multiple regression is used
SAR	Simultaneous autoregressive models (SAR is also used as an abbreviation for Spatial Autoregression). A form of regression model including adjustments for spatial autocorrelation. Many variants of SAR model have been devised
CAR	Conditional autoregressive models — as per SAR, a form of regression model including adjustments for spatial autocorrelation. Differs from SAR in the specification of the inverse covariance matrix. In this model the expected value of the response variable is regarded as being conditional on the recorded values at all other locations
Logistic	Logistic regression applies where the response variable is binary, i.e. of type 1/0 or Yes/No. Typically this involves use of the logit transform (Table 1.4), with linear regression being conducted on the transformed data. Variants on the basic binary model are available for response variables that represent more than two categories, which may or may not be ordered
Poisson	Poisson regression applies where the response variable is a count (e.g. crime incidents, cases of a disease) rather than a continuous variable. This model may also be applied to standardised counts or "rates", such as disease incidence per capita, species of tree per square kilometre. It assumes the response variable has a Poisson distribution whose expected value (mean) is dependent on one or more predictor variables. Typically the log of the expected value is assumed to have a linear relationship with the predictor variables. An excess of zeros in many sample datasets may present problems when attempting to apply this form of regression
Ecological	The term ecological regression does not relate directly to the subject of ecology, but to the application of regression methods to data that are aggregated to zones (lattices), as is often the case with census datasets and information collected by administrative districts. The related issue of the so-called *ecological fallacy* (referred to in Section 3.2.3.1) concerns the difficulty of making inferences about the nature of individuals within an aggregated region on the basis of statistics (data values, parameters, relationships) that apply to the aggregated data
Hedonic	The term hedonic regression is used in economics, especially in real estate (property) economics, to estimate demand or prices as a combination of separate components, each of which may be treated as if it had its own market or price. In the context of regression these separate components are often treated as the independent variables in the modelling process
Analysis of variance	Applies if all of the predictors are either qualitative or classified into a (small) number of distinct groups. Analysis of variance methods are often used to analyse the significance of alternative regression models under the Normality assumption for the distribution of errors
Analysis of covariance	Applies if some of the predictors are qualitative and some are quantitative. Analysis of covariance methods are also widely applied in spatial modelling, where the covariance of observations at pairs of locations is examined

A second information theoretic measure, which places greater weight on the number of parameters used, is often also provided. This is known as the Bayesian Information Criterion or BIC, or sometimes as the Swartz Criterion. It has the simpler form:

$$BIC = -2\ln(L) + k\ln(n)$$

5.6.2 Simple regression and trend surface modelling

Simple regression techniques have been widely applied in spatial analysis for a very long time. The general form of model employed can be represented by the expression:

$$y = f(x_1, x_2, \mathbf{w})$$

where: y is the (possibly transformed) observed value of some continuous dependent variable y at location (x_1, x_2); $f()$ is a general function to be defined, but often a linear expression in the coefficients; and $\mathbf{w}$ is a vector of attributes measured at (x_1, x_2). For example, if $f()$ is simply modelled as a linear function of the coordinates we can write

$$y = \beta_0 + \beta_1 x_1 + \beta_2 x_2$$

Model performance can be assessed using a number of statistics that are discussed in standard statistical text books. One such statistic, which will be referred to in our subsequent discussion, is the correlation coefficient or R^2 statistic. This statistic records the proportion of variation in the data that is explained by the model. A modification this statistic, adjusted R^2, takes into account the complexity of the model in terms of the number of variables that are specified.

With a set of n observations made at different locations in the study area, $y_i(x_{1i}, x_{2i})$, a series of n equations can be set up and solved for the $p=3$ unknown parameters using OLS. This is simply a linear trend or best fit linear surface through the point set, $\{y_i\}$. A set of differences

between the observed values and the best fit surface can then be computed:

$$\varepsilon_i = \hat{y}_i - y_i$$

where the y hat values (the estimated values) are obtained by evaluating the trend surface at each point i. Assuming the value of n is reasonably large and provides a representative coverage of the complete set of locations in the study area, these residuals may be treated as samples from a continuous surface and mapped using simple contouring or filled contouring procedures. If, on the other hand, the sample points represent centroids of zones (pre-defined regions or generated zones, e.g. Voronoi polygons) the visualisation of residuals is generally applied to these zones as a choropleth map. Plotting the residuals using a combination of the fitted surface and stick/point display in 3D may also be helpful but is less common within GIS packages.

Whichever approach is adopted the fitting of a trend surface and mapping of residuals is a form of exploratory spatial data analysis (ESDA - see further, Section 5.2) and is supported in many GIS and related packages. Within GIS packages such residual maps may be overlaid on topographic or thematic maps with a pre-specified transparency (e.g. 50%) enabling spatial associations between large positive or negative residuals to be visually compared to potential explanatory factors.

With higher order polynomials or alternative functions (e.g. trigonometric series) the fit of the trend surface to the sample data *may* be improved (the sum of the squared residuals reduced) but the overall utility of the procedure may well become questionable. This may occur for many reasons including the increased difficulty in interpreting the fitted surface (e.g. in terms of process), and increased volatility of the surface close to and immediately beyond the convex hull of the sample points and/or in sparsely sampled regions. Higher order polynomial trend surfaces are usually limited to second order equations, although in some applications 3rd order or above have been shown to be useful — for example, Lichstein *et al.* (2002) fit a full 3rd order polynomial trend equation to the spatial coordinates of bird species sightings. They then proceed to analyse the distribution of bird species by modelling habitat + trend (large scale spatial pattern) + spatial autocorrelation (small scale spatial pattern) in order to identify separate factors that might be used to explain observed distribution patterns by species. Note that any analysis of trends in this way will result in a fitted surface whose form is heavily dependent upon the set of specific locations at which measured values of *y* are available and the scale factors operating for the problem being examined. Frequently data locations are not the result of selections made by the analyst but are pre-determined survey locations that may reflect convenience for data gathering (e.g. proximity to a forest path, access to quarry sites) rather than an ideal sampling frame for analytic purposes.

The most common measure of the quality of fit of a trend surface or similar regression is the *coefficient of determination*, which is the squared coefficient of (multiple) correlation, R^2. This is simply a function of the squared residuals with standardisation being achieved using the sum of squared deviations of observations from their overall mean:

$$R^2 = 1 - \frac{\sum \varepsilon_i^2}{\sum (y_i - \bar{y})^2}, R^2 \in [0,1]$$

Under appropriate conditions, e.g. $\varepsilon \sim N(0, \sigma^2 I)$ the significance of this coefficient can be evaluated in a meaningful way, with values close to 1 generally being highly significant. Frequently, however, the statistical significance of the measure cannot be determined due to distribution conditions not being met, and the value obtained should be seen as no more than an indicator of goodness of fit.

Trend surface and residuals analysis is essentially a form of non-statistical ESDA — it

is not necessary to state how significant the fit of the model is or what confidence limits might be placed around the values of the estimated coefficients and the fitted surface. For inferential analysis the standard pre-conditions that apply are extremely unlikely to be met in the context of spatial analysis. Conditions that may apply for statistical purposes include:

1. the set $\{y_i\}$ is comprised of independent (uncorrelated) observable random variables. This set does not have to be drawn from a Normal distribution but it is often helpful or necessary that they are at least asymptotically Normal (possibly after transformation)

2. the set $\{\varepsilon_i\}$ is comprised of independently and identically distributed (*iid*) unobservable random variables with mean 0 and constant variance, σ^2, where σ^2 is not a function of β or **x**

3. the set $\{\varepsilon_i\}$ is Normally distributed — this condition is not a pre-requisite but is a requirement for inference on finite sets (as previously noted, this condition is directly connected to the first bullet point above)

4. the model applied is appropriate, complete and global. This assumption incorporates the assumption that the independent variables, **x**, are themselves uncorrelated, and the parameters β are global constants

Tests for the Normality of the observations and the residuals are commonplace (the first and third conditions above), and a variety of tests for heteroskedasticity (resulting in failure to meet the second condition) have been implemented in a several GIS-related packages. Note that heteroskedasticity may or may not be associated with positive spatial autocorrelation.

As discussed in Section 5.5, testing for (global) spatial autocorrelation using the Moran *I* statistic is a useful starting point for such analyses. A spatial weights matrix, **W**, is required in order to generate the Moran I statistic. As shown earlier this statistic has the form:

$$I = \frac{1}{p} \frac{\sum\sum w_{ij}(y_i - \bar{y})(y_j - \bar{y})}{\sum(y_i - \bar{y})^2}, \text{ where}$$

$$p = \sum\sum w_{ij} / n$$

This measure is the spatial equivalent of the standard product moment correlation coefficient. The weighting terms, w_{ij}, are defined either from patterns of zone adjacency or using some form of distance-decay function. A value of Moran's $I=1$ indicates perfect positive spatial autocorrelation, whilst a value of I very close to 0 indicates no spatial autocorrelation. A value such as $I=0.2$ indicates some positive spatial autocorrelation, although this may or may not be significant — with large spatial datasets small positive I-values are very often technically identified as significant, although the value of this interpretation may not be great.

Assuming that significant spatial autocorrelation is identified several options exist for continued analysis using regression methods:

- proceed with the analysis ignoring the observed spatial autocorrelation (particularly if it is small) accepting that the drawing of confidence intervals and determination of significance levels will not be possible, or if computed may be misleading

- drop the requirement that the parameters β are global constants and allow these to vary with location. Effectively this involves fitting a series of regression surfaces to the dataset in such a way as to ensure a continuous prediction surface is created. One procedure for achieving this objective is known as Geographical Weighted

Regression (GWR), and is described in Section 5.6.3; or

- include additional elements into the regression model that take explicit account of the observed pattern of spatial autocorrelation. This may be achieved in a number of ways, several of which are described in Section 5.6.4

5.6.3 Geographically Weighted Regression (GWR)

GWR is the term introduced by Fotheringham, Charlton and Brunsdon (1997, 2002) to describe a family of regression models in which the coefficients, β, are allowed to vary spatially. GWR uses the coordinates of each sample point or zone centroid, t_i, as a target point for a form of spatially weighted least squares regression (for some models the target points can be separately defined, e.g. as grid intersection points, rather than observed data points). The result is a model of the form:

$$y = X\beta(t) + \varepsilon$$

The coefficients $\beta(t)$ are determined by examining the set of points within a well-defined neighbourhood of each of the sample points. This neighbourhood is essentially a circle, radius r, around each data point (anisotropic modelling is not currently supported). However, if r is treated as a fixed value in which all points are regarded as of equal importance it could include every point (for r large) or alternatively no other points (for r very small). Instead of using a fixed value for r it is replaced by a distance-decay function, $f(d)$, as described earlier (Section 4.4.6). This function may be finite or infinite, much as with kernel density estimation (Section 4.3.4). The functions utilised in the GWR software package are of the form:

$f(d) = e^{-d^2/2h^2}$, or $f(d) = e^{-d/h}$, or

$$f(d) = \left(1 - \frac{d^2}{h^2}\right)^2, d < r$$

$f(d) = 0$ otherwise

In these functions the parameter, h, also known as the *bandwidth*, is the key factor determining the way in which the weighting schemes operate. A small bandwidth results in very rapid distance decay, whereas a larger value will result in a smoother weighting scheme. This parameter may be defined manually or alternatively by some form of adaptive method such as cross-validation minimisation (e.g. jack-knifing — see further, Section 6.7.2.2) or minimisation of the Akaike Information Criterion (AIC). The use of a kernel function also raises the possibility of generating additional descriptive statistics, as have been described earlier.

Using a selected kernel function and bandwidth, h, a diagonal weighting matrix, $\mathbf{W}(t)$, may be defined for every sample point, t, with off-diagonal elements being 0. The parameters $\beta(t)$ for this point can then be determined using the standard solution for weighted least squares regression:

$$\hat{\beta}(t) = \left(\mathbf{X}\mathbf{W}(t)\mathbf{X}^T\right)^{-1}\mathbf{X}^T\mathbf{W}(t)\mathbf{y}$$

Or letting $\mathbf{D} = \left(\mathbf{X}\mathbf{W}(t)\mathbf{X}^T\right)^{-1}\mathbf{X}^T\mathbf{W}(t)$

$\hat{\beta}(t) = \mathbf{D}\mathbf{y}$, and $\text{var}\left(\hat{\beta}(t)\right) = \mathbf{D}\mathbf{D}^T\sigma^2$

The standard errors of the parameter estimates can be computed as the square root of these variances and used in t-tests to obtain estimates of the significance of the individual components. In this model the variance component, σ^2, is defined by the normalised residual sum of squares (RSS) divided by the degrees of freedom. The latter are defined by the number of parameters, p, in a global model, or the *effective number of parameters*

in the GWR model. This value is approximated by the authors as the trace of a matrix $\mathbf{S}$, $\text{tr}(\mathbf{S})$ (the sum of the diagonal elements of $\mathbf{S}$) defined by the relation:

$$\hat{\mathbf{y}} = \mathbf{S}\mathbf{y}$$

A set of such equations is solved for all points, t. The fit of the model may be examined in the usual manner, although it is to be expected that the fit in terms of variance explanation will almost always be an improvement over global methods, if only because there are far more parameters fitted to the dataset. For this reason comparisons should be made on additional criteria, for example the AIC measure which takes account of the model complexity. As with conventional regression, the modelled surface and (standardised) residuals may be mapped for exploratory purposes, but additionally the parameters $\beta(t)$ and their estimated standard errors may also be mapped since these also vary spatially. Within GWR standardised residuals are determined as the sum of squared residuals, $\varepsilon\varepsilon^T$, divided by the degrees of freedom, $n\text{-tr}(\mathbf{S})$. The authors recommend examining any values for these residuals >3.

To illustrate this process we shall use an example dataset comprising educational attainment by county in the state of Georgia, USA (the dependent or response variable, Table 5.9). This dataset lists the percentage of University graduates by county together with a range of social data that might act as independent variables to be used in predicting the dependent variable. The data have been assigned to a set of 159 point locations (county centroids) and show an overall average of 19% of the population recorded as being graduates, with an average per county of 10.9% (i.e. not population weighted). The range by county is from 4.2% to 37.5%.

Table 5.9 shows the predictor variables and the global regression parameters estimated by OLS, which collectively account for around 63% of the variance. Also shown are the GWR parameter estimates, expressed as a range of values that have been computed. In the diagnostics section of this table note the drop in residual sum of squares, the increase in the adjusted R^2, and a modest fall in the AICc statistic. The authors suggest that a fall of 3 or more in the AICc value warrants examination as demonstrating a meaningful improvement in model fit. Note that differences in the method of calculation of the AIC statistic can easily result in differences of greater than 3, so caution is required when comparing alternative software packages on this measure.

As mentioned above, the standardised residuals from the GWR predictions can be mapped in order to identify any prediction outliers. Figure 5-28A shows this mapped dataset, with the dark blue and red (or darkest grey) counties illustrated being those with the highest and lowest deviations. These counties may then be examined to try and ascertain if there are any special characteristics of these cases that might explain the large residuals. By definition the GWR modelled parameters falling within the range shown in Table 5.9 include a value for every county. Hence each parameter can also be mapped, as shown in Figure 5-28B. In this case the map highlights a distinct pattern of variation, with higher values in the north and lower values in the south.

The same kind of GWR analysis can be carried out on count data, using Poisson regression (GWPR), and on binary data, using Logistic regression (GWLR). The GWR V3 program supports both models. The standard Poisson and Logistic regression models described in Table 5.8 are utilised, but with the coefficients $\beta(t)$ varying with location, t, as before. As an example, GWPR has been applied to counts of disease incidence amongst a particular age/sex grouping, recorded by health district. In such models an *offset* value is applied to the model based on a matching count variable, such as the total number of people in that district in the selected age/sex cohort. For example, Nakaya *et al.* (2005) applied GWPR to mortality rates in Tokyo. The dependent variable in this instance was based on the standardised mortality ratio (SMR) for

each of 262 municipality zones. The SMR is defined as the observed number of deaths, O_i, in a specified time period (e.g. 1990) in a given zone i, divided by the expected number, E_i, for that zone based on national or regional mortality rates (i.e. by demographic grouping). In the GWPR model the O_i became the response variable and the E_i provided the offset values. The study was able to examine relationships between the dependent variable and a number of independent variables at the local level, highlighting particular relationships that global models may not have identified. In fact the best model the authors were able to produce included a mix of global and local parameter estimates, with the proportion of older people (64+) and of house-owners taken as globals and the proportion of professional and technical people and the proportion of unemployed people being allowed to vary regionally.

For Logistic GWR one might have true presence/absence data, or recoded continuous data based on some critical threshold value. For example, with the Georgia dataset the dependent variable could be recoded as 1 if the percentage of graduates is above the state average and 0 otherwise. This is rather an artificial example, but recoding of this type is often applied in decision-making — for example coding land as contaminated (1) if, say, the average measured cadmium level in the soil exceeds a certain number of parts per million (ppm) and not contaminated (0) with respect to this trace element if below this threshold level.

Both Poisson and Logistic GWR require model fitting using a technique known as iteratively reweighted least squares (IRLS). The analysis is carried out in much the same manner as previously described, but the computation of the Akaike Information Criterion (AIC) and AICc differs from the OLS expressions. The ready availability of GWR V3 software, supporting Gaussian, Poisson and Logistic models, together with a companion book and materials, has resulted in an upsurge in interest in the technique. This includes its consideration by spatial econometricians, medical statisticians and ecologists amongst others. It has the attraction of accepting the non-stationarity of most spatial datasets (see further, Section 6.7.1.8), and proceeding to create models with improved information characteristics and amenable to further exploratory analysis.

Table 5.9 Georgia dataset — global regression estimates and diagnostics

Predictor variables	Global parameter estimate	GWR parameter estimates
Total population, β_1	0.24×10^{-4}	0.14 to 0.28×10^{-4}
% rural, β_2	-0.044	-0.06 to -0.03
% elderly, β_3	-0.06 (not signif.)	-0.26 to -0.06
% foreign born, β_4	1.26	0.51 to 2.42
% poverty, β_5	-0.15	-0.20 to -0.00
% black, β_6	0.022 (not signif.)	-0.04 to 0.08
Intercept, β_0	14.78	12.62 to 16.49
Diagnostics		
Residual SS	1816	1506
Adjusted R^2	0.63	0.68
Effective parameters	7	12.81
AICc	855.4	839.2

Figure 5-28 Georgia educational attainment: GWR residuals map, Gaussian adaptive kernel

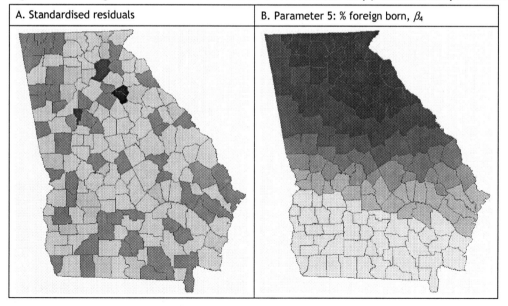

A. Standardised residuals	B. Parameter 5: % foreign born, β_4

5.6.4 Spatial autoregressive and Bayesian modelling

5.6.4.1 Spatial autoregressive modelling

Many of the techniques that are briefly described in this final subsection originate from time series analysis and were subsequently developed from the mid-1950s within the discipline known as spatial statistics. They have been applied and substantially extended in the last 25 years, notably by econometricians, geographers and medical statisticians. Additional disciplines that have made extensive use of these techniques include the actuarial, ecological and environmental sciences. Detailed discussion of the methods and underlying theory may be found in Cressie (1993), Bailey and Gatrell (1995), Anselin (1988), Anselin and Bera (1998), Anselin (2002) and Haining (2003). The procedures have been implemented in the SpaceStat, S-Plus, MATLab Spatial Statistics Toolbox (Pace *et al.*), WinBUGS and GeoDa packages, amongst others. A number of these have been specifically developed to deal with large (and often sparse) matrix difficulties that arise with detailed regional and national datasets.

A pure spatial autoregressive (SAR) model simply consists of a spatially lagged version of the dependent variable, **y**:

$$y = \rho \mathbf{W} y + \varepsilon$$

As can be seen this is similar to a standard linear regression model where the first term is constructed from a predefined n by n spatial weighting matrix, **W**, applied to the observed variable, **y**, together with a spatial autoregression parameter, ρ, which typically has to be estimated from the data. The spatial weights matrix, **W**, is almost always standardised such that its rows sum to 1. For an individual observation the equation is simply:

$$y_i = \rho \sum_j w_{ij} y_j + \varepsilon_i$$

Note the similarity of this model with a simple time series autoregressive model (from which it is derived):

$$y_t = \rho y_{t-1} + \varepsilon_t$$

Since the dependent variable, **y**, appears on both sides of the expression:

$$y = \rho \mathbf{W} y + \varepsilon$$

it can be re-arranged to solve for **y**:

$$y = (\mathbf{I} - \rho \mathbf{W})^{-1} + \varepsilon$$

from which we can obtain an expression for the variance of **y** as:

$$\text{var}(y) = E(yy^{\mathsf{T}}) = (\mathbf{I} - \rho \mathbf{W})^{-1} E(\varepsilon \varepsilon^{\mathsf{T}})(\mathbf{I} - \rho \mathbf{W}^{\mathsf{T}})^{-1}$$

hence

$$\text{var}(y) = E(yy^{\mathsf{T}}) = (\mathbf{I} - \rho \mathbf{W})^{-1} \mathbf{C} (\mathbf{I} - \rho \mathbf{W}^{\mathsf{T}})^{-1}$$

where **C** is the variance-covariance matrix. This derivation has made no distributional assumptions regarding the response variable or the errors. Furthermore the matrix $\rho \mathbf{W}$ does not have to be symmetric. The equivalent result for the conditional autoregressive (CAR) model, which we discuss later in this subsection, is:

$$\text{var}(y) = (\mathbf{I} - \rho \mathbf{W})^{-1} \mathbf{I} \sigma^2$$

If we add to the pure SAR model additional predictor variables, **x**, we have a *mixed regressive spatial autoregressive* model (*mrsa*), which is defined as follows:

$$y = \mathbf{X}\mathbf{B} + \rho \mathbf{W} y + \varepsilon$$

As can be seen this is the same as a standard linear regression model with the addition of the SAR component. The design of this kind of mixed model specifically incorporates spatial autocorrelation whilst including the influence of other (aspatial) predictor variables. The objective of this revised approach is to obtain a significant improvement over a standard OLS model. The level of improvement will depend on how well the revised model represents or explains the source data, and to an extent this will vary depending on the detailed form of the weighting matrix, **W**.

Theoretical analyses have shown that this type of model can be derived from a variety of different processes, including direct processes such as spatial diffusion, certain forms of spatial interaction (including spillover and gravity or potential-type process models), and indirect processes such as resource distribution. This lack of a well-defined link between process and form is commonplace in spatial analysis, and is well-documented in fields such as point set clustering and fractal analysis. That is also applies here, in spatial regression modelling, should come as no surprise.

A second approach to SAR modelling is known as the spatial error model. This model is applied when there appears to be significant spatial autocorrelation, but tests for spatial lag effects do not suggests that inclusion of the latter would provide a significant improvement. A decision diagram for selecting the appropriate model based on a set of additional diagnostics (Lagrangian multiplier test statistics) is included in the GeoDa tutorial materials. The spatial error model (from GeoDa) is defined as:

$$\mathbf{y} = \mathbf{X}\boldsymbol{\beta} + \boldsymbol{\varepsilon}, \text{ where}$$
$$\boldsymbol{\varepsilon} = \lambda \mathbf{W}\boldsymbol{\varepsilon} + \mathbf{u}$$

Hence the basic model is as per the standard linear model, but now the error term is assumed to be made up of a spatially weighted vector, $\lambda \mathbf{W}\boldsymbol{\varepsilon}$, and a vector of *iid* errors, **u**.

The Georgia educational attainment dataset used to illustrate GWR can be analysed in a

similar manner using SAR methods. If this is conducted within GeoDa the OLS results match those within GWR (although the AIC values differ slightly owing to the differences in the detailed expressions applied). However, to apply an SAR model a spatial weights matrix is required. In the following example we have set the spatial weights to be defined by simple rook's move contiguity (adjacent edges), and then examined the GeoDa diagnostics to determine which form of SAR regression model seems most appropriate to apply. In this instance the spatial error model was identified as the most appropriate and the regression re-run using this model. The results are summarised in Table 5.10, which is simply an extended version of the Table 5.9, including the new SAR parameter estimates. Although the RSS value is not as low as with GWR, the model is intrinsically far simpler and enables a more global view of the relationship between variables. There is an argument for utilising both global OLS/SAR and GWR approaches when analysing datasets of this type, since they provide different perspectives on the data, and different insights into the use of such data for predictive purposes.

Given the error term:

$$\boldsymbol{\varepsilon} = \lambda \mathbf{W}\boldsymbol{\varepsilon} + \mathbf{u}$$

and observing that also:

$$\boldsymbol{\varepsilon} = \mathbf{y} - \mathbf{X}\boldsymbol{\beta}$$

we have:

$$\mathbf{y} = \mathbf{X}\boldsymbol{\beta} + \lambda \mathbf{W}(\mathbf{y} - \mathbf{X}\boldsymbol{\beta}) + \mathbf{u}, \text{ or}$$
$$\mathbf{y} = \mathbf{X}\boldsymbol{\beta} + \lambda \mathbf{W}\mathbf{y} - \lambda \mathbf{W}\mathbf{X}\boldsymbol{\beta} + \mathbf{u}$$

Hence this expression models the dependent variable **y** as a combination of a general (global) linear trend component, $\mathbf{X}\boldsymbol{\beta}$, plus a pure SAR component, $\lambda \mathbf{W}\mathbf{y}$, minus a neighbouring trend component, $\lambda \mathbf{W}\mathbf{X}\boldsymbol{\beta}$, plus a set of *iid* random errors, **u**. Comparing this to the *mrsa* model above:

$$y = X\beta + \rho Wy + \varepsilon$$

we see that the spatial error model can be viewed as a form of mixed spatial lag model

with an additional autoregressive component, the neighbouring trend, $\lambda WX\beta$.

Table 5.10 Georgia dataset — SAR comparative regression estimates and diagnostics

Predictor variables	Global parameter estimate	SAR-E parameter estimates	GWR parameter estimates
Total population, β_1	0.24×10^{-4}	0.24×10^{-4}	0.14 to 0.28×10^{-4}
% rural, β_2	-0.044	-0.046	-0.06 to -0.03
% elderly, β_3	-0.06*	-0.099*	-0.26 to -0.06
% foreign born, β_4	1.26	1.196	0.51 to 2.42
% poverty, β_5	-0.15	-0.145	-0.20 to -0.00
% black, β_6	0.022*	0.013*	-0.04 to 0.08
Intercept, β_0	14.78	15.46	12.62 to 16.49
lambda, λ		0.313	
Diagnostics			
Residual SS	1816	1708	1506
Adjusted R^2	0.63	0.67	0.68
Effective parameters	7	7	12.81
AIC / AICc	855.4	846.0	839.2

* not significant

The type of model can be generalised still further (Haining, 2003, p355), for example as:

$$y = \alpha I + X\beta + \rho Wy + WX\delta + (I - \phi W)^{-1} \varepsilon$$

where the scalars α, ρ, and ϕ, and the vectors β and δ are all parameters to be estimated, and the final term represents an SAR on the errors. Clearly one could proceed from the generalised model to the particular, or vice versa. Likewise one could progressively increase or decrease the set of explanatory variables in the model. Given the considerable complexity of spatial phenomena, Haining suggests a data-driven approach to statistical modelling, which can be seen as fitting comfortably within the Data and Analysis components of the PPDAC framework described in Section 3.2.1 and Figure 3-4. His approach commences with ESDA, proceeds to

model specification for the current data, and then progresses to an iterative cycle of selection and implementation of parameter estimation, assessment of model fit and re-specification where necessary.

5.6.4.2 Conditional autoregressive and Bayesian modelling

A somewhat different conceptual model, which may also provide similar results, is known as conditional autoregressive modelling (CAR). In the study by Lichstein *et al.* (2002), cited earlier, they chose to use CAR rather than SAR modelling, following the recommendation of Cressie (1993) and because they felt it to be more appropriate for their study. They found no real difference in the results obtained with the CAR model from those achieved using SAR modelling.

The standard or proper CAR model for the expectation of a specific observation, y_i, is of the form:

$$E\left(y_i \mid all\ y_{j \neq i}\right) = \mu_i + \rho \sum_{j \neq i} w_{ij}\left(y_j - \mu_j\right)$$

where μ_i is the expected value at i, and ρ is a spatial autocorrelation parameter that determines the size and nature (positive or negative) of the spatial neighbourhood effect. The summation term in this expression is simply the weighted sum of the mean adjusted values at all other locations j — this may or may not be a reasonable assumption for a particular problem under consideration. Note also that SAR weighting schemes (W_s) can be converted to conditional or CAR schemes (W_c) through a matrix relation of the form:

$$W_c = W_s + W_s^T - W_s W_s^T$$

although the reverse is not generally possible.

In the standard CAR model spatial weights are often computed using some form of distance decay function. The range of this function may be unbounded or set to a value beyond which the weights are taken as 0. This range might be determined from some *a priori* knowledge relating to the problem at hand, or perhaps estimated from a semivariogram or correlogram (see further, Section 6.7.1.15). Also, as noted earlier, in the CAR model the covariance matrix is of the form:

$$var(y) = (I - \rho W)^{-1} M$$

and if the conditional variances of **y** are assumed constant this simplifies to:

$$var(y) = (I - \rho W)^{-1} I \sigma^2$$

Requirements on the specification of the weighting matrix, **W**, and conditional variance matrix, **M**, include: (i) **M** is an n by n diagonal matrix with $m_{ii} > 0$; (ii) to ensure symmetry of the variance-covariance matrix $w_{ij} m_{ji} = w_{ji} m_{ij}$; and (iii) $0 > \rho > \rho_{max}$ (typically) where ρ_{max} is

determined from the largest eigenvalues of $M^{-1/2} W M^{1/2}$.

A range of CAR models are supported by the GeoBUGS extension to the WinBUGS package. This software is specifically designed to support Bayesian rather than frequentist statistical modelling, and uses computationally intensive techniques (Markov Chain Monte Carlo or MCMC simulation with Gibbs sampling) to obtain the fitted parameter estimates and confidence intervals. Haining (2003) discusses the use of such Bayesian models, in which additional (prior) information (for example, national or regional crime survey data) is used to strengthen the modelling process and reduce bias in local estimates. The Bayesian approach treats the unknown parameters (e.g. the vector β) as a set of random variables, just like the data, to which may be associated prior distributions. The prior guesses for these parameters are then combined with the likelihood of the observed data to obtain posterior distributions for the parameters, from which inferential analysis proceeds. Essentially this provides a broader range of modelling approaches than pure (classical) frequentist analysis, and has been shown to result in substantial improvements over using simple rate data such as SMRs. See for example, Yasui *et al.* (2000) for a fuller discussion of this question.

In the so-called *proper* CAR model (WinBUGS function *car.proper*) the variance-covariance matrix is positive definite. The example values given in the WinBUGS manual for **M** and **W** based on expected counts, E_i, are of the form:

$m_{ii} = 1 / E_i$

$w_{ij} = (E_j / E_i)^{1/2}$ for neighbouring areas i, j or

$w_{ij} = 0$ otherwise

This particular example relates to the Sudden Infant Death Syndrome (SIDS) data described in Section 4.3.3 and in Cressie and Chan (1989) and more recently revisited by Berke (2004). Here the definition of *neighbouring area* was not based on adjacency but on distance between county seats ($d<30$ miles), a value determined from an examination of an experimental variogram (an estimate of a variogram based on sample data). The specific model applied in this case was actually of the form:

$$w_{ij} = \left\{ C(k)d_{ij}^{-k} \right\} \left(E_j / E_i \right)^{1/2}$$

where the term in curly brackets is a distance decay function, with k selected as 0, 1 or 2, and $C(k)$ is a constant of proportionality to ensure results are easily compared across different values of k. In this study the authors chose $k=1$ as this provided the best results when considered from a likelihood perspective, hence their weights were of the form:

$$w_{ij} = \left(E_j / E_i \right)^{1/2} 1/d_{ij}$$

Edge effects in this model are quite significant, since over a third of counties lie on the State boundary and clearly States do not represent closed systems for many (most) applications.

In this example the 'proper' (or autoGaussian) model fitted for this dataset was not applied to the full raw dataset, but to a Freeman-Tukey variance-stabilising square root transform of the data (see Table 1.4 and Freeman and Tukey, 1950) with Anson County omitted as an outlier. This county is the one picked up as an outlier in Figure 4-31, the Excess risk rate map for SIDS data.

Cressie and Chan had looked for non-spatial explanatory variables based on population density, percentage urban, number of hospital beds per 100,000 population, median family income and non-white live-birth rate. They then extended their analysis to include spatial patterns, but even after doing so could not adequately explain the observed variations in the data for this period, or for the subsequent 5 year period. It remains the case that the causes of SIDS are not fully understood, but medical research has shown that the

placement of children on their back when sleeping, the use of pacifiers (dummies) and avoidance of overheating, all help to reduce the risks involved substantially. It is reasonable to suggest that the spatial variations observed and their changes over time might have been, in part, a reflection of cultural and social factors (such as advice given to mothers by local medical staff). These factors were not explicitly picked up by the non-spatial explanatory variables. Although such factors may be related to race-specific customs, it is likely that the spatial variations observed and modelled *may* have reflected variations in these advisory and behavioural factors. Certainly it would have warranted a very close examination of such factors in counties with unusually high and low death rates in each time period.

An *intrinsic* version of the CAR model (IAR or ICAR) is also supported, in which the variance-covariance matrix is not positive definite, but is semi-definite (WinBUGS functions *car.normal* and the robust variant *car.l1*). The intrinsic version (applied initially in an image processing context) is based on pairwise differences between the observed values (similar to the computations used in variogram analysis, from which it originates — see Matheron (1973) and Künsch (1987) for a detailed mathematical treatment) and is now a more popular choice of CAR model for many researchers. Intrinsic models are a generalisation of the standard conditional autoregressive models to support certain types of non-stationarity. The example values given in the WinBUGS manual for **M** and **W** for the intrinsic CAR model, based on Besag *et al.*, (1991) and Besag and Kooperberg (1995) are of the form:

$$m_{ii} = 1/n_i$$

where n_i is the number of areas adjacent to i, and

$$w_{ij} = 1 \text{ for neighbouring areas } i,j \ i \neq j \text{ or}$$

$$w_{ij} = 0 \text{ otherwise}$$

The use of simple 1/0 weighting schemes for SAR or CAR models is not really appropriate for finite irregular lattices, and frequently a row-adjusted scheme of the form $\mathbf{W}^*=\{w_{ij}^*\}$ is used, where $w_{ij}^*=w_{ij}/w_{i.}$ (often written within this field as w_{ij}/w_{i+}). Hence the expected conditional means, for example, refer to an average rather than a summation. The symmetry requirement for CAR models cited earlier, i.e. $w_{ij}m_{ji}=w_{ji}m_{ij}$ implies that the conditional variances should be proportional to $1/w_{i+}$.

Although widely used, Wall (2004) has pointed out significant weaknesses in the spatial interpretation of such weighting schemes when applied in SAR and CAR models of this type. She recommends the use of geostatistical models as an alternative or additional approach, especially when attempting to understand the spatial structure of lattice (zonal) datasets.

Having fitted the chosen CAR or SAR model to the sample data, the residuals may be examined by mapping and/or by using the Moran I correlogram, $I(h)$, to identify any remaining patterns. If the residuals appear to show little or no spatial pattern it supports the view that the fitted model provides a good representation of the observed spatial patterns. However, as noted earlier, different models with fundamentally different interpretations may provide equally good fits to the data, hence drawing inferences from such models is difficult. Detailed examinations of the likely processes that apply for the particular dataset under consideration are vital for such analyses.

In the examples cited in this subsection, the response variable, $\mathbf{y}$, has been assumed to be continuous. As with GWR, autoregressive models have been developed to handle discrete and binary data, for example autoLogistic and autoPoisson models — see Haining (2003, Chapters 9 and 10) for more details. Haining (2003, p.367 et seq) provides examples of the use of WinBUGS for Bayesian autoregressive modelling of burglaries in Sheffield, UK, by ward (Binomial logistic model) and children excluded from school (Poisson model). He includes sample code and data for these examples, together with maps of the results and provisional interpretations.

5.6.5 Spatial filtering models

Conceptually the spatial lag models and spatial error models are substantially different, yet the results from their use on a given dataset may be similar. Furthermore, both are related to a process known as *spatial differencing* which is a form of *spatial filtering*. This is similar to simple differencing methods applied in time series analysis, but instead of analysing values at each time slot, y_t, one examines the differences between sequential time slots (first differences), y_t-y_{t-1}. In the spatial context the expression:

$$\mathbf{y} = \mathbf{X\beta} + \varepsilon$$

can be written with a first-order spatially autocorrelated difference as:

$$\mathbf{y} - \rho\mathbf{Wy} = \mathbf{X\beta} - \rho\mathbf{WX\beta} + \varepsilon, \text{ or}$$
$$\mathbf{y}(\mathbf{I} - \rho\mathbf{W}) = (\mathbf{I} - \rho\mathbf{W})\mathbf{X\beta} + \varepsilon, \text{ hence}$$
$$\mathbf{y} = \mathbf{X\beta} + (\mathbf{I} - \rho\mathbf{W})^{-1}\varepsilon$$

Thus a model with a spatially autoregressive error term is a form of standard regression equation (OLS) on spatially filtered data. The term

$$(\mathbf{I} - \rho\mathbf{W})^{-1}$$

is the spatial filter in this instance. Similarly the mixed spatial lag model, *mrsa*, described earlier:

$$\mathbf{y} = \rho\mathbf{Wy} + \mathbf{X\beta} + \varepsilon$$

can be written with a first-order spatially autocorrelated difference as:

$$\mathbf{y} - \rho\mathbf{Wy} = \mathbf{X\beta} + \varepsilon, \text{ or}$$
$$\mathbf{y}(\mathbf{I} - \rho\mathbf{W}) = \mathbf{X\beta} + \varepsilon$$

Hence, once again the model can be seen as a form of spatial filter where in this case only the dependent variable is filtered.

A somewhat different approach has been pioneered by Getis and Griffith. They apply spatial filtering to potential independent variables in a regression equation using one of two procedures: (i) using the G_i statistic described in Section 5.5.3 to transform the source data (Getis method); or (ii) decomposition of the Moran I statistic into orthogonal and uncorrelated map pattern components (Griffith). The approach of Getis is the most straightforward to describe and apply, since it involves a simple transformation of the original data values.

Let $G_i(d)$ be the local Getis and Ord spatial autocorrelation statistic applying to observation (point) i:

$$G_i(d) = \frac{\sum_j w_{ij}(d)x_j}{\sum_j x_j}, i \neq j$$

where d is the estimated range of observed spatial autocorrelation. This expression is the Observed value of the statistic for location i. The Expected value of the statistic for location i is of the form $W_i/(n-1)$. The filtering transformation of the observed data values x_i applied is then:

$$x_i^* = x_i \frac{E_i}{O_i}, \text{ i.e.}$$

$$x_i^* = x_i \left(\frac{W_i}{n-1}\right)\bigg/ G_i(d)$$

where W_i is the sum of the binary weights, w_{ij}, within a distance d of i, and n is the number of observations. The original dataset can then be viewed as being comprised of a spatial component, $x_{sp} = \{x_i - x_i^*\}$, and a non-spatial (filtered) component, $x_f = \{x_i^*\}$. These two components are then used in standard linear OLS as dependent variables and analysis proceeds as described earlier. Hence for each

filtered dependent variable there will be two additional terms in the regression equation.

Getis and Griffith (2002) applied unfiltered OLS, SAR, Getis filtered OLS and Griffith filtered OLS to sample US state expenditure data (observation taken as applying to state centroids). For OLS, SAR and the Getis filtering the coefficient of determination was found to be 75%, whilst with the Griffith method a slightly higher value of 80% was obtained. The authors conclude that both filtering methods provide a more appropriate specification of the regression model, facilitating meaningful interpretation of all of the regression coefficients, with the advantage of not requiring formal specification of an appropriate SAR model. Overall the Griffith approach appears somewhat more powerful, but the Getis approach is simpler to apply, although it does require that the filtered variables are positive and have a natural origin (i.e. are positive ratio-scale variables).

6 Surface and Field Analysis

6.1 Modelling Surfaces

6.1.1 Test datasets

The analysis of physical surfaces and fields is a topic that is best explored using tangible examples. To this end, this part of the Guide uses a number of common datasets to illustrate the various forms of analysis that may be readily carried out. Our first dataset comprises eight sub-tiles from the GB Ordnance Survey 10m digital elevation model (DEM), taken from tiles TQ81, TQ82, TQ91 and TQ92. Detailed analysis is conducted on one of the tiles, TQ81NE. Each DEM tile contains approximately 250,000 cells, and the mosaic-ed (combined) 8-tile region is thus roughly 2 million cells. A 3D visualisation of tile TQ81NE is shown in Figure 6-1. This was created in Surfer. Comments on artefacts associated with this dataset appear later in this chapter. The data files referred to may be obtained from the Ordnance Survey directly or via the EDINA data download service (for academic use only – see the "Web links" Section at the end of this Guide for details).

Each main tile (e.g. TQ81) is made up of 4 sub-tiles (in this example TQ81NE, NW, SE and SW). These sub-tiles are 5000mx5000m, and the TQ81NE tile has a modest maximum height of around 70m. The tiles consist of a number of ridges, separated by relatively broad very flat valleys, and border on the English Channel.

Figure 6-1 East Sussex test surface, OS TQ81NE

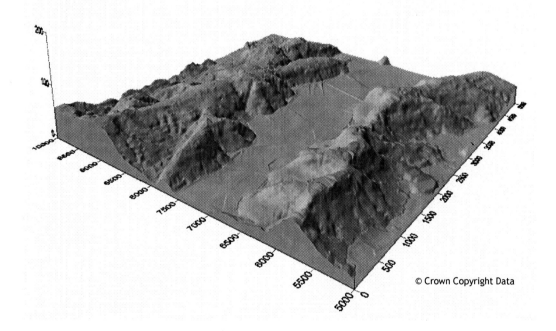

© Crown Copyright Data

A separate dataset representing a much more rugged landscape is also used, both in the present chapter (notably in the section covering interpolation methods) and elsewhere in this Guide. This covers part of the Pentland Hills area to the south of Edinburgh in Scotland.

Figure 6-2 Pentland Hills test surface, OS NT04

A. Spot heights in NT04 tile

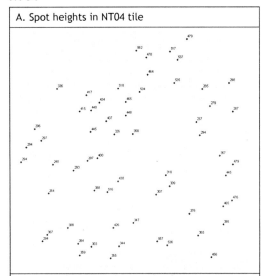

B. Relief map: 25m DEM for NT04 tile

© Crown Copyright Data

The principal OS tile used in this case is NT04. It has already been used in our discussion of Delaunay triangulations and Voronoi polygons (Section 4.2.14.3).

6.1.2 Surfaces and fields

The test datasets described in Section 6.1.1 are representations of physical surfaces. For every location in the set $\{x,y\}$ there is a single scalar value z representing the height of the surface at that point. There is no reason however to restrict the set of z-values to be heights — they could equally well be some other variable, such as the cost of land or the level of a soil trace element. The range of values that $\{z\}$ may take would typically be the set of (positive) real numbers, although only integer values may be recorded depending on the dataset in question.

In some instances a number of z-values are associated with each (x,y) location. This is the case, for example, with multi-spectral remote sensing data, where separate spectral bands are coded for each pixel of an image (as described in Section 4.3.3). In this case each band can be considered as a distinct surface, and may well be stored as separate layers within an image folder. Frequently such data are integer-coded colour values, so have a well-defined and limited range, e.g. [0,255]. Another example is the case of spatial datasets (surfaces) that are recorded over a number of time periods — for example atmospheric pressure or temperature at a specified altitude, recorded at hourly intervals.

In all of these examples the recorded information is a single interval or ratio-scaled value. Spatial datasets of this type are considered as surfaces or (scalar) *fields* within GIS. Datasets in which multiple values for the same variable are recorded for a single location and time (e.g. geological borehole data) and/or have a directional component (e.g. wind speed and direction), are typically excluded from the set of objects described as surfaces or fields, or are re-cast where possible to fit such a model (e.g. as multiple layers). The term *vector field* is used to refer to fields that include both a magnitude and a directional value at every point.

Having defined the kind of datasets that qualify, the question of how such information is obtained and stored becomes central to the process of analysing and interpreting such data. A very large proportion of data that describes the physical world (terrain, vegetation, land use etc.) is obtained from one of two sources (possibly both): national mapping agencies and sample surveys.

National agencies provide details of topography; land use; geology etc.; derived from a mixture of terrestrial, aerial and satellite surveys. Traditionally such data were analysed and then recorded on paper maps, although this has now been largely superseded by digital encoding and storage, with paper maps as one of a range of possible output forms. For terrain data, output consisted largely of contour maps with spot heights, and the great majority of current digital terrain datasets, including those available in grid or raster format, have been derived from such contour maps. One result of this process has been to make available grid datasets or *digital elevation models* (DEMs) for many areas of the world. However, since these have been programmatically generated there is a tendency for artefacts to appear, such as small ridges, troughs and hollows, and these can be seen even in the most carefully produced national datasets, including the GB Ordnance Survey and United States Geological Survey (USGS) datasets. For example, the test dataset for the OS tile TQ91SW includes an area of sea, which the file records as having a range of values from 0 to -3m! The same issue applies to conversion utilities built-in to many GIS packages — converting input contour datasets to grid format, or grid data to contours or TIN format.

Another principal source of field-like data is from sample surveys. In this case one or more sample values are known for a set of distinct (point-like) locations. These might be soil samples, radioactivity data, meteorological station data or some other variable that is known to be spatially continuous. Values at unsampled locations, typically defined as a fine grid, are estimated from the measured data using a variety of interpolation and prediction methods. Hence again grid data may be subject to artefacts and additional estimated or unknown errors of modelling.

In each of these examples the surface or grid can be thought of as a single valued function $z=f(x,y)$. Mathematically derived functions that apply to an entire region or to sub-sections of regions may be used to create grid datasets. For example the surface shown in Figure 6-3 is a representation obtained by fitting a linear regression surface of the form $z=ax+by+c$ to the set of spot heights provided by the GB Ordnance Survey for tile NT04.

Figure 6-3 Linear regression surface fit to NT04 spot heights

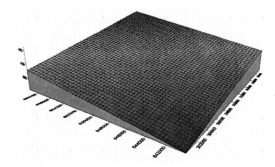

More complex mathematical models are frequently used to generate and analyse surface datasets. Many of those available in modern GIS packages and related software (notably geostatistical packages) are described in the subsections that follow.

6.1.3 Raster models

A raster, grid or image file may be thought of as a rectangular array of identical square cells, each of which has an associated value (which may be zero or a code for "missing value"). The array therefore comprises a number of rows and columns. In some instances (as in mathematics) rows run from 1 to n from the top and from 1 to m from the left, whilst in others (including standard geographic referencing) the rows may be in the reverse order, i.e. the first row is actually the bottom of the array, or most southerly position, with subsequent rows representing successive steps to the north (northings). Columns in this model correspond to east-west location, and may be referred to as eastings. In Cartesian coordinates eastings correspond to x-values and northings to y-values. The contents of cells are then considered as field values or surface heights, and conventionally referred to by the letter z. A raster grid is thus an array of n rows and m columns $\{x,y,z\}$. Where multiple bands or values exist for each (x,y) location a series of triples $\{x,y,z_1\}$, $\{x,y,z_2\}$, $\{x,y,z_3\}$ ... are recorded, and typically treated as a set of separate (associated) surfaces. Collections of grid datasets that are adjacent are often called tiles, and separate tiles may be processed independently or may first require combining into a single larger grid before analysis takes place. This process is known as creating a *mosaic* in some GIS packages, or a combination operation in others. When such datasets are combined there may be user-selectable rules on how any small overlaps of edges should be treated — typically these default to a First model, i.e. the values in the first grid to be processed take precedence, but other options might include averaging, taking the largest or smallest values etc. The advantage of combining grids as a mosaic is that analysis is conducted on the entire dataset (region) of interest, taking into account all of the recorded values, and minimising edge effects; the main disadvantage is that such datasets can become extremely large and thus very cumbersome and slow to process.

We can draw up the neighbourhood of a single cell of a grid structure in (at least) two separate ways, depending on the referencing system applied (Figure 6-4).

Figure 6-4 Raster file neighbourhoods

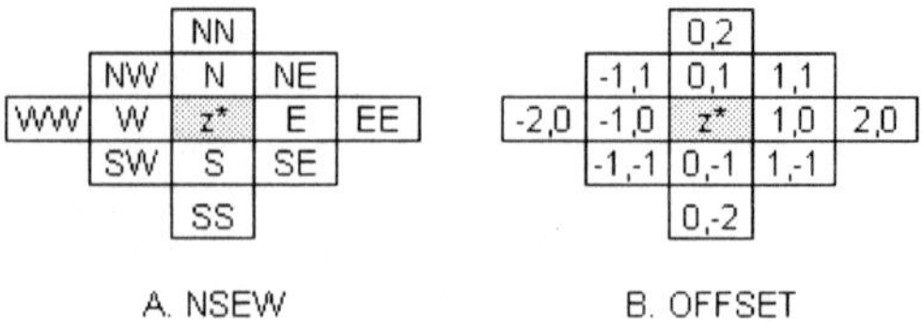

A. NSEW   B. OFFSET

Here we have illustrated referencing for cells using compass-type notation and using numerical offset notation. Both may be extended to cells further away from z^*, but the numerical system is simpler as it remains in the same format (2 values) as more distant cells are included.

Using either of the notational forms we can define operations on raster grids. For example, the rate of change in the x-direction and y-direction about z^* (the first-order partial derivatives) can be estimated as:

$$\frac{\partial z}{\partial x} \approx \frac{z_E - z_W}{2\Delta x}, \frac{\partial z}{\partial y} \approx \frac{z_N - z_S}{2\Delta y} \text{ or}$$

$$\frac{\partial z}{\partial x} \approx \frac{z_{1,0} - z_{-1,0}}{2\Delta x}, \frac{\partial z}{\partial y} \approx \frac{z_{0,1} - z_{0,-1}}{2\Delta y}$$

where the Δx and Δy values are determined by the grid resolution. These are the formulas used by the Surfer and Idrisi packages, amongst others. Basic second order differentials, required for curvature calculations, follow a similar format:

$$\frac{\partial^2 z}{\partial x^2} \approx \frac{z_E - 2z^* + z_W}{\Delta x^2},$$

$$\frac{\partial^2 z}{\partial y^2} \approx \frac{z_N - 2z^* + z_S}{\Delta y^2},$$

$$\frac{\partial^2 z}{\partial x \partial y} \approx \frac{z_{NE} - z_{NW} - z_{SE} + z_{SW}}{4\Delta x \Delta y}$$

An alternative approximation, first introduced by Horn (1981) uses all 8 immediate neighbours of z^*. ArcGIS is an example of a GIS package that uses this method for some of its calculations: for the x-direction the directly east and west positions are weighted 2 whilst the remaining 4 directions (NE, NW, SE, SW) are weighted 1; for the y-direction the directly north and south points are weighted 2 and the remainder weighted 1. The result is a slightly more complicated pair of formulas for the first-order differentials that have been found to be better than many alternatives, especially for rough surfaces. Using these approximations both slope and aspect can be estimated for each point of a grid (see further, Section 6.2.1). The formulas are based on standard finite difference expressions:

$$\frac{\partial z}{\partial x} \approx \frac{(z_{1,1} + 2z_{1,0} + z_{1,-1}) - (z_{-1,1} + 2z_{-1,0} + z_{-1,-1})}{8\Delta x}$$
$$\frac{\partial z}{\partial y} \approx \frac{(z_{1,1} + 2z_{0,1} + z_{-1,1}) - (z_{1,-1} + 2z_{0,-1} + z_{-1,-1})}{8\Delta y}$$

Surface curvature measures (including plan, profile and tangential curvature) require second order derivatives, and again these may use relatively simple formulas or more complex variants. ArcGIS uses a method of estimation based on fitting a partial fourth-order polynomial to the 3x3 window in which z^* sits, whilst Surfer applies simpler differential approximations. Both approaches are described in more detail in Section 6.2. Landserf uses a different method from those described above for all of its main surface analysis facilities. The method is based on a quadratic surface approximation at each grid cell, using the immediate 8-cell neighbourhood (i.e. a 3x3 window) as standard. It also provides the option to use a larger window size and a distance decay function for weighting cells that are more distant (such functions are described in Section 4.4.6).The quadratic fitted surface is of the form:

$$z=ax^2+by^2+cxy+dx+ey+f$$

where a, b, c, d, e and f are constants to be estimated. The fitted quadratic can be analytically differentiated, the components equated to 0, and the slope, aspect and curvature components derived directly from

this process. The result is a set of very simple expressions in terms of the constants for the main surface characteristics: aspect is simply computed as:

$$A=\tan^{-1}(e/d),$$

and slope as:

$$S_t=\tan^{-1}(e^2+d^2)$$

It should be noted that the question of determining the coefficients remains, and since there are only 6 of these and potentially 9 cell values to utilise, a level of redundancy exists permitting a range of choices or procedures for coefficient selection, with least squares fitting being one such choice.

An alternative, devised by Zevenbergen and Thorne (1987), is to fit a 9 parameter function that is of quartic (4th) order but only quadratic in x and y:

$$z=ax^2y^2+bx^2y+cxy^2+dx^2+ey^2+fxy+gx+hy+i$$

This function has the advantage that all its parameters are uniquely determined from the values of the cells in the 3x3 window, and the function passes exactly through each data point, including the central point. The coefficients a, b, c and i are not used in computations of slope, aspect or curvature, so may be ignored for these measures. The coefficient $i=z^*$, the value at the central point of the 3x3 window ($z_{0,0}$). The coefficients g and h turn out to be identical to the simple expressions used by Surfer and others as estimates of first order derivatives, and these determine both slope and aspect computations so the method reduces to simple finite differences for these measures. The expressions for plan and profile curvature are more complex, and use all of the coefficients d-h. This quartic model is the one used by ArcGIS as described above, and in the RiverTools facility Extract|Finite difference grid. The terms d-h are also determined in exactly the same manner as for the second-order derivatives used by Surfer, as defined above. The derivation and use of these expressions is discussed in more detail in Moore *et al.* (1991, 1993), Wood (1996) and Zevenbergen and Thorne (1987). Useful analyses of alternative algorithms and the associated issues of errors and uncertainty analysis are provided by Jones (1998), Zhang and Goodchild (2002, Chapter 5) and Zhou and Liu (2004).

Surfaces represented by grids suffer a number of disadvantages. These include:

- very large requirements for data storage, processing and display — the test grid TQ81NE that will be used for illustrative purposes here is 1.4Mbytes as an NTF file, 1.7Mbytes as an ASCII or Surfer grid, and less than 150kbytes as an ArcGIS TIN file and only 33kbytes as a 500-vertex Landserf TIN file (albeit with some degree of information loss). Image files are often far larger — a single USGS High Resolution Orthoimage covering approximately 4.5kmx4.5km in colour at 0.3m resolution requires 660Mbytes of memory to load (40Mbytes when compressed with no information loss)

- a fixed grid size and orientation that does not reflect variations in underlying surface complexity

- lack of clarity regarding linear and point-like features

On the other hand grids are extremely convenient for data manipulation, combination and display, which explains their widespread popularity.

6.1.4 Vector models

The most common vector-based models of surfaces are contours and TINs. Source data may be provided in either of these formats, or may be converted to these formats from DEM files. Typically they are far more compact than DEMs, but less suitable for analytical methods, and TINs may result in a significant loss of information unless the underlying surface is of triangular form. TINs are widely used in engineering modelling (they are the principal form of surface model in Autodesk's Land Desktop package for example). Since DEMs may well have been generated from contour datasets, reversing the process may not result in the same pattern of contours, depending on the algorithms used in each case. Contour creation is described in Section 6.5.3.

Figure 6-5A-C shows the test area source DEM together with derived contour and TIN representations, created using the Landserf package. The TIN example is based on a program selected set of 500 vertices, resulting in approximately 1000 triangular regions. The representation is not unique, and a similar computation using another package (e.g. TNTMips, which uses a similar method that it calls adaptive densification) may result in a different pattern of triangles. Note that representation of flat regions with the TIN model can be problematic. The generated models may be used to re-generate an approximation to the source DEM as shown in Figure 6-5D — the differences between the TIN-generated DEM and the original DEM can be plotted as a grid, as shown in Figure 6-5E. The darkest reds (greys) are the areas of maximum "error" — in this case up to 10m. The error map can be improved (the quality of the TIN representation enhanced) by increasing the number of vertices. Note that this is not the same as the difference between the TIN and the true surface, which in principle can be computed everywhere.

Figure 6-5 Vector models of TQ81NE

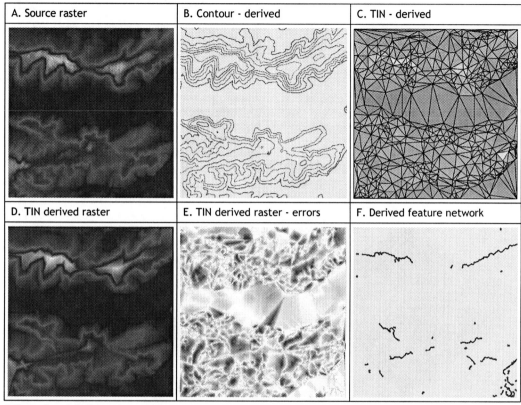

| A. Source raster | B. Contour - derived | C. TIN - derived |
| D. TIN derived raster | E. TIN derived raster - errors | F. Derived feature network |

© Crown Copyright Data

TIN creation from DEMs benefits from a level of feature extraction in order to produce the most efficient description of the surface (i.e. with minimum surface error and inclusion of key surface features, such as peaks, pits and ridge lines). However, such facilities are rarely provided in TIN-generation algorithms as noted earlier (Section 4.2.14.2). Pure feature-based vector models, which are essentially networks of polylines or curves linking peaks, pits and passes or saddles along ridges or channels are sometimes computable as output from GIS analysis tools, but are not provided as a standard form of vector representation (see Rana and Wood (2000) and Rana (2004a) for further discussion of this topic). Figure 6-5F shows feature network extraction from the source DEM. The level of network connectivity, and hence surface representation, obtained by

such methods will vary depending on the values of user-definable parameters that are set, and on the nature of the surface itself. The factors are in addition to the computational problems we have alluded to.

6.1.5 Mathematical models

We have previously noted that a surface or grid can be thought of as a single valued function $z=f(x,y)$. Mathematical models of surfaces typically assume that the function $f(x,y)$ has certain distinctive properties: (i) single valued — i.e. there is only one z value for each (x,y) pair; (ii) continuous — in broad terms this means that a value exists for every (x,y) pair; and (iii) twice differentiable — first and second differentials can be computed at (almost) all locations. The mathematical surface model is then a single function or a piecewise function that satisfies these requirements. Piecewise functions are continuous functions over some interval or region (e.g. a 3x3 grid of cells) but which have breaklines at the edges of the region. Piecewise continuous functions are widely used in surface modelling and interpolation, where particular conditions are set on behaviour of the function at the edges (for example that the gradient matches that of the adjacent regions).

Single function mathematical models are typically used to approximate real-world surfaces, often using least squares methods to provide best fit parameter estimation. As such they are described as approximating or non-exact models. Piecewise models are typically exact, but inexact or smoothing variants of such models are also provided within some software packages. In most cases models used are algebraic functions with powers limited to linear, quadratic or cubic expressions in x and y. Example surfaces of orders 1-3 are shown in Figure 6-6 — each of these examples was generated as a grid using the Surfer Grid menu, Functions option.

Figure 6-6 First, second and third order mathematical surfaces

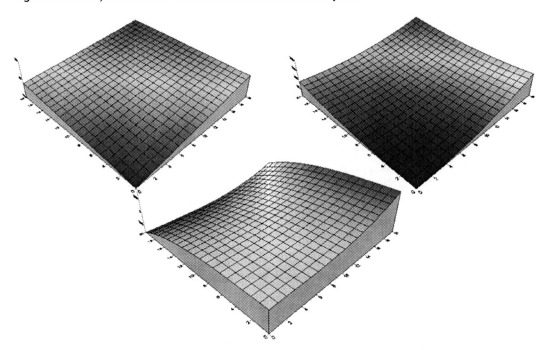

Some software tools talk about quadric or quartic functions (i.e. in 4th powers), but these are typically limited to terms of power 2 in x and y, i.e. expressions like ax^2y^2. Quintic functions (i.e. in 5th powers) are also used occasionally, for example TNTMips provides the option for quintic polynomial fit to individual triangles in a TIN network (deriving the parameters from the triangle itself and its immediate neighbours). This is then used in interpolation within the selected triangles. Quintic functions, with all 21 parameters, are the minimum degree of polynomials needed to ensure zero, first and second-order continuity across triangle edges.

In principle, mathematical models, which are scale-independent compact representations of surfaces, could be used in place of alternative vector and grid models. At present this idea, which has been described as *Map Calculus* by Haklay (2004, 2006), has not been implemented in commercial GIS packages. Haklay provides a simple demonstration of this approach using Manifold, and it is possible that such models will be incorporated into GIS packages in the future.

6.1.6 Statistical and fractal models

Several GIS packages facilitate the creation of grids or surfaces using statistical functions rather than just mathematical functions. The generated surfaces may then be combined with existing surfaces or mathematically defined surfaces to create new grids which may be used for a variety of purposes: uncertainty modelling; providing reference grids for comparison or residuals analysis purposes; for the generation of idealised terrains; or for statistical analyses. ArcGIS provides a function for creating raster files whose entries are uniform random values in the range [0,1] or Normally distributed random z-values with mean=0 and standard deviation=1, i.e. $z \sim N(0,1)$. The resulting grids can be modified (e.g. multiplied by a constant, adding an offset or trend) when generated or subsequent to generation. GRASS (r.random.surface function) provides similar functionality, but also facilitates a user-specified level of spatial dependence, implemented using a simple distance decay function with filtering. Figure 6-7A and B illustrates the ArcGIS functionality, where a 250x250 grid with cell size 1 unit has been generated. Each cell is given a value independently as a random value from the chosen distribution. In the random uniform case all values will lie in the range [0,1] whilst in the Normal case values may take any value, but will typically fall within the bounds [-5,+5], with just over 95% of values (the paler colours) falling in the range [-2,2].

Figure 6-7 Random and fractal grids

A. Random Uniform [0,1]

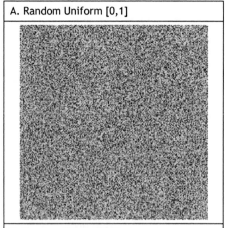

B. Random Normal(0,1)

C. Ridged multi-fractal

Figure 6-7C shows a markedly different pattern, which has been generated by a hybrid random process designed to yield a surface with known fractal dimension, $D_C>2$, and surface characteristics that are more terrain-like than purely random. GIS software packages do not appear to provide models for artificial terrain generation, although this is an area that may be developed in future. This particular example was created using the Fractal Terrains package, which provides a range of fractal generation methods, from simple Brownian motion to various ridged multi-fractal procedures. In principle such methods could be implemented within many of the popular grid-oriented GIS packages using scripting or object-based programming, but there appear to be few examples of this at present. For more details on this area see Hanrahan *et al.* (2003, notably Chapter 16).

A slightly different approach, based on random perturbation of mathematical surfaces, is illustrated in Figure 6-8. In this instance a simple linear surface has been modified by random perturbation of each cell. In the upper example a 5x5 grid and a random uniform function has been used; in the lower case a 20x20 grid and a random Normal function has been applied. Both examples were generated using the Grid|Functions facility in Surfer, followed by using the Map facility for the 3D surface visualisation and wireframe overlay. The wireframe lines in this case are simply straight line connections between the grid values taken in *x*- and *y*-directions separately.

Figure 6-8 Pseudo-random surfaces

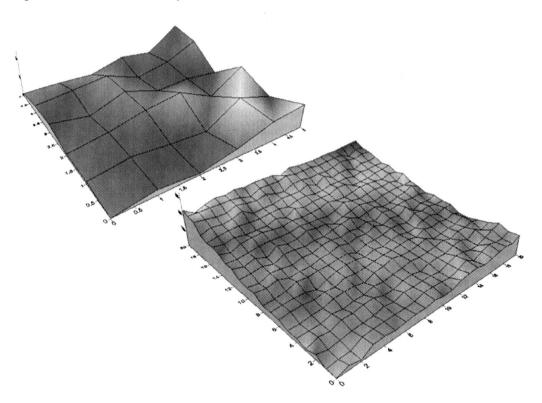

6.2 Surface Geometry

6.2.1 Gradient, slope and aspect

The term gradient refers to a vector quantity, i.e. an object that has both magnitude and direction. The magnitude or size of the gradient is the slope, whilst the direction in which the maximum value of this magnitude occurs is known as the aspect. Slope is one of the most widely used of surface attributes so understanding how it is generated and what alternatives exist is extremely important.

6.2.1.1 Slope

The common notion of the slope of a surface or terrain, t, is the amount of rise (e.g. change in elevation) over run in some direction, usually the direction for which the rise over run is greatest, i.e. the steepest path up or down the surface. In practice we would specify some distance over which the slope is to be calculated, such as the amount of rise over a 10 metre distance *in the plane*. If this was 2 metres, for example, the rise over run would be $S=2{:}10$ or 1:5 or 0.2 or 20%. This ratio is actually the tangent function: $S=\tan(S_t)$ where S_t is the terrain slope in radians. If the run were taken as the surface distance rather than the plane distance then we would have $S=\sin(S_t)$. For gentle slopes (less than 1:4) this ambiguity in the definition of slope is not serious, but in general the tan() version is preferred.

The normal use of the term slope within GIS packages is the tan() value S_t, usually reported in degrees from the horizontal (11.3° in this example). To convert between the two one would use an expression such as:

$S_t=\text{degrees}(\tan^{-1}(S))= 360\tan^{-1}(S)/2\pi$

For an analytical surface, $F(x,y)$, slope is defined as the magnitude of the first derivative of the surface function:

$$S = \sqrt{\left(\frac{\partial F}{\partial x}\right)^2 + \left(\frac{\partial F}{\partial y}\right)^2}$$

This formula is not based on a rise-over-run calculation over a fixed interval, but rather assumes that a plane surface can be placed at any point on the surface $F(x,y)$ in such a way that it only just touches the surface — it is a tangential plane and relies on the notion of infinitesimally small distances. It is closely related to the formula for Euclidean distance, and simply shows how much the surface F rises with a small fixed increment in x and y. However, surfaces within GIS are rarely if ever represented by analytic functions — typically they will be modelled as TINs or grids, with a finite resolution. Hence slope calculations will use approximations to the formula above depending on the surface model used, which is itself an approximate representation of the true surface. In almost all instances software packages apply grid-based computation, with TINs being first converted to grid format. Strictly speaking such conversion is not necessary, since TINs can provide interpolated surface values at every point, and use of a suitable (generally non-linear) function could then provide slope values directly.

The slope of a line crossing the surface (e.g. a road or pipeline) is dependent on the direction of the line relative to the underlying surface — it is always less than or equal to the slope of the surface. If the direction in question is an angle α we need to adjust the gradient vector by taking a dot product with a unit vector in the direction of interest. This enables slope values to be computed in any path direction. Such values are known as *directional derivatives* (see further, subsection 6.2.3). If the differential components for gradient and curvature have been calculated for a surface (e.g. a grid) then directional derivatives can be computed with very little further computation.

Using the grid model described in Section 6.1.3 we have surface values represented by triples $\{x,y,z\}$, and can compare this with the mathematical formulation above to see that $z(x,y) \equiv F(x,y)$. We have written this as an equivalence, but it should not be assumed that this is strictly true — the set of values $z(x,y)$ is a particular representation of a non-mathematical surface or field at a pre-defined set of locations.

With a grid model slope can be determined at any particular location z^* by simple approximation methods using the rook's move or NSEW values for z at that point:

$$S = \sqrt{\left(\frac{z_E - z_W}{2\Delta x}\right)^2 + \left(\frac{z_N - z_S}{2\Delta y}\right)^2}$$

and S_t computed as before. This is the approach used by Surfer, as described in Section 6.1.3. Within ArcGIS the 8-cell method due to Horn (1981), described in Section 6.1.3 is used.

Methods based on fitting a plane surface to the unweighted 8-cell neighbourhood of z^* and using its slope as the surface slope estimate have been found to be less satisfactory. Even less acceptable are slopes computed by examining the maximum difference in cell values in the 8-cell neighbourhood in the x- and y-directions and across the diagonals (suitably adjusted for cell size and diagonal length where appropriate).

Vertical Mapper for MapInfo uses yet another approach. It computes the slope and aspect of the 8 triangles defined by the mid-points of the 3x3 window centred on z^* (Figure 6-9). It then averages these to provide its estimates for slope and aspect at the point z^*. The result is a measure that is less susceptible to directional bias, providing a full range of directions (infinitely many are possible). The 8 individual triangles of this local neighbourhood each have distinct slope values and aspect (directions). These individual directions can be examined to see which is the steepest, and this slope value used in determining drainage flows across the surface. This approach, with the added element of assigning flows to two

adjacent "downslope" cells, provides the basis for the method described by Tarboton (1997) as the D-infinity drainage method (see further, Section 6.4). ERDAS Imagine uses yet another method, based on the average of the 3 east-west and 3 north-south differences for the estimates of dz/dx and dz/dy and then utilises these as above for slope and aspect calculation.

Figure 6-9 D-infinity slope computation

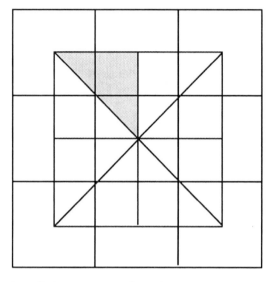

It will be apparent that these computations are extremely sensitive to scale. Larger grid intervals will result in smoothing of slope values, which may be desirable in some situations — in others it will be problematic. The availability of suitable datasets and the purposes to which the slope computations are to be put should be the basis of selection of method and resolution. Some of these issues are illustrated in Figure 6-10. Here the jagged black line represents the surface we are analysing, which has some periodicity in its form. A grid resolution represented by the vertical green (pale grey) lines would always result in a slope estimate that was fairly constant and steep, whilst halving the interval (doubling the resolution) would give a rapidly alternating picture of the slopes.

Figure 6-10 Gradient and sampling resolution

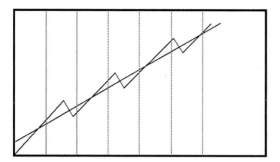

However, it may be that for the purposes of analysis a broader picture is required, as illustrated by the straight line, which would be generated if the underlying grid were smoothed (e.g. by a local averaging process) or if a much larger grid spacing were used.

Figure 6-11B shows the result of computing slope values in degrees for the source map displayed in Figure 6-11A. The red and yellow (lightest grey) areas are those of highest values (up to 65 degrees). The source raster has a resolution of 10 metres, and as the resolution is reduced/cell sizes for analysis are made larger, the maximum slope computed from this particular dataset falls to roughly 50 degrees with a 20m resolution and under 30 degrees with a 100m resolution. Slope computation is also affected to a greater or lesser degree by grid orientation, since it relies on grid values in particular directions.

6.2.1.2 Aspect

Aspect is defined as the directional component of the gradient vector and is the direction of maximum gradient of the surface at a given point. As with slope, aspect is calculated from estimates of the partial derivatives:

$$\frac{\partial z}{\partial x}, \text{ and } \frac{\partial z}{\partial y}$$

Figure 6-11 Slope computation output

A. Source terrain map

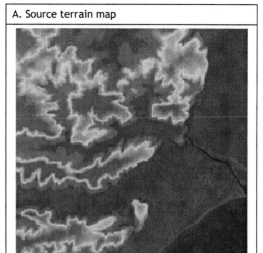

B. Slope map

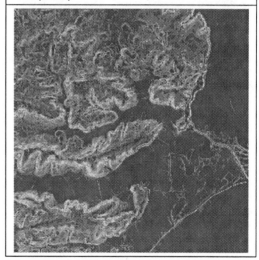

© Crown Copyright Data

Aspect is computed in degrees from due north, i.e. as an azimuth in degrees not radians. The expression required is:

$$A = 270 - \frac{360}{2\pi} \tan^{-1}\left(\frac{\partial z}{\partial x}, \frac{\partial z}{\partial y}\right)$$

where the inverse tan function applies to both components — this function is often described, and implemented in software, using the notation atan2(a,b). Since the function is equivalent to atan(b/a) for $a>0$ the standard inverse tan function may be used if preferred for $a>0$.

As with the calculation of slope, Surfer uses simple finite differences for aspect calculation whereas ArcGIS uses the more complex 8-point formulas of Horn. This does not avoid problems of directional bias resulting from the computational procedure, grid structure and/or errors in or rounding of grid values. Figure 6-12A shows the frequency distribution produced by the ArcGIS Aspect facility when applied to the terrain shown in Figure 6-11A. Strong peaks in this distribution can be observed at approximately 45° intervals, which of course does not reflect the true frequency distribution of landscape aspect. Landserf analysis of aspect for a subset of the region covered by Figure 6-11A is shown in Figure 6-12B (with zero values excluded). There are still discernable peaks, especially in the 8 major compass-point directions, but the computations appears to be smoother overall. Computations of aspect suffer from edge effects, resulting in zero or undefined values for boundary cells. Zero values are also generated wherever there are flat areas, and these can dominate frequency diagrams and associated analyses. Different packages treat flat areas in different ways, and these should be checked before relying too heavily on the results generated. Exclusion of zeros may be desirable in any further analysis of such data, and checks on the treatment of 360° should also be made, although these do not seem to be a particular problem with most packages.

Figure 6-12 Frequency distribution of aspect values

A. ArcGIS Spatial Analyst: Aspect

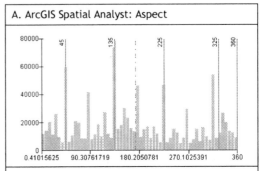

B. Landserf: Analyse|Surface parameter|Aspect

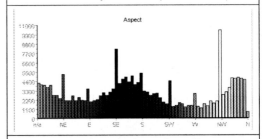

C. Landserf Aspect data — Rose diagram (MATLab), Log freq.

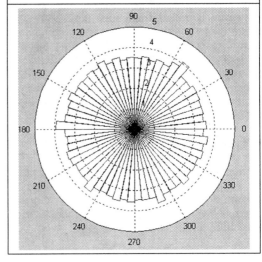

In Figure 6-12C we have generated a circular frequency histogram of the set of aspect values again with zeros excluded. This is a more appropriate way of examining the data but is not available in most GIS packages. In this case the diagram was generated using the rose() function in MATLab with 60 intervals. Note that 0° and 360° in Figure 6-12C are in the usual position for mathematical analysis, i.e. horizontal, and the $\log_{10}$ of frequencies rather than frequencies themselves have been plotted in order to reduce the effect of isolated peaks at certain frequencies.

Figure 6-13 Aspect computation output

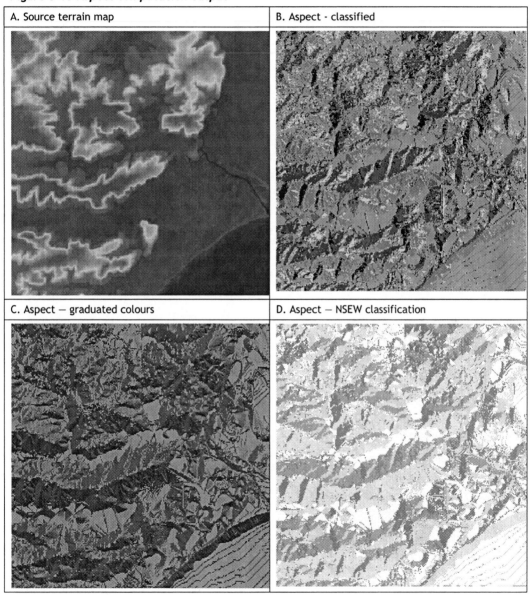

A. Source terrain map	B. Aspect - classified
C. Aspect — graduated colours	D. Aspect — NSEW classification

© Crown Copyright Data

Clearly there are some problems with this computational process, since areas of sea to the lower right of the output (Figure 6-13A-C) are shown as having variable aspect – these are linear artefacts or ghost lines, which arise due to artefacts in the underlying source DEM. If a graduated colour scale is used to represent this data, as in Figure 6-13C, the pattern appears clearer, with high values shown in purple (darker grey) in this example, but the circular nature of the data means that this representation is misleading. A more appropriate visualisation is one in which categories and colours reflect the circular nature of the data, as in Figure 6-13D. Here the range 325°-45° (broadly north-facing) has been coloured pale orange (lightest grey), east-facing yellow, west-facing blue (darkest grey) and southerly pale green.

6.2.2 Profiles and curvature

6.2.2.1 Profiles and cross-sections

Many GIS packages provide tools for examining the profile of a surface along a selected straight line or series of straight line segments (a polyline). The form of the profile reflects: the number of observations taken along the selected line (with only two sampled points, at the start and end, the profile will be a straight line showing the slope between the two points, and as the number of observations is increased the surface will be represented in greater detail); the direction(s) of the selected line segments; and the nature of the surface itself, as represented in the surface model being utilised. Figure 6-14 shows a profile from south to north of the test surface TQ81NE along its westerly edge (the edge shaded grey in Figure 6-1), in this case produced using Landserf – similar functionality is provided in many packages (e.g. TAS, ArcGIS).

The vertical extent is clearly exaggerated, but the profile shown does provide an indication of the slope and curvature of the surface *along the selected line*. The important point to note here is that the selected line does not represent the direction of maximum gradient, so values for slope and curvature apply solely to the direction chosen and will differ from calculations of slope and curvature for the surface as a whole.

Figure 6-14 Profile of NS transect, TQ81NE

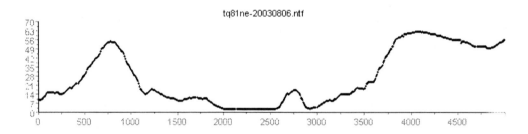

In addition to single profile computation multiple-slice profiling can be performed using some packages, such as TNTMips. Figure 6-15 illustrates this procedure using the TQ81NE dataset once more, with a selection of slices taken at 50 grid cell intervals (500m intervals as this is a 10m DEM). The profile averages for the *entire* input raster object are shown as the baselines against which the curve-filling is generated. In this example we have rotated the output (profiles plus relief map backdrop) through 90 degrees for clarity of viewing.

Figure 6-15 Multiple profile computation

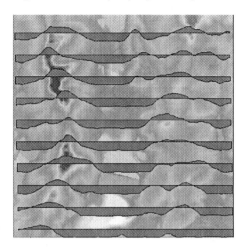

6.2.2.2 Curvature and morphometric analysis

As with slope, curvature values depend upon the line or plane along which such calculations are made. There are several alternative measures of surface curvature. The three most frequently provided within GIS software are profile curvature, plan curvature and tangential curvature. Additional terms and measures include longitudinal curvature, cross-sectional curvature, maximum and minimum curvature, and mean curvature.

In order to clarify these various terms we have drawn some illustrative details on part of test tile TQ81NE (Figure 6-16).

Figure 6-16 Surface morphology

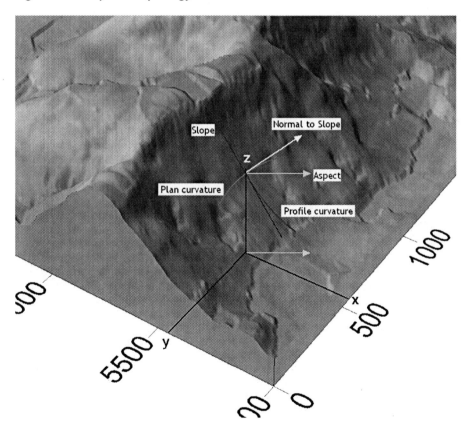

At the sample location $(x,y)=(600,5400)$ a straight line has been drawn to indicate the slope and arrows indicates the direction in which the slope has been computed (the aspect). At this point a vertical plane (i.e. a plane drawn in the Z-direction) could be constructed that is oriented in the direction of the aspect and passes through (x,y,z). It is the plane in which gravitational processes are maximised, so is of great interest in problems of hydrology, soil movement and geomorphological analyses. A plane constructed at right angles (orthogonal) to this vertical plane, i.e. a plane passing through the target point and slicing the surface horizontally (parallel to the XY-plane) maps out a line on the surface that is effectively a contour at that point. This second plane is the one for which gravitational effects are minimised. *Profile and plan convexity* (or curvature) is defined in terms of these two planes (see further, subsections 6.2.2.3 and 6.2.2.4). A slice through the surface could also be drawn in the YZ-plane, also passing through (x,y,z). This would again generate a line where it intersected with the surface, and the somewhat confusingly named *tangential curvature* is defined in terms of this arrangement.

To provide an additional level of complexity and confusion to the above terms, there are a further set of curvature values that correspond to those used in the mathematical analysis of surfaces (see for example, do Carmo (1976) and/or the Mathworld web site for more details). The slope and aspect of a surface at any given point are computed by imagining that a plane can be drawn that just touches the surface. The direction this tangential plane faces (not to be confused with the YZ plane described above) defines the aspect, and the slope is the magnitude of the gradient in this direction. But the tangential plane is not aligned with the XY-plane, the YZ-plane or the ZX-plane. It sits at an angle to all three. If a pole were constructed attached to this tangential plane at the point (x,y,z) at right angles to the surface it would point in a direction known as the Normal to the surface (as shown in Figure 6-16, white arrow). As with

the three coordinate planes XY, XZ, and YZ we can now construct three planes that are orthogonal to each other based on the tangential plane and the surface Normal. One of these three is the tangential plane itself, the second is a plane through the surface Normal and in the direction of the aspect, and the last is a plane at right angles to both of these. These last two planes will cut the surface at an angle that is neither horizontal nor vertical. Where they cut the surface they will generate a line, whose curvature can be calculated. The Normal plane produce a line whose curvature at (x,y,z) is known as the *longitudinal curvature*, and the plane at right angles to the Normal generates a line with curvature at (x,y,z) known as the *cross-sectional curvature* (see further, subsection 6.2.2.6). Perhaps the simplest way of understanding the difference between the mathematical expressions of curvature and those commonly used in landscape analysis is to compare lines of latitude on a globe, which are formed by the intersection of a horizontal plane with the surface of a sphere, with great circles, which are formed by the intersection of a plane Normal to the surface, passing through the centre of the sphere. The latter (corresponding to the mathematical view of a tangent plane and Normal) will always be larger circles, so have a greater radius of curvature, r, and thus a lower curvature value than the former, which correspond to the contour or landscape perspective.

Curvature is conventionally described using the Greek letter kappa, κ, so we use this in conjunction with subscripts to define each form of curvature. The main expressions provided are based on those provided by Surfer, which in turn are based on Moore *et al.* (1991, 1993). For quadratic surface modelling as used in Landserf, expressions are based on Evans (1979) cited in Wood (1996). GIS packages, such as ArcGIS, may multiply curvature values by -100 for convenience, as recommended by Zeverbergen and Thorne (1987), to give values in the approximate range $[-1,1]$ with a sign that ensures that positive curvature equates to convex forms and negative curvature equates to concave forms. The value $r=1/\kappa$ is the radius of curvature, hence κ can be viewed as directly related to the size of a circle or a sphere that just touches the surface at sampled points.

Wood (1996) uses the signs of the measured slope, and the curvature measures: cross-sectional, maximum and minimum curvature, to characterise surface features, i.e. to help identify peaks, ridges, passes, planar regions, channels and pits. The simplified relationships he defines are shown in Table 6.1, where blanks indicate the value or sign of the measure has no relevance, and if a second line of conditions is included it relates to the immediate 3x3 neighbourhood of the target point. Thus these various curvature measures have applications in both process modelling and feature extraction, and as such may be applied not only to physical surfaces but in some cases, other types of surface data or fields. Readers are referred to Rana and Wood (2000) and Rana (2004a) for a fuller discussion of this area and the problems associated with trying to carry out such analyses on real-world surfaces.

Table 6.1 Morphometric features – a simplified classification

Feature	Slope	Curvature		
		Cross-sectional	Maximum	Minimum
Peak	0		+	+
Ridge	0		+	0
	+	+		
Pass	0		+	-
Plane	0		0	0
	+	0		
Channel	0		0	-
	+	-		
Pit	0		-	-

after Wood (1996, Table 5.3). + and − indicate positive or negative curvature, 0 indicates no curvature, blank indicates not relevant

6.2.2.3 Profile curvature

The labelled red line in Figure 6-16 indicates the *profile curvature* which is the shape of the surface in the immediate neighbourhood of the sample point contained within the vertical plane described in Section 6.2.2.2. It represents the rate of change of the slope at that point in the vertical plane, and is negative if the shape is concave, positive if the shape is convex and zero if there is no slope. Profile curvature is defined as:

$$\kappa_{pr} = \frac{\dfrac{\partial^2 z}{\partial x^2}\left(\dfrac{\partial z}{\partial x}\right)^2 + 2\dfrac{\partial^2 z}{\partial x \partial y}\dfrac{\partial z}{\partial x}\dfrac{\partial z}{\partial y} + \dfrac{\partial^2 z}{\partial y^2}\left(\dfrac{\partial z}{\partial y}\right)^2}{pq^{3/2}},$$

$$p = \left(\frac{\partial z}{\partial x}\right)^2 + \left(\frac{\partial z}{\partial y}\right)^2, q = 1 + p$$

Surfer implements the above expression using the approximations provided in Section 6.1.3. Both ArcGIS and Landserf utilise mathematical models fitted to the surface using the 3x3 window surrounding the target point z^*, in order to obtain estimates of profile curvature. These result in expressions that are defined in terms of the coefficients of the surface functions provided earlier.

For Landserf (quadratic surface model) we have:

$$\kappa_{pr} = \frac{-200\left(ad^2 + be^2 + cde\right)}{\left(e^2 + d^2\right)\left(1 + d^2 + e^2\right)^{3/2}}$$

For ArcGIS (fourth order surface model) we have:

$$\kappa_{pr} = \frac{200\left(dg^2 + eh^2 + fgh\right)}{\left(g^2 + h^2\right)}$$

where the coefficients *d-h* are computed in the manner described in Section 6.1.3.

6.2.2.4 Plan curvature

The *plan curvature* is the shape of the surface viewed as if a horizontal plane has sliced through the surface at the target point (the labelled green line in Figure 6-16). It is essentially the curvature of a contour line at height z and location (x,y). The differential expression for plan curvature is similar to that for profile curvature:

$$\kappa_{pl} = \frac{\frac{\partial^2 z}{\partial x^2}\left(\frac{\partial z}{\partial x}\right)^2 - 2\frac{\partial^2 z}{\partial x \partial y}\frac{\partial z}{\partial x}\frac{\partial z}{\partial y} + \frac{\partial^2 z}{\partial y^2}\left(\frac{\partial z}{\partial y}\right)^2}{p^{3/2}}$$

$$p = \left(\frac{\partial z}{\partial x}\right)^2 + \left(\frac{\partial z}{\partial y}\right)^2$$

Surfer implements the above expression using the approximations provided in Section 6.1.3. Both ArcGIS and Landserf utilise mathematical models fitted to the surface using the 3x3 window surrounding the target point z^* in order to obtain estimates of profile curvature. These result in expressions that are defined in terms of the coefficients of the surface functions provided earlier.

For Landserf (quadratic surface model) we have:

$$\kappa_{pl} = \frac{200\left(bd^2 + ae^2 - cde\right)}{\left(e^2 + d^2\right)^{3/2}}$$

For ArcGIS (fourth order surface model) we have a similar looking formula, although the coefficient values will differ from those of the quadratic model:

$$\kappa_{pl} = \frac{-200\left(dh^2 + eg^2 - fgh\right)}{\left(g^2 + h^2\right)}$$

The coefficients d-h, which are different for the quadratic and quartic models, are computed in the manner described in Section 6.1.3.

Note that ArcGIS also defines a general term which it describes as *curvature*, defined as the difference between the plan and profile curvatures, which simplifies to $\kappa = -200(d+e)$

6.2.2.5 Tangential curvature

Tangential curvature is again closely related to plan and profile curvature, and also to the slope, S_t. The differential formula is:

$$\kappa_{tg} = \frac{\frac{\partial^2 z}{\partial x^2}\left(\frac{\partial z}{\partial x}\right)^2 - 2\frac{\partial^2 z}{\partial x \partial y}\frac{\partial z}{\partial x}\frac{\partial z}{\partial y} + \frac{\partial^2 z}{\partial y^2}\left(\frac{\partial z}{\partial y}\right)^2}{pq^{1/2}}$$

$$= \kappa_{pl}\left(\frac{p}{q}\right)^{1/2}, \text{ where}$$

$$p = \left(\frac{\partial z}{\partial x}\right)^2 + \left(\frac{\partial z}{\partial y}\right)^2, \text{ and } q = 1+p$$

Surfer implements the above expression using the approximations provided in Section 6.1.3. ArcGIS and Landserf do not provide this function.

6.2.2.6 Longitudinal and cross-sectional curvature

Landserf provides these additional curvature measures based upon approximation of the surface using a local quadratic. The formulas used for each are as follows, where the letters *a-e* refer to the coefficients of the quadratic form described in Section 6.1.3:

$$\kappa_{lon} = \frac{-200\left(ad^2 + be^2 + cde\right)}{\left(e^2 + d^2\right)}$$

$$\kappa_{cro} = \frac{-200\left(bd^2 + ae^2 - cde\right)}{\left(e^2 + d^2\right)}$$

6.2.2.7 Mean, maximum and minimum curvature

Mean curvature is a term used in a variety of ways. In some instances it is defined as the difference between the plan and profile curvature, or as a measure of the average curvature for specific cells (the average of the maximum and minimum curvature for that cell), or even the mean value over all cells in a grid. Likewise maximum and minimum (profile) curvatures can be defined in both local and global terms. Within Landserf local formulas for the mean, maximum and minimum values are provided based on the quadratic model used throughout the package. These again are drawn from Evans (1979) as cited by Wood (1996) and are as follows:

$$\kappa_{min} = -a - b + \sqrt{(a-b)^2 + c^2}$$

$$\kappa_{max} = -a - b - \sqrt{(a-b)^2 + c^2}$$

Hence

$$\kappa_{mean} = (\kappa_{max} + \kappa_{min})/2$$

6.2.3 Directional derivatives

As noted earlier, once the components for gradient and curvature calculations have been estimated (first and second differentials) the computation of similar functions for other directions, α, is relatively straightforward. In these computations a step in the chosen direction is defined as having length ds, so the first directional derivative is obtained as follows:

$$\frac{dz}{ds} = \frac{\partial z}{\partial x}\cos(\alpha) + \frac{\partial z}{\partial y}\sin(\alpha)$$

If $\alpha=0$ or a multiple of π this reduces to the x-component of the vector, whilst if $\alpha=\pi/2$ or a similar multiple it reduces to the y-component (although the sign may change). In a similar manner the second derivative is defined as:

$$\frac{d^2z}{ds^2} = \frac{\partial^2 z}{\partial x^2}\cos^2(\alpha) + 2\frac{\partial^2 z}{\partial x \partial y}\cos(\alpha)\sin(\alpha)$$
$$+ \frac{\partial^2 z}{\partial y^2}\sin^2(\alpha)$$

Surfer provides computation of both of these derivatives based on a user-defined angle input. This angle applies to every grid cell, so will almost always differ from the values obtained by the kind of terrain modelling computations described in Section 6.2.2. Curvature computations based on the mathematical view of tangential curvature are also provided. Collectively these computations show how the surface changes in terms of gradient and curvature *in a particular constant direction*.

If the selected direction is a major compass point (NSEW) one of the sin or cos terms will be zero and the other =+1 or -1. The first directional derivative for the directions due north or due east would then be:

East: $\frac{\partial z}{\partial x} \approx \frac{z_E - z_W}{2\Delta x}$, North: $\frac{\partial z}{\partial y} \approx \frac{z_N - z_S}{2\Delta y}$

These simple expressions are those described in Section 6.1.3 and also, in the context of filtering, in Section 4.6.2.

6.2.4 Paths on surfaces

A path crossing a surface may be modelled in various ways. Simple cross-sectional profiles or transects, as we have previously seen, are typically modelled by sampling the line at a given frequency, or constructing the profile directly from the height of cells representing the surface across which the line passes. In general paths on surfaces are assumed to lie exactly on the surface and are modelled as polylines, smooth curves or sets of adjacent cells in the plane. Strictly speaking a curve crossing a surface is a form of space (3D) curve rather than a plane curve, and could be modelled as such. In practice the planar form is modelled and smoothed where necessary, with horizontal and vertical profiles treated separately. A path, P, may be modelled using incremental distance, t, along its length as a parameter: P={x(t),y(t)} or for a space curve P={x(t),y(t),z(t)}. As t increases from 0 (the start point) to T (the end point, defined by the total length of the path) the set of points {x(t),y(t),z(t)} map out the precise route of the path.

Incremental path slope, direction and curvature in plan or profile views may be computed using simple finite difference expressions based on the point set, or by calculating the relevant differentials using smooth curve approximations to the point set. For example, using spline curve modelling, as described in de Smith (2006), paths may be constructed across surfaces that meet specific engineering design requirements in terms of the minimum radius of curvature acceptable in vertical and plan profiles. Such paths will not necessarily remain on the surface, but will pass above or below the original surface at various points. Construction of these paths then requires a degree of cut-and-fill and/or bridging and tunnelling.

Path curvature, κ, at point t, is defined by the expression shown below — see for example, do Carmo (1976, Ch.1) or the Mathworld website entry for curvature for more details:

$$\kappa(t) = \left\| \gamma'(t) \times \gamma''(t) \right\| / \left\| \gamma'(t) \right\|^3$$

where $\gamma(t)$ is the parameterised space curve, $\gamma'(t)$ and $\gamma''(t)$ are the first and second differentials of the curve, $\times$ is the 3D vector cross product and $\|x\|$ is the Euclidean Norm of x. This expression may be written in the form we have provided earlier as:

$$\kappa(t) = \frac{\dfrac{\partial x}{\partial t}\dfrac{\partial^2 y}{\partial t^2} - \dfrac{\partial y}{\partial t}\dfrac{\partial^2 x}{\partial t^2}}{\left(\left(\dfrac{\partial x}{\partial t}\right)^2 + \left(\dfrac{\partial y}{\partial t}\right)^2 \right)^{3/2}}$$

As noted earlier, the radius of curvature is simply $r=1/\kappa$. Sample computation of κ for use within MATLab is described in de Boor (2003, p. 2-31).

Vertical profile path smoothing using spline functions is illustrated in Figure 6-17. This shows the profile of an initial path (dotted line) across a surface that was designed to meet maximum slope constraints but did not meet horizontal curvature requirements. The vertical profile of the horizontally smoothed version of this path is shown as a solid line, which shows greater variation in profile than the original path. The new path has then been smoothed again (bold line) to meet vertical curvature and slope constraints.

Figure 6-17 Path smoothing — vertical profile

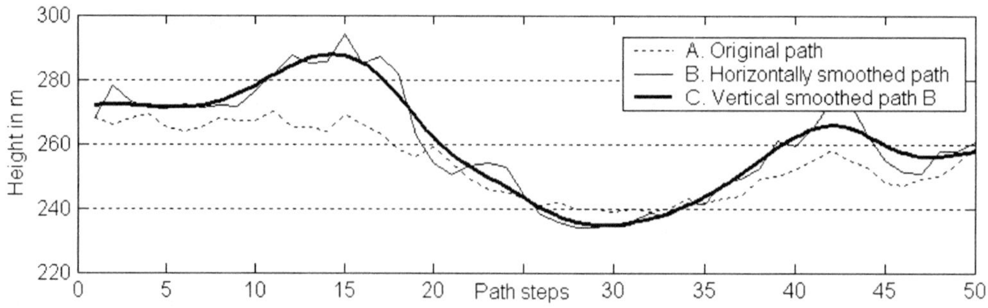

6.2.5 Surface smoothing

Surfaces may be smoothed in a variety of different ways. Many of the interpolation methods described in Sections 6.6 and 6.7 may be described as surface smoothing functions, since they provide estimates that do not pass exactly through the point sets provided. More generally, however, the term smoothing for surfaces refers to procedures that amend an existing grid. This is achieved through one of three main methods: (i) resolution increase (this may be the main objective of the procedure); (ii) grid recalculation, which if the grid has fewer rows and columns than the original is sometimes described as thinning; and (iii) filtering or kernel smoothing, involving operations on NxN windows or kernels. Smoothing operations that are terrain-specific (e.g. filling pits in a surface) are discussed separately in Section 6.2.6.

Smoothing a grid by resolution increase involves increasing the number of nodes or cells and then fitting an exact smoothing interpolator through these points and the original set. Typically the function used will be a bicubic spline. Figure 6-18A shows a low resolution grid (with 6x6 nodes — the same as that illustrated in Figure 6-8, upper diagram). In Figure 6-18B this surface has been spline smoothed to provide a 16x16 grid that passes exactly through the original point set, but is also a much smoother surface model.

Smoothing a grid by recalculation involves defining the number of grid rows and columns required, and then computing the result by interpolation to this resolution whilst again honouring the original point set. If a recalculated grid has the same number of rows and columns as the original grid no recalculation is carried out. If it is defined as the same resolution as in case (i), the result is an identical grid to that shown in Figure 6-18B.

Applying a filter or kernel to a grid can result in a wide range of effects, one of which is smoothing. Figure 6-18C illustrates this process for the same source grid, using a 3x3 kernel in which the central weight is 4, the NSEW

position weights are 2, and the diagonal weights 1 (as described in Section 4.6.2). The total weighting is thus 16, so the result is divided by 16 after each grid position has been adjusted using this kernel.

Figure 6-18 Grid smoothing

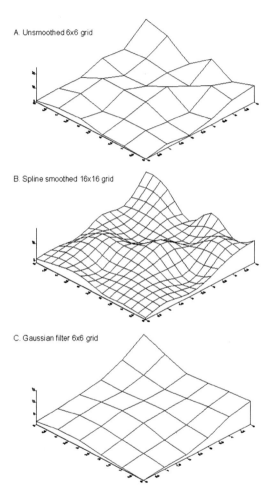

A. Unsmoothed 6x6 grid

B. Spline smoothed 16x16 grid

C. Gaussian filter 6x6 grid

This is a form of simple weighted average, in this case often described as a Gaussian kernel owing to the general symmetric, centre-weighted kernel form applied. This procedure is very similar to operations that are standard in image processing, such as edge detection and de-blurring.

6.2.6 Pit filling

Grid models of terrains often contain individual cells or groups of cells that are surrounded by cells with larger (higher) values. Pits are uncommon features of natural terrains, except in karst landscapes and some types of desert, so in many instances observed pits arise from errors in data capture and subsequent modelling of the surface. Pit filling is thus primarily of form of error correction procedure.

For hydrological analysis pits can be assumed to fill with water during flow and it is often convenient to remove them prior to analysis. This is an application-specific form of smoothing, but may be applied to any grid file, assuming that the result is meaningful for the problem at hand. GIS software such as Landserf, MFWorks, PCRaster, GRASS, ArcGIS and Idrisi all provide such functionality as standard (e.g. see the *lddcreatedem* and related functions in PCRaster, the Fill operations in MFWorks, the specialised procedures provided in ANUDEM and similar functions in other packages). Some of these implementations involve simple pit removal working on the assumption that such pits are likely to be minor errors in modelling the landscape, whilst many adopt a broader view of the hydrology. These programs attempt to distinguish between errors or artefacts, and true hydrological depressions. They then seek to remove selected pits (generally by filling) using an iterative process in order to correctly generate local drainage directions and from these, a completely connected drainage network. Several programs, including GRASS and Idrisi implement the algorithm developed by Jenson and Domingue (1988), whilst ArcGIS utilises their work and also the work of Tarboton *et al.* (1991) and Hutchinson (1989).

Implementations vary depending upon the number of cells to replace, the conditions applied and the value used for replacement. The simplest condition is that all 8-neighbour cells must contain values greater than a single central cell and the replaced value is the minimum of those surrounding the central cell.

Output will consist of a new, pit-less grid file and in some cases, a grid file showing the removed pits and their values. Variants of this operation include masked fill in which a pre-defined sub-region is subject to pit removal, and depression removal in which multi-cell pits (depressions) are removed based on hydrological flow and the depth of fill required. ANUDEM applies a more complex procedure, as described briefly in Section 6.6.14

6.2.7 Volumetric analysis

Surface-related volume calculations are usually performed on: (i) a single grid (which we call here the "upper" surface) with respect to a reference plane (i.e. z=constant, which we call the "lower" surface), rather as per the surface area computations described in Section 4.2.3.1; or (ii) on a pair of grids. In the latter case the difference in volumes is computed assuming that the first grid represents the upper surface and the second represents the lower surface, even though the two may intersect. The result of the computations is an estimated positive volume (upper surface volume that is above the lower surface) and an estimated negative volume (where the lower surface is above the upper surface). If the upper surface is always above the lower then all values will be positive. Some packages require an additional reference surface to be defined, e.g. as an absolute base value.

For example, using our test surface TQ81NE, with a reference plane of z=0, the entire surface is above sea level so the total volume reported is entirely positive with a value of roughly 570 million cubic metres, i.e. around $0.57km^3$. Note that this figure includes adjustment to allow for a 10m x 10m grid size. Using a reference plane at z=30m the positive (solid matter) volume is around $0.13km^3$ and the negative (air space) around $0.3km^3$. At 22.5m the two volumes are approximately equalised at $c.0.2km^3$, i.e. cut=fill if one were trying to level out this region using the available materials.

Volumes are calculated by numerical integration procedures, similar to those described earlier for determining the area of a polygon. With very small grid sizes the accuracy of volume estimation is, in theory, better, but this will depend on the way in which the surface is modelled. Surface roughness affects the results, especially where the surface is complex in form, the differences between upper and lower surface are small, and the computational method simplistic. The simplest procedure (which appears to be that adopted by ArcGIS) involves use of the trapezoidal rule (see also, Section 4.2.1), first in the x-direction across all grid nodes or cells, for the first row in the grid, and then for each subsequent row. This creates a set of slices with areas $A_1, A_2, ... A_n$. The slices times the grid resolution in the y-direction provides an estimate of volume. The first and last elements in each row may be weighted 1, with all other elements weighted 2, and then the sums divided by 2 to adjust for double counting. Although this provides a fairly crude and clearly biased estimate of the surface form at the top of each DEM cell, it is quite effective — the difference between this approach and more complex procedures, such as the extended Simpson's rule and variants of this which are provided in Surfer for example, is minimal in the test example we have cited (a tiny fraction of a percent).

Engineering-oriented packages, such as Autodesk's Land Desktop product, and civil engineering extensions to this, offer a wider range of options including volume estimation from surface point data and cross-sectional slices. ArcGIS includes volume computation from TIN surface representations, which one would presume is exact with respect to this surface model. Mathematical packages, such as MATLab perform surface integration in much the same way as described above, although the input is typically a mathematical function rather than a terrain model. This enables the numerical integration procedure to meet precise tolerances (accuracy levels) based on using successively finer steps (i.e. similar to utilising an arbitrarily fine grid model).

6.3 Visibility

The visibility of points in a landscape from one or more locations has many applications. These include studies of scenic quality, sound reduction, urban design, civil and military observation needs and telecommunications planning, amongst others. Many GIS packages offer visibility analysis, with varying degrees of functionality. Two principal functions are provided: line of sight, which is essentially a point-to-point operation; and viewshed, which is typically a point or point set to surface operation. Packages such as ArcGIS with Spatial or 3DAnalyst and MapInfo with Vertical Profiler provide a broad set of facilities, whilst programs such as TNTMips, Idrisi and Manifold provide a subset of these functions. Typically these operations are performed on grid files, but a similar, vector-based procedure known as Isovist analysis has been developed by Rana (2004b). This includes an add-on "Isovist Analyst" for ArcView (see further, Section 6.3.3).

For large distances the curvature of the Earth and optical refraction may be important, and packages often provide options to automatically adjust for these factors. For radio-wave modelling, Earth curvature tends to have an enlarged effect, which is generally modelled as 4/3 times the basic Earth curvature effect (possibly adjusted for different wavelengths). Other radio-wave modelling factors, such as Fresnel and non-linear signal decay effects (attenuation, interference, reflection, refraction) are not provided for, but may be incorporated as application-specific models using the programming facilities of the package.

An additional and in many ways more important issue is the question of offsets from the surface. Lines of sight are rarely if ever made from the surface itself — the observer's height and/or a structure (such as a radio mast) provide the vertical position from which computations should be made (the same question may arise for target points, such as communication or observation towers). A further and very important factor requiring consideration is the impact of surface uncertainty and obstructions (such as vegetation, road embankments, signs and buildings) on such computations. Lines of sight and viewsheds are highly susceptible to such factors, principally because of the low angles involved. For example, with zero offset from a surface, visibility is equivalent to shining a powerful light beam from the observation point and scanning around in a full circle. Any small rise or obstruction, such as a 1 metre high mound of earth or a bush, especially if close to the observation point, will cast a very large shadow behind it. Scaling this problem up to include larger objects over a range of distances and variable curvature of the landscape highlights the need to view results with caution. Raising and lowering the offset height of the observation point(s), moving the observation points and inspecting the results is one way to evaluate the sensitivity of results to such factors. Another is to increment and/or decrement the surface where known obstructions are understood to exist (e.g. by adding an estimated vegetation height and/or buildings height raster to a DEM).

6.3.1 Viewsheds and RF propagation

Viewsheds are regions of visibility observable from one or more observation points. Typically the inputs for viewshed analysis are:

- a surface raster or set of raster files that have been joined into a single file by a mosaic operation

- one or more observation points. In vector GIS these are typically points, sets of points or polylines, whilst in raster GIS these are generally a second raster containing non-zero values for viewpoint cells (thus may be any number, representing points, lines or polygons). For convenience we shall define the number of such observation points as m; and

- offset values for the observation points. The scope of the analysis may be limited by selecting: (a) a maximum (and in some cases, minimum) range for computations; and (b) horizontal and/or vertical limits (upper and lower) for angular computation

Executing the computation produces an output raster with the same extent as the input raster and a typical coding scheme of 0 for not visible, and $1 \geq n \geq m$ for a count of the frequency with which the cell in question may be observed from the m source points. Mapping this raster for values that equal m will identify those parts of the landscape visible from all observation points (in Manifold this is a separate selectable option). Several GIS packages use the term Radiate rather than Viewshed for this type of analysis (e.g. MapCalc, MFWorks). Because of the sensitivity of viewshed computations (algorithmic and surface detail uncertainties) some packages and algorithms provide a level of uncertainty built into the procedure, for example dividing the regions into: (i) almost certainly visible; (ii) possibly visible; (iii) possibly invisible; and (iv) almost certainly invisible.

As an example we shall consider the case of preliminary selection of sites for mobile radio masts. The first step is to obtain a DEM for the study region, with the finest resolution available. In this case GB Ordnance Survey 10m DEMS were used. The second step is to identify potential locations for masts based on areas that are known to be poorly serviced or not serviced at all. The mast height is an important factor, as may be issues of access, land ownership, power availability, security, proximity to schools and houses etc. All of these are readily accounted for within GIS packages. Figure 6-19 illustrates this process for a sample region covering 10kmx10km in Southern England (the lower left corner is the TQ81NE tile we have used elsewhere).

Figure 6-19 Viewshed computation

A. Topography

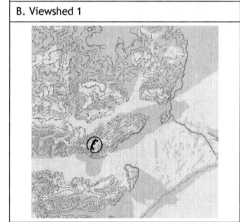

B. Viewshed 1

C. Viewshed 2

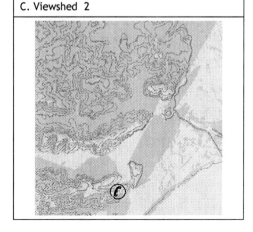

Figure 6-19A shows the topography of the region, with the dark blue (darkest grey) area in the lower right corner being the sea. Figure 6-19B and Figure 6-19C show two different viewshed rasters, based on 10m high masts located where indicated, with the pink areas (darker grey) being non-visible locations and green areas (lighter grey) as visible. The different "shadows" cast by local topographic features are clearly illustrated.

This analysis is simplistic for radio transmission analysis since it does not take into account the characteristics of radio wave propagation. Visibility may not be sufficient to determine signal quality. Other factors to be considered include: the frequency of the transmitter; the power of the transmitter (or gain); and the terrain surface in the neighbourhood of the transmission path, all of which affect the strength of the signal at the receiving site (the signal loss or attenuation). For example, trees and buildings that rise above the land surface but are not in the direct line of sight may produce signal attenuation and a zone of clearance around/above such obstacles is required in order to obtain a satisfactory received signal. Special purpose software packages exist for performing such computations, many including mapping and profile analysis facilities, and a number using GIS format files (e.g. .SHP, USGS DEM) for input and optionally output. Examples of such software include Nokia's NETACT, ATDI's HerTZ Mapper and Softwright's Terrain Analysis Package (TAP). Software built upon GIS platforms includes programs such as Cellular Expert which runs as an ArcGIS application (Figure 6-20, ArcScene visualisation) and RCC's ComSiteDesign suite.

Figure 6-20 3D Urban radio wave propagation modelling using Cellular Expert and ArcGIS

Figure 6-21 Radio frequency viewshed

A. Topography, Swansea area, South Wales	B. Predicted signal strength Viewshed

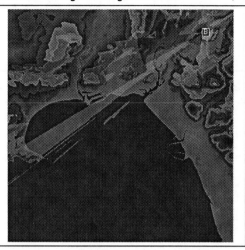

Figure 6-21 illustrates this process using the HerTZ Mapper software for their sample DEM in South Wales. A test site for a mast has been placed at the location marked B. This was specified as an 8m high mast, with transmission frequency 1800MHz and 34dB gain (typical of GSM mobile phone masts designed for small cell operations). Figure 6-21B shows the predicted signal strength based on a Bullington-Fresnel propagation model, with colours indicating the difference signal strengths. As noted earlier, these kinds of computations can be developed as programmed extensions to mainstream GIS packages, such as ArcGIS.

In addition to software providers and consultancies, a number of telecommunications operators and service providers have adopted this approach and produced their own internal systems built on GIS platforms (e.g. the US operator, Teligent, again using ArcGIS software). A significant advantage of this approach is that a vast range of fundamental and associated data are available and managed within GIS frameworks, enabling pure engineering modelling to be combined with customer and service area analysis, facilities management, and environmental and planning considerations.

6.3.2 Line of sight

Line of sight computations provide point-to-point information on visibility by means of: (a) mapped lines — variable colouring of a line drawn on the map between the source and target points; (b) tabulated data for the selected transect; and (c) profiles. Lines of sight may be computed from a single point to multiple destinations at regular angular intervals, providing a simplified picture of overall visibility from an observation point. However, for a more detailed picture complete viewsheds should be computed.

Using the example provided in Figure 6-19A we might examine the line of sight from the proposed mobile phone mast in Figure 6-19B to a sample point on the coast. The result is illustrated in Figure 6-22, where the source point has an offset of 10m. Locations highlighted in red (darker grey) are not visible from the source or target along this line. Again, as discussed in Section 6.3.1, with telecommunications modelling specific line-of-sight profile analysis tools are required. These include the ability to insert surface *clutter* such as trees, water areas and buildings, and to observe the effect of these on the signal strength and the Fresnel zones.

Some modelling packages, such as MATLab, provide a number of functions for line of sight and viewshed computation that can be combined in a variety of ways and displayed as 3D visualisations. This approach has the advantage of flexibility of both display and access to the computational details and results. An example is illustrated in Figure 6-23, where visible areas are shown in dark blue (dark grey), visible lines of sight in yellow (white) and invisible line profiles in dotted red (pale grey). The computations are based on an offset from the surface of 20 metres.

Figure 6-22 Line of sight analysis

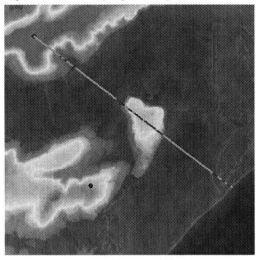

Figure 6-23 Viewsheds and lines of sight on a synthetic (Gaussian) surface

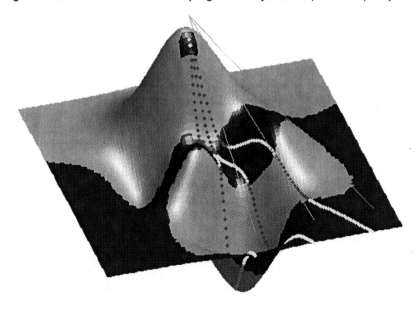

6.3.3 Isovist analysis

Analysis of visibility within urban landscapes (external and/or internal visibility) raises additional problems. With buildings and streets represented as 2D polygons, rather than 3D grids, a first level of visibility analysis can be carried out, with applications including questions such as where to place CCTV cameras or security guards in order to ensure all points are visible from the selected locations. This in turn involves an NP-hard optimisation problem (see further Section 7.1.4), that of determining the minimum number of such points to ensure complete coverage. The problem also extends in a natural manner to 3D environments. Rana (2004b) has developed efficient heuristic procedures for solving such problems based on a combination of systematic evaluation of visibility from a fine grid of initial observer positions, and then reducing this set by a ranking rule (Ranking and Overlap Elimination, or ROPE procedure). By stochastically iterating this procedure using randomly selected observer positions (S-ROPE) the final count of observer points may be reduced and thus approach optimality. An ArcView add-on is available to perform this operation. The procedure is illustrated in Figure 6-24 for an area of central London (Aldwych). A sample isovist region is shown in green (pale grey street areas), based on one of the initial test observer locations shown as a red dot symbol. The blue dotted box symbols indicate the 27 locations identified by the program using the S-ROPE algorithm with around 8000 starting points as the (approximately) optimal positions for surveillance cameras, ensuring 100% coverage. Restricting the initial solution set to medial axis regions determined by simple distance transform methods (as described in Section 4.4.2.2) shows an immediate further improvement of these results (Rana, 2006). It is interesting to consider how difficult it would be for architects, surveillance specialists or traffic monitoring units to perform such analysis manually.

Figure 6-24 Isovist analysis, Street network, central London

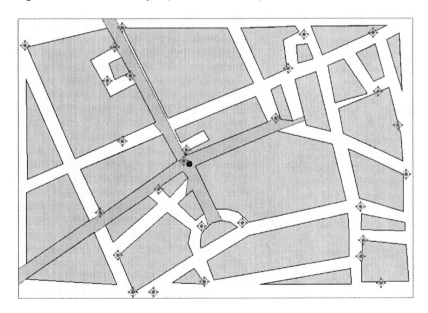

6.4 Watersheds and Drainage

6.4.1 Overview of watersheds and drainage

Many GIS programs support analysis of surface water hydrology and hydrological processes. In all cases the data input is a grid file of surface elevations (a DEM file) plus optional seed or *pour points*, and output is a number of vector and grid files. Programs with a focus on grid processing such as PCRaster, Idrisi, GRASS, TNTMips, ANUDEM and MFWorks provide a wide range of facilities for this type of analysis. Vector-oriented programs with strong grid support, such as ArcGIS and Manifold (V6.5 onwards) also provide support for such analyses. The central notion is that of a uniform pattern of raindrops falling throughout the study region, creating water flows from location to location down the locally steepest path, combining with other paths and accumulating the total of upstream flows in the process. Where flows in adjacent cells are away from each other it suggests that a local watershed exists. As streams combine larger and larger regions which drain into these streams are determined, and these regions are described as *stream basins* with boundaries known as *watersheds*.

6.4.2 Drainage modelling

There are many assumptions made in the basic drainage model, but it has proven to be of considerable value. These assumptions include: (i) uniform precipitation; (ii) flows take place entirely across surfaces, which they do not alter, and are unaffected by absorption (notably different soil or rock types) or groundwater; (iii) flows grow as a linear function with distance and are not altered by the slope values, just by the direction; (iv) there are no barriers to flow; and (v) the study region is complete and meaningful in the context of the analysis. Several GIS packages provide facilities and options that allow many of these assumptions to be relaxed or

explicitly modelled. The facilities for generating streams and watersheds within most GIS packages are separated into several functions, whilst in TNTMips and Manifold integrated tools are provided to carry out many of the procedures required. Within this Section we shall utilise the terminology and procedures applied in TNTMips, but these are very similar to those used by all packages. A good description of the methods involved is given in Burrough and McDonnell (1998, Ch.8).

The core steps in this kind of analysis are:

- Creation of an amended initial DEM with pits or depressions removed, as described in Section 6.2.6. This ensures that flows will be continuous across the surface, depending on the parameters selected (in some cases the GIS program may force fill pits until continuous flows can be generated)

- Systematic examination of the entire grid to identify flow directions. This is similar in concept to the creation of an *aspect* raster (some programs, such as GRASS, utilise an aspect grid as input). Most GIS implementations use the so-called *D8 algorithm*, which involves examining a 3x3 window of cells for the locally steepest direction (rise over run) and then coding each cell to indicate which one direction (of the 8 available) the flow is to follow (e.g. coded as 1,2,...8 or 1,2,4,...128). Some packages, such as RiverTools from RIVIX and the TAUDEM add-in for ArcGIS from Tarboton, implement the Tarboton (1997) D-infinity algorithm (Section 6.4.3 and Figure 6-25). RiverTools also implements an enhanced procedure known as the Mass Flux Method (MFM). Both D-infinity and MFM provide improved modelling since they handle divergent as well as convergent flows, but nonetheless they retain grid-related directional bias. The output from these procedures is often described as an *ldd* or local drainage direction grid. In some GIS implementations the 3x3 window is expanded if all surrounding cells have the same slope. In this case the neighbourhood is enlarged until a unique direction can be chosen. GRASS and several other packages also consider all possible directions, avoiding problems of significant directional bias (zigzag and parallel lines)

- Identification of flat areas and local extrema (peaks and pits)

- Computation of hypothetical stream paths and accumulated flows. This is typically carried out by accumulating the number of uphill cells (counting rook's moves as 1 unit and diagonal flows as longer — either 1.414... units or approximated as 3/2 units for speed of computation by working in integer arithmetic for most of the calculations) to obtain an accumulated value per cell. There may be a minimum value required or user-specified before accumulations are considered to constitute a stream, and hence be included in generated stream networks. PCRaster, for example, includes an extensive set of options relating to flow patterns

- Merging streams into topologically consistent networks and calculating simple stream-order statistics, e.g. Strahler, Horton and Shreve order values — see Haggett and Chorley (1969) or similar texts for more details on branching networks

- Identification of the upstream limits of different stream systems, in order to determine watersheds and compute watershed-related statistics

- Identification of basins within these watersheds, typically representing the watersheds of secondary streams (first level tributaries) from the main stream within the watershed

- Optional specification of points within the grid file that may be examined for the downstream impact of introduced flows, e.g. spillovers or pollutants entering the system, and/or the upstream flows that may impact upon a downstream location

- Computation of statistics relating to the various elements generated, for example details of stream lengths, basin and watershed areas, membership lists (streams within watersheds), estimation of sediment transport amounts etc. and computation of estimated soil loss using the Universal Soil Loss Equation (USLE)

6.4.3 D-infinity model

The D-infinity flow assignment model is illustrated in Figure 6-25. Flows are assigned to two cells or pixels (coloured here in yellow/pale grey) based on the direction of maximum gradient of the 8 triangular facets illustrated. The proportions assigned to each pixel are determined by the two ratios:

$$p_1 = \frac{\alpha_1}{\alpha_1 + \alpha_2}, p_2 = \frac{\alpha_2}{\alpha_1 + \alpha_2}$$

where α is defined in radians or degrees, as measured as in mathematics (i.e. from due east).

Figure 6-25 D-Infinity flow assignment

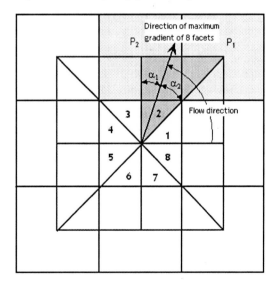

In the D-infinity model, if the steepest slope vector angle illustrated lies outside of a facet (a special case that may well occur) then the direction vector is taken along the steepest edge of the facet.

6.4.4 Drainage modelling case study

In the following discussion we use the NT04 test DEM to illustrate the procedures involved. If a relatively flat and/or poorly bounded region is selected, with integer height values, the output from hydrological modelling may be ambiguous, so here we have selected a more mountainous test dataset.

6.4.4.1 Flow accumulation

Figure 6-26A shows the NT04 source DEM as a hillshaded relief map. This version of the relief has been pre-processed to remove small pits in the surface, as described in Section 6.4.2. Using this amended grid file as input, a local drainage direction (ldd) grid has been generated (Figure 6-26B), which is similar to, but not precisely the same as, the aspect map in this case. This is because it is based on the simple D8 algorithm rather than more complex computations used in aspect determination. With the ldd map as input, flow accumulation modelling can proceed. Figure 6-26C shows a close-up of a part of the NT04 grid after the accumulation process has been run. Lighter pixels are those with higher accumulation values, and the pixelated nature of the accumulations and the boundary between accumulation zones can be clearly seen.

6.4.4.2 Stream network construction

The flow accumulation grid provides the basis for stream network construction, typically marking all accumulation values greater than some pre-defined figure (e.g. 10) as being streams or potential streams. This process is illustrated in Figure 6-27A. Note that an assignment rule (e.g. proximity) is required for flat regions. Figure 6-27B shows those regions of the source DEM (prior to pit-filling) that may warrant closer attention.

Figure 6-26 Flow direction and accumulation

A. Source DEM after pit removal

B. ldd grid — D8 algorithm

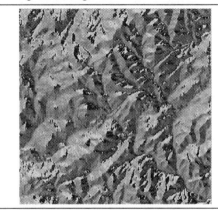

C. Flow accumulation — selected area

Figure 6-27 Stream identification

A. Streams

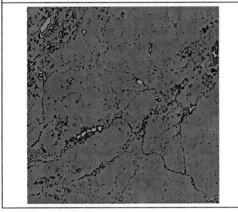

B. Flat areas and extrema

© Crown Copyright Data

Figure 6-28 Watersheds and basins

A. Watersheds

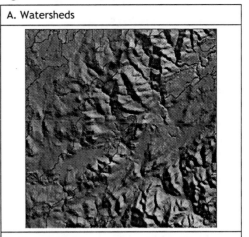

B. Stream basins

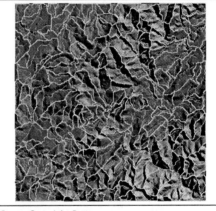

© Crown Copyright Data

This may be because their values indicate that they are either flat (shown here in yellow/pale grey) or local extrema (maxima in red, minima in blue, saddles in green). Each stream is identified by colouring the appropriate pixels; in this case without making allowance for accumulation value in the number of pixels coloured (i.e. unlike Figure 6-26C).

6.4.4.3   Stream basin construction

Surrounding collections of streams are the upstream limits of flow, which identify the watersheds, as shown in Figure 6-27A and without the stream overlay in Figure 6-28A.

Figure 6-28B illustrates the construction of stream basins. These are shown here as yellow/white polygons, and are subdivisions of the main watershed polygons. The stream branching patterns shown and the identification of stream basins are both determined by flow threshold parameters, e.g. by setting the parameter lower a larger number of smaller stream basins will be identified, whilst increasing this parameter will reduce the number of such basins within each watershed.

6.5 Gridding, Interpolation and Contouring

6.5.1 Overview of gridding and interpolation

The term gridding refers to the process of generating a grid of data values from an existing set (generally of the form {x,y,z} triples) using two-dimensional (spatial) interpolation methods. These interpolation methods assume that the attributes being modelled, z, are continuous or piecewise continuous across the study region, i.e. that z=f(x,y) exists and is single-valued for all (x,y) except where regions are explicitly masked off (i.e. excluded from modelling).

Interpolation methods often assume data points are correct (exact), but they may assume they are subject to error (generally of known or estimated extent). The models used may be exact (go precisely through the sample data points) or inexact (approximate the values at the data points). If the data points are relatively sparse and irregular, interpolation needs to be more sophisticated (subtle) than for dense, regularly spaced data. However, regularly spaced data may be subject to bias due to intrinsic frequencies in the data (spacing and/or directional effects). An interesting example cited by Burrough and McDonnell (1998) is that of soil samples taken in a field that had previously been ploughed, with values recorded being affected by the furrow patterns. Sparse datasets are typical of many applications in which data are costly or time-consuming to collect, for example using boreholes, meteorological stations, soil sampling or radiation measures.

There have been a many studies that compare the effectiveness of alternative interpolation techniques, using a wide range of different test datasets and conditions. Overall it has been found that a number of well-defined factors have a major influence on the quality of interpolation: data measurement accuracy; data density; data distribution; and spatial variability. These are fairly predictable findings, but prior examination of each of these elements may assist in choosing the most appropriate technique for the problem at hand and/or be used in guiding sampling of new or supplementary datasets. Interpolation quality can often be substantially improved through the use of ancillary information, such as remote sensing data or additional environmental information (e.g. location of stream networks).

Having obtained the best possible dataset, within budget and time constraints, achieving the maximum usage and value is very important — hence the need for interpolation procedures that assist in estimating values at unsampled locations. More generally, spatial interpolation is required:

- to convert a sample of data points to a complete coverage (set of values) for the study region

- to convert from one level of data resolution or orientation to another (resampling)

- to convert from one representation of a continuous surface to another, e.g. TIN to grid or contour to grid

Grid files may be pre-generated (as for example, with elevation grid datasets from national mapping agencies), or may be generated from an input dataset of {x,y,z} form. In the latter case values of z are estimated for unsampled points, typically a square grid of pre-specified resolution. This generated grid may then be analysed in order to identify its statistical attributes, its approximation or fit to the input dataset and/or other information available for validation of its quality. It may then be mapped in a variety of forms, e.g. contour, filled contour, shaded relief, wireframe, perspective 3D surface. Grids may also be generated in a number of other ways: from remote sensing datasets (e.g. hyperspectral images); as a result of conversion of vector datasets (e.g. contour to grid conversion); as representations of mathematical functions (e.g. as a output from a mathematical expression, with or without random or fractal perturbations); or as a result of Map Algebra, resampling and/or overlay operations.

Grid generation using local interpolation functions, of whatever type, is typically a process based on weighted averages of values at nearby points. The assumption is that each grid cell or intersection is likely to be similar to other values in its neighbourhood. Most such models are of the general form:

$$z_j = \sum_{i=1}^{n} \lambda_i z_i$$

where z_j is the z-value to be estimated for location j, the λ_i are a set of estimated weights (proportional contributions that sum to 1) and the z_i are the known (measured) values at points (x_i, y_i). Assuming that $\lambda_i > 0$ for all i, then

$\min\{z_i\} \le z_j \le \max\{z_i\}$

i.e. z_j will always lie within the range of observed data values. If one imagines z_j being a point on a small hilltop surrounded by observations that are below it such estimators will generate a flat surface in the neighbourhood of z_j at the height of or slightly below the maximum observed value.

The size and shape of the search neighbourhood is usually selectable. The size is taken either as a static value (e.g. a specific radius) or is determined in an adaptive manner (e.g. requiring a minimum number of observations to be used in any calculation). The shape of the neighbourhood or search area may be circular, elliptical, or segmented variants of these forms. A number of packages facilitate the inclusion of known breaklines, exclusion zones and/or missing values as part of the gridding process. Most GIS packages provide a wide range of facilities for gridding, many of which are described in Section 6.5.2. Surfer provides amongst the most extensive set

of options and offers suggestions as to their selection and use. Recommendations based on those of Surfer and similar GIS-related facilities are summarised in Table 6.2.

Generating a grid from a (global) function does not require separate interpolation, merely definition of the maximum and minimum values for the x and y components and either the grid increment (size) or number of rows and columns to be generated. An example of such a function might be:

$z = ax + by + c + 100*\text{rnd}()$

where a, b and c are constants and rnd() is a random number generator providing numbers in the range [-1,1]. Another example is the Gaussian surface plotted in Figure 6-23.

6.5.2 Gridding and interpolation methods

6.5.2.1 Comparison of sample gridding and interpolation methods

Examples of each of 12 methods for generating the grids described in Table 6.2 are plotted in Figure 6-29C-N. The source data in this case is a set of 62 spot heights from the GB Ordnance Survey NT04 tile, which covers part of the Pentland Hills area to the south of Edinburgh. Also shown is the source OS vector file of contours for this region at 20m contour intervals. Each of the plotted maps has been generated with a 20m contour interval using the methods listed and taking their default parameter settings as provided within the Surfer package.

If the input $\{x,y,z\}$ dataset is already a complete or nearly complete set of grid values, little or no interpolation is required — simple nearest-neighbour methods are typically used for filling in missing values in such cases.

There are several key points to note about these maps:

- all the derived contour maps shown have been generated by first creating a grid file using the interpolation procedure in question, and then contouring applied using linear grid interpolation (as described in the Section 6.5.3)

- with a highly variable surface such as this, 62 data points are insufficient to re-create the source data, despite the relatively small size of the region in question

- two of the interpolation methods (natural neighbour and triangulation plus linear interpolation) limit their extent to the convex hull of the input point set

- some methods, including moving averages, nearest-neighbour and triangulation produce very unsatisfactory-looking results with this dataset

- several methods generate very similar looking results, notably natural neighbour and Kriging, and minimum curvature and modified Shepard (see further, Sections 6.6 and 6.7)

- interpolation methods based on profiles tend to produce much better results, since these use contour data as their primary input, from which they generate a grid

Table 6.2 Gridding and interpolation methods

Method	Speed	Type	Comments
Inverse distance weighting (IDW)	Fast	Exact, unless smoothing factor specified	Tends to generate bull's eye patterns. Simple and effective with dense data. No extrapolation. All interpolated values between data points lie within the range of the data point values and hence may not approximate valleys and peaks well
Natural neighbour	Fast	Exact	A weighted average of neighbouring observations using weights determined by Voronoi polygon concepts. Good for dense datasets. Typically implementations do not provide extrapolation
Nearest-neighbour	Fast	Exact	Most useful for almost complete datasets (e.g. grids with missing values). Does not provide extrapolation
Kriging - Geostatistical models (stochastic)	Slow/ Medium	Exact if no nugget (assumed measurement error)	Very flexible range of methods based on modelling variograms. Can provide extrapolation and prediction error estimates. Some controversy over aspects of the statistical modelling and inference. Speed not substantially affected by increasing number of data points. Good results may be achieved with <250 data points
Conditional simulation	Slow	Exact	Flexible range of techniques that use a fitted variogram as a starting point. Effective as a means of reproducing statistical variation across a surface and obtaining pseudo-confidence intervals
Radial basis	Slow/ Medium	Exact if no smoothing value specified	Uses a range of kernel functions, similar to variogram models in Kriging. Flexible, similar in results to Kriging but without addition assumptions regarding statistical properties of the input data points
Modified Shepard	Fast	Exact, unless smoothing factor specified	Similar to inverse distance, modified using local least squares estimation. Generates fewer artefacts and can provide extrapolation

Method	Speed	Type	Comments
Triangulation with linear interpolation	Fast	Exact	A Delaunay triangulation based procedure. Requires a medium-large number of data point to generate acceptable results.
Triangulation with spline interpolation	Fast	Exact	A Delaunay triangulation based procedure, with bicubic spline fitting rather than plane surface fitting. Generates very smooth surfaces. Widely used in computer-aided design
Profiling	Fast	Exact	A procedure that converts contour data to grid format. Similar to linear interpolation methods used in generating contour data (see subsection 6.5.2.2). See also, Topogrid, subsection 6.6.14
Minimum curvature	Medium	Exact/Smoothing	Generates very smooth surfaces that exactly fit the dataset
Spline functions	Fast	Exact (smoothing possible)	Available as a distinct procedure and incorporated into a number of other methods. Bicubic and biharmonic splines are commonly provided
Local polynomial	Fast	Smoothing	Most applicable to datasets that are locally smooth
Polynomial regression	Fast	Smoothing	Provides a trend surface fit to the data points. Most effective for analysing 1st-order (linear) and 2nd-order (quadratic) patterns, and residuals analysis/trend removal. Can suffer from edge effects, depending on the data. See also, Section 5.6
Moving average	Fast	Smoothing	Uses averages based on a user-defined search ellipse. Requires a medium-large number of data points to generate acceptable results
Topogrid/Topo to Raster	Slow/ Medium	Not specified	Based on iterative finite difference methods. Interpolates a hydrologically "correct" grid from a set of point, line and polygon data, based on procedures developed by Hutchinson (1988, 1989, 1996). Requires contour vector data as input. Available in ArcGIS based on Hutchinson's ANUDEM program

In subsection 6.5.2.2 a series of contour plots are provided, in each case based on the set of GB Ordnance Survey spot heights shown in Figure 6-29A. The first plot, Figure 6-29B, shows the GB Ordnance Survey contour map for the same region, as a base for comparison purposes. The subsequent 12 contour maps show the results obtained using a variety of interpolation techniques that have been applied to the source spot heights. These are each discussed in more detail in Sections 6.6 and 6.7.

6.5.2.2   Contour plots of sample gridding and interpolation methods

Figure 6-29 Contour plots for alternative interpolation methods — generated with Surfer 8

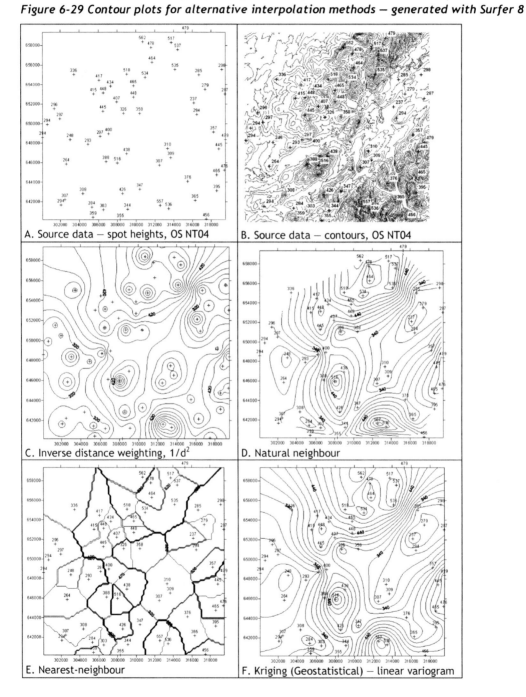

A. Source data — spot heights, OS NT04

B. Source data — contours, OS NT04

C. Inverse distance weighting, $1/d^2$

D. Natural neighbour

E. Nearest-neighbour

F. Kriging (Geostatistical) — linear variogram

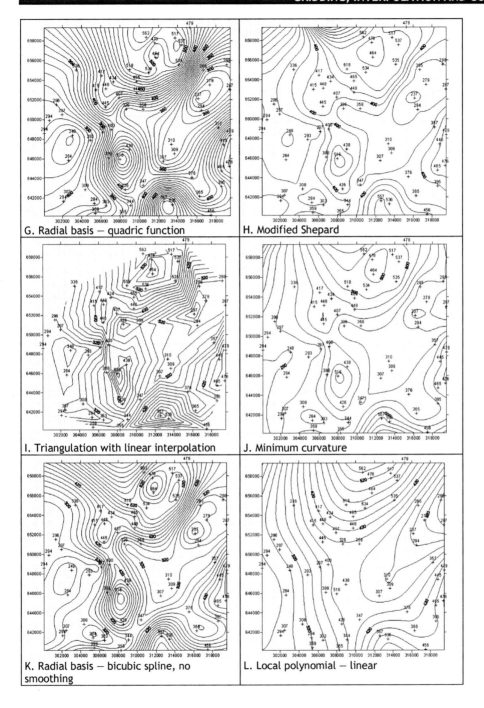

G. Radial basis — quadric function

H. Modified Shepard

I. Triangulation with linear interpolation

J. Minimum curvature

K. Radial basis — bicubic spline, no smoothing

L. Local polynomial — linear

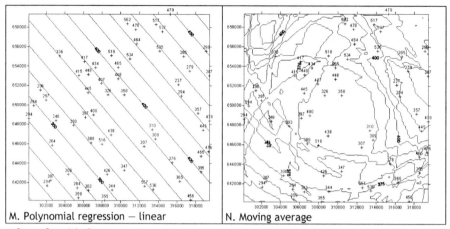

M. Polynomial regression — linear

N. Moving average

© Crown Copyright Data

6.5.3 Contouring

Contour maps of surfaces (topographic or other) can be derived from grid, DEM or TIN representations, or may be sourced from vector datasets such as those produced by national mapping agencies. Contours are lines of equal value on a surface and are thus level sets — lines that may be generated by flooding a surface to a specified level or by slicing a surface horizontally. Where contours are generated from grid data they tend to be interpolated linear segments or smoothed segments derived from linear interpolation. If grid data are viewed as a set of grid lines rather than cells, the line intersections or grid nodes have values. A straight line segment of constant value (e.g. 480m) can be constructed from pairs of NS and EW nodes by simple linear interpolation (Figure 6-30). The result is a polyline with vertices at each grid line intersection and a constant value (e.g. elevation estimate).

Figure 6-30 Linear interpolation of contours

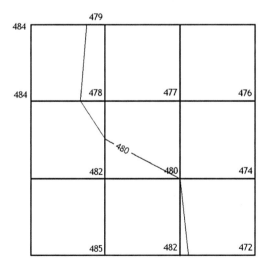

The polylines so generated may then be smoothed in a variety of ways — the most common being to fit a spline function to the straight line segments and interpolate a denser set of points that may be plotted to provide a smooth set of contours. Note that independently smoothed contours can result in separate contours crossing, so the procedure needs to be carried out with care. Alternatively input grids may be resampled to provide a finer resolution input or pre-smoothed using mathematical or Map Algebra operations with contouring conducted on the resultant revised dataset. For TIN datasets linear interpolation is typically the only choice, with post-computation polyline smoothing as an option.

Figure 6-31 illustrates the standard process using the sample TQ81NE region represented as a DEM and using TNTMips software. As noted earlier, there appear to be artefacts (linear forms) represented in the flat regions of the source data. Figure 6-31B shows contours generated using simple linear interpolation, with a close-up of a section of the surface provided in Figure 6-31C. The angular form of some line elements is reduced in Figure 6-31D using the iterative thresholding procedure provided as a standard option in TNTMips. This is essentially a form of bicubic spline smoothing. A similar result, often with even smoother curves, is obtained if the input file is pre-smoothed using simple weighted average or Gaussian smoothing over a sample window, e.g. a 3x3 or 5x5 window, and then linear interpolation applied. ArcGIS recommends using pre-smoothing of grids with a 3x3 weighting matrix and the Map Algebra operation Focal Sum (Weight option) with a custom weight of 0.96 for the central cell and the remaining 8 cells weighted to 0.05.

Figure 6-31 Contour computation output

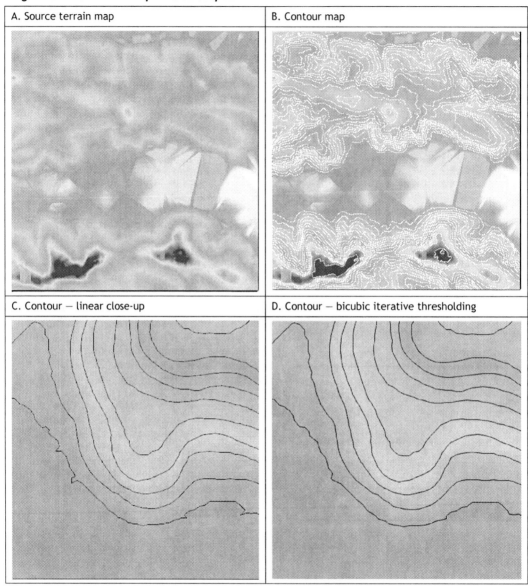

| A. Source terrain map | B. Contour map |
| C. Contour — linear close-up | D. Contour — bicubic iterative thresholding |

© Crown Copyright Data

6.6 Deterministic Interpolation Methods

In Section 6.5 we noted that interpolation typically involves generating a fine rectangular grid covering the study region, and then estimation of the surface value or height for every grid intersection or cell. The estimation process involves the use of a simple linear expression in order to compute grid values:

$$z_j = \sum_{i=1}^{n} \lambda_i z_i$$

where z_j is the z-value to be estimated for location j, the λ_i are a set of estimated weights and the z_i are the known (measured) values at points (x_i, y_i). As z_j is a simple weighted average an additional constraint is required ensuring that the sum of the weights adds up to 1:

$$\sum_{i=1}^{n} \lambda_i = 1$$

The interpolation problem is to determine the optimum weights to be used. If λ_i=0 for all i except for the measured point closest to the grid intersection then this would represent a form of nearest-neighbour interpolation. If all n points in the dataset were used and weighted equally every point would have weights $1/n$ and would be given the same z-value. In many cases, as per Tobler's First Law, measured points closer to z_j are more likely to be similar to z_j than those further afield and hence warrant weighting more strongly than observations that are a long way away. Indeed, there may be little to gain from including points that lie beyond a given radius, or more than $m<n$ points away, or points that lie in certain directions.

Each of the methods described has its own approach to computation of these weights. In each case the weights are determined by the choice of model, algorithm and user-defined parameters. This differs from the approach described in Section 6.7, where the structure of the input data is first analysed and on the basis of this a model and preferred set of parameters are identified — i.e. these are not directly selected by the user. Despite this apparent distinction, all methods may be subjected to a procedure known as cross-validation, which can assist in the choice of model and parameters. The simplest form of cross-validation is systematic point removal and estimation. The procedure is as follows:

- the model and parameters are chosen

- one by one each of the original source points is removed from the dataset and the model is used to estimate the value at that particular point. The differences between the input (true) value and the estimated value is calculated and stored

- statistics are then prepared providing measures such as the mean error, mean absolute error, RMSE, the maximum and minimum errors, and whereabouts these occur

- the user then may modify the model parameters or even select another model, repeating the process until the error estimates are at an acceptable level

- the gridding process then proceeds and the data are mapped

There are variants of this procedure (for example selection of a random subset of the source data points for simultaneous removal — a technique known as jack-knifing), and various alternative cross-validation methods. The latter include:

- re-sampling — obtaining additional data

- detailed modelling of related datasets (e.g. easy-to-measure variables) — examples might include distance from a feature such as a river or point-source of pollution, or measurement of a highly correlated variable, such as using light emission at night as a surrogate for human activity leading to increased CO_2 emissions

- detailed modelling incorporating related datasets — look at possible stratification, modelling non-stationarity, boundaries/ faults

- comparison with independent data of the same variate (e.g. aerial photographs)

6.6.1 Inverse distance weighting (IDW)

Inverse distance weighting models work on the premise that observations further away should have their contributions diminished according to how far away they are. The simplest model involves dividing each of the observations by the distance it is from the target point raised to a power α:

$$z_j = k_j \sum_{i=1}^{n} \frac{1}{d_{ij}^{\alpha}} z_i$$

The value k_j in this expression is an adjustment to ensure that the weights add up to 1. If the parameter $\alpha=1$ we have:

$$k_j = \sum_{i=1}^{n} \frac{1}{d_{ij}}$$

Many GIS packages provide this kind of inverse distance model for interpolation, as it is simple to implement and to understand. Often the model is generalised in a number of ways:

- a faster rate of distance decay may be provided, by including a power function of distance, $\alpha>1$, rather than simple linear distance. While any α value convenient for a given application may be used, common practice is to use distance ($\alpha=1$) or distance squared ($\alpha=2$)

- since it is possible that the grid intersection could coincide with a data point (especially likely at region corners or edges if MBRs have been used on the

original point set), an explicit check or adjustment to the expression is needed to avoid computational errors (overflow). Typically the adjacent point weight is set to 1 (its value is copied) and the remaining weights are set to 0

- if a user-selectable adjustment to the minimum inter-point distance is specified, this can result in smoothed rather than exact interpolation. This may be a simple incremental amount added to the distance, t, or an adjusted distance value such as $d_{ij}^{*} = \sqrt{(d_{ij}^{2}+t^{2})}$

- as with all methods, additional controls may be applied or available: limiting the number of points included; specifying the search directions and search shape; and limiting computations by excluding pre-defined regions, breaklines or faults

Figure 6-32 and Figure 6-33 provide illustrations of the method applied to the test data for OS NT04. The surface plot shows how simple IDW with no smoothing and power 2 distance decay results in dips and peaks around the data points but is otherwise relatively smooth in appearance.

Figure 6-32 IDW as surface plot

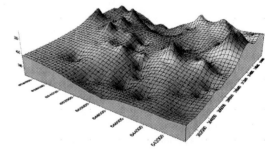

Figure 6-33A shows the source data (spot heights and contours) with Figure 6-33B and Figure 6-33C illustrating the surface obtained using parameters of $\alpha=1$ and $\alpha=2$. Both exhibit the familiar bull's eye effect of standard IDW. Figure 6-33D is markedly different and seems much closer to the source contours. In this case we have selected $\alpha=3$, a smoothing factor of $t=2$, and an anisotropy (directional bias) of 45 degrees using an elliptical search region with a ratio of 2:1. The selection of these values was made after limited experimentation using simple cross-validation and comparison with additional information. Some GIS products, such as the ArcGIS Geostatistical Analyst (but not the ArcGIS Spatial Analyst) attempt to select an optimal value for α

automatically. The method adopted varies, but typically involves cross-validating the data for incremental values of α, for example $\alpha=0(0.25)4$, and selecting the value (or locally interpolated value) that yields the minimum RMSE value for the surface as a whole. In the example illustrated ArcGIS estimates $\alpha=2.96$ and RMSE at approximately 5 metres, although the maximum absolute error is around 100 metres.

Figure 6-33 Contour plots for alternative IDW methods, OS NT04

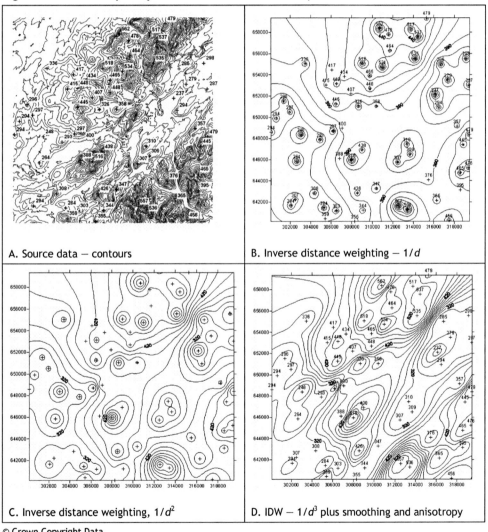

A. Source data — contours

B. Inverse distance weighting — $1/d$

C. Inverse distance weighting, $1/d^2$

D. IDW — $1/d^3$ plus smoothing and anisotropy

© Crown Copyright Data

6.6.2 Natural neighbour

Natural neighbour interpolation creates weights for each of the input points based on their assumed "area of influence". These areas are determined by the generation of Voronoi polygons around each input point. In principle every grid intersection created would be in one of these polygons and could be assigned the value of the point around which the polygon has been created. This would result in a step-like surface of patches, as illustrated in Figure 6-34.

Figure 6-34 Simple Voronoi polygon assignment

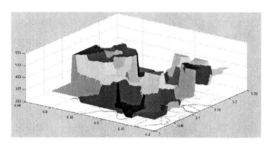

This is the kind of result that is obtained from Nearest neighbour interpolation. A far more effective approach involves a development of this idea, as described below. The end result is a smooth surface with discontinuities at the input points (Figure 6-29D).

The first step in the process is to create a Delaunay triangulation of the input data points (Figure 6-35A) as a preliminary stage in the creation of Voronoi polygons. The blue (darker) line identifies the immediate or *natural neighbours* for a sample input point which in this example has a spot height of 310m. The second stage is to generate a set of Voronoi polygons for the study region (Figure 6-35B). Each of the points, *j*, in the highlighted area has its own polygon, with known area A_j, and each of these polygons shares a common border with the sample point. In order to determine the estimated value at a sample point P the point P is temporarily added to the set and Voronoi polygons re-computed (Figure

6-35C). Adding point P has resulted in a new Voronoi polygon and redefinition of those immediately surrounding it. This new polygon has an area A_p. Effectively this new point has "borrowed" some of the area of influence from each of the nearby points. This can be seen in Figure 6-35D, where the new region has been overlaid on the original set. There are $k=5$ original polygons that the new polygon has borrowed area from. Let us call these borrowed areas A_{ip}, $i=1,...k$ then the total area of P's Voronoi polygon is:

$$A_p = \sum_{i=1}^{k} A_{ip}$$

and thus the proportion borrowed from each of the original points is:

$$\lambda_i = \frac{A_{ip}}{\sum_{i=1}^{k} A_{ip}}$$

These proportions are the weights used to compute the estimated value at P, based on the standard linear weighting equation:

$$z_p = \sum_{i=1}^{k} \lambda_i z_i$$

If the point P coincides with one of the existing points its area of overlap with that point would be 100%, hence its weight would be 1, as required. If the Voronoi polygon for P does not overlap a region the weight associated with that region is 0. One of the main advantages of this method of interpolation is that it requires no decision-making regarding the number of points to use, the radius or direction of search, or any other parameters. It is a surprisingly effective and straightforward technique that is widely used in many disciplines other than GIS.

The standard implementation of this method provides interpolated values for the convex hull of the input point set. However, one or two software packages extend this to enable limited extrapolation to the boundary of a rectangular region (the MBR). This is achieved by using a slightly enlarged MBR (e.g. increased by 10%). Within this enlarged region temporary points are added to the original set at each corner (and/or along the region boundary). z-values for these temporary points are estimated by simple IDW or a similar technique, and then the Voronoi regions computed for all points. The procedure continues as before, but on completion the temporary points are discarded and the MBR re-drawn based on the original point set.

Figure 6-35 Contour plots for Natural Neighbour method, OS NT04

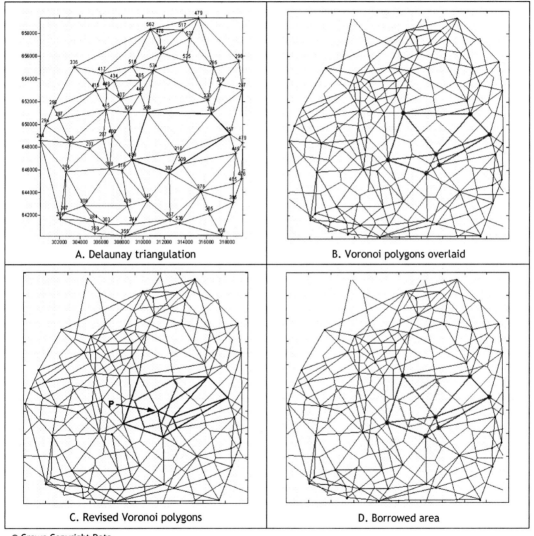

A. Delaunay triangulation

B. Voronoi polygons overlaid

C. Revised Voronoi polygons

D. Borrowed area

6.6.3 Nearest-neighbour

This is a simple technique whereby each grid intersection is assigned the value of the nearest point in the input set, as illustrated in Figure 6-29E. The result is a stepped surface if applied to data such as the NT04 spot heights, but this is not the kind of situation in which the method is employed. The commonest applications are for: (i) evenly or near-evenly spaced input data that need to be converted to a grid file with similar spatial frequency; (ii) resampling grids or images, especially where no change to pixel/grid values is desirable; (iii) where there is a need to substitute missing values in a grid where most cells are complete; and (iv) where there is concern over the continuity of the underlying field and/or expectation that steps in the surface provide the most appropriate representation of the underlying data.

6.6.4 Radial basis and spline functions

Radial basis interpolation is the name given to a large family of exact interpolators. In many ways the methods applied are similar to those used in geostatistical interpolation (see Section 6.7), but without the benefit of prior analysis of variograms. On the other hand they do not make any assumptions regarding the input data points (other than they are not co-linear) and provide excellent interpolators for a wide range of data.

A great deal of research has been conducted into the quality of these interpolators, across many disciplines. For terrain modelling and Earth Sciences generally the so-called multiquadric function has been found particularly effective, as have thin plate splines. The simplest variant of this method, without smoothing (i.e. as an exact interpolator) can be viewed as a weighted linear function of distance (or inverse distance) from grid point to data point, plus a "bias" factor, m. The model is of the form:

$$z_p = \sum_{i=1}^{n} w_i \phi(r_i) + m$$

or the equivalent model, using the untransformed data values and data weights λ_i:

$$z_p = \sum_{i=1}^{n} \lambda_i z_i$$

In these expressions z_p is the estimated value for the surface at grid point p; $\phi(r_i)$ is the radial basis function selected, with r_i being the radial distance from point p to the ith data point; the weights w_i and λ_i and the bias value m (or Lagrangian multiplier) are estimated from the data points. This requires solving a system of n linear equations. Using the second of the two models above the procedure is then essentially the same as for Ordinary Kriging (see further, Section 6.7.2). For clarity we outline this latter procedure below as a series of steps using matrix notation:

- Compute the $n \times n$ matrix, D, of inter-point distances between all (x,y) pairs in the source dataset (or a selected subset of these)

- Apply the chosen radial basis function, $\phi()$, to each distance in D, to produce a new array Φ

- Augment Φ with a unit column vector and a unit row vector, plus a single entry 0 in position $(n+1),(n+1)$. Call this augmented matrix A — see below for an illustration

- Compute the column vector r of distances from the grid point, p, to each of the source data points used to create D

- Apply the chosen radial basis function to each distance in r, to produce a column vector ϕ and then create the $(n+1)$ column vector c as ϕ plus a single 1 as the last entry

- Compute the matrix product $b = A^{-1}c$. This provides the set of n weights to be used in the calculation of the estimated value at p, plus the Lagrangian value m using the linear equation cited earlier in this Section. The value m is not used in calculating the estimate at p in this formulation

In matrix form the system of linear equations being solved is of the form:

$$\begin{bmatrix} \Phi & 1 \\ 1' & 0 \end{bmatrix} \begin{bmatrix} \lambda \\ m \end{bmatrix} = \begin{bmatrix} \varphi \\ 1 \end{bmatrix}$$

i.e. **Ab=c** hence **b=A⁻¹c**.

A variety of different radial basis functions may be used. A selection of those commonly available follows. Where more than one functional form is cited by separate GIS software suppliers the name of the package has been included. The parameter c in these expressions (not to be confused with the vector **c** above), which may be 0, determines the amount of smoothing. For an example implementation using a number of alternative Radial Basis functions on track (transect-like) data see Carlson and Foley (1991, 1992).

Multiquadric:

$$\phi_1(r) = \sqrt{r^2 + c^2}$$

If the application of this function is restricted to some range, a, and we write $h = r/a$, then

$$\phi_1(h) = \sqrt{h^2 + c^2}$$

is approximately linear over the range $[c,1]$ and approximately c over the range $[0,c]$.

Inverse multiquadric:

$$\phi_2(r) = \frac{1}{\sqrt{r^2 + c^2}}$$

Thin plate spline:

$$\phi_3(r) = c^2 r^2 \ln(cr) \text{ (ArcGIS)}$$

$$\phi_3(r) = (c^2 + r^2)\ln(c^2 + r^2) \text{ (Surfer)}$$

Multilog:

$$\phi_4(r) = \ln(c^2 + r^2)$$

Natural cubic spline:

$$\phi_5(r) = (c^2 + r^2)^{3/2} \text{ (Surfer)}$$

Spline with tension:

$$\phi_6(r) = \ln(cr/2) + I_0(cr) + \gamma \text{ (ArcGIS)}$$

where $I_0()$ is the modified Bessel function and $\gamma = 0.5771...$ is Euler's constant. As noted earlier in this Guide (Table 1.4) the modified Bessel function of order 0 is given by:

$$I_0(cr) = \sum_{i=0}^{\infty} \frac{(-1)^i (cr/2)^{2i}}{(i!)^2}$$

Completely regularised spline function:

$$\phi_7(r) = \ln(cr/2)^2 + E_1(cr)^2 + \gamma \text{ (ArcGIS)}$$

where $E_1()$ is the exponential integral function and γ is Euler's constant, as before. The exponential integral function is given by:

$$E_1(x) = \int_1^{\infty} \frac{e^{-tx}}{t} dt$$

In each of the above models the smoothing parameter c remains to be determined. ArcGIS Geostatistical Analyst seeks to optimise c by computing the RMSE of the prediction versus the data point values using simple cross-validation. Surfer selects a default value based on the diagonal length of the MBR of the data points and the number of points this MBR contains. This may then be user modified to meet the user's requirements.

6.6.5 Modified Shepard

There are many variations on Shepard's original interpolation method, which is essentially the standard inverse distance (IDW) procedure. One variant introduced by Shepard involves the use of two separate powers: a lower value (generally 2) for nearby data points and a higher value (generally 4) for points further away. Another variant, implemented in some packages, adjusts the weights based on how far away the furthest point (in the entire set or within a given radius) is to be found. If this distance is R then a revised IDW formula is:

$$z_j = k_j \sum_{i=1}^{n} \left(\frac{R - d_{ij}}{R d_{ij}} \right)^{\alpha} z_i$$

The value k_j in this expression is:

$$k_j = \sum_{i=1}^{n} \left(\frac{R - d_{ij}}{R d_{ij}} \right)^{\alpha}$$

This is the form used in the Groundwater Modelling System (GMS) of Environmental Modelling Systems Inc.

Surfer uses a more complex procedure based on a local quadratic polynomial fit in the neighbourhood of each data point. This is sometimes called the modified quadratic Shepard's method — for an example implementation see Carlson and Foley (1991, 1992). The procedure then continues as an inverse distance model using surface values obtained from the fitted quadratic surface rather than the original data points. The result is an IDW-type interpolator without the strong bull's-eye effect that simple IDW can suffer from. The method can be exact or approximate, depending on whether a smoothing factor is specified (see Figure 6-29H).

6.6.6 Triangulation with linear interpolation

The standard form of interpolation using triangulation of the data points is a widely available exact method. The Delaunay triangulation of the point set is first computed with the z-values of the vertices determining the tilt of the triangles. Interpolation is then simply a matter of identifying the value at each grid node by linearly interpolating within the relevant triangle.

Linear interpolation may be achieved, for example, using matrix determinants in a similar manner to those used in Section 4.2.3.1. A plane surface through three points $\{x_i, y_i, z_i\}$, $i=1,2,3$ has a formula in determinant form of:

$$\begin{vmatrix} x & y & z & 1 \\ x_1 & y_1 & z_1 & 1 \\ x_2 & y_2 & z_2 & 1 \\ x_3 & y_3 & z_3 & 1 \end{vmatrix} = 0$$

For example, consider a TIN element defined by the three coordinates: $(0,10,10)$, $(10,0,20)$, $(0,0,5)$. Points in the planar triangle defined by the three (x,y) coordinates (including the vertices) may be interpolated from the above plane surface. In this case evaluating the determinant gives $z = 5 + 3x/2 + y/2$, from which all values may be computed.

6.6.7 Triangulation with spline-like interpolation

A number of other procedures exist for interpolation of smooth surfaces from TIN representations. Many of these have their origins in computer-aided geometric design (CAGD), with engineering and visualisation applications. Some software packages, such as GMS, include CAGD interpolation methods, notably Clough-Tocher procedures. These involve fitting a piecewise two-dimensional cubic (i.e. a bicubic) function to each triangular patch, rather as per piecewise cubic spline interpolation on a regular grid. The fitted function may be regarded as being of the form:

$$f(x,y)=a_1+a_2x+a_3y+a_4x^2+a_5y^2+a_6xy+a_7x^2y+a_8xy^2$$

$$+a_9x^3+a_{10}y^3$$

The vertices of each triangle provide 3 values, $z=f(x,y)$, and the first-order partial differentials estimated at each vertex $z_y=f_y(x,y)$ and $z_x=f_x(x,y)$ provide a further 6, giving a total of 9 values from which the constants in the complete bicubic expression may be determined (Figure 6-36). However, there are 10 such constants, and at least one additional value is needed to uniquely determine the fitted function. In practice the original triangle is subdivided into three sub-triangles using the centroid of the main triangle. Each sub-triangle includes one edge from the original triangle and the first-order derivative of the Normal to the surface at the midpoint of this edge is used to provide the final parameter. Separate bicubic patches are then fitted to each of the three subtriangles, which collectively provide a continuous spline-like surface approximation at all points. This procedure is exact, fast and smooth, with interpolated values inside the triangle having values that may be above or below the maximum and minimum values of the vertex z-values.

Figure 6-36 Clough-Tocher TIN interpolation

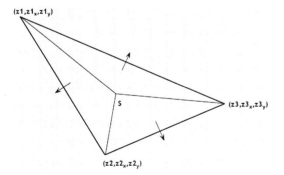

(z1,z1$_x$,z1$_y$)

s

(z3,z3$_x$,z3$_y$)

(z2,z2$_x$,z2$_y$)

6.6.8 Rectangular or bi-linear interpolation

With a reasonably dense and uniform arrangement of data points (or with a grid image file) a number of simple and fast interpolation procedures can be performed. A common procedure for non-uniform data involves dividing the area around a grid point to be interpolated with a segmented circle enclosing a number of neighbouring data points. The nearest data point in each of four segments or sectors of the circle is then selected and a 4-sided polygon defined connecting these points. This polygon can be used to define a weighted average for the grid point, either directly (e.g. using simple inverse distance weighting) or indirectly by assigning an adjusted value to each of the 4 grid cells that surround the sample point and taking the average of these. In either case a square or rectangle is determined which again may be used to assign a linearly interpolated value to the central grid point. The cell-based version of this procedure is widely used in image processing, including re-scaling/resampling operations.

6.6.9 Profiling

This is a surface fitting procedure that utilises elevation contours to create an interpolated grid surface. The standard form, implemented in TNTMips for example, is a simple linear interpolation based on the nearest pair of contour lines. Nearest is defined as the closest to the selected grid point in any one of 8 directions (i.e. four direction pairs). The resultant grid will provide an exact interpolation at contour locations but will be discontinuous across contour lines. This procedure will flatten ridges and "fill" valleys unless a non-linear interpolant is used. For terrain surfaces procedures that correctly model the hydrological structure of the landscape are preferable, e.g. Topogrid-type interpolation (see Section 6.6.14).

6.6.10 Polynomial regression

Polynomial regression provides interpolation by approximating the source data points by a global polynomial expression. The surfaces are typically simple polynomial functions of order 1, 2 or 3, fitted to the point set by ordinary least squares. Such surfaces are described and illustrated in Sections 5.6, 6.1.4 and 6.5.1). Typically global polynomial surface fitting is not used as a means of interpolation, but rather as a form of trend analysis, with further analysis being carried out on the residuals obtained when the fitted surface values are subtracted from those of the source. Linear fitting and the computation of residuals is also the first stage in the minimum curvature interpolation method, described in Section 6.6.11.

6.6.11 Minimum curvature

Minimum curvature is similar to bi-cubic spline fitting, in that the surface is modelled assuming that a smooth elastic-like membrane is used to approximate the surface. However, the interpolation is not exact, but close to exact, and is designed to ensure that the amount of surface curvature is as small as possible. This is achieved through a multi-stage process. As mentioned in Section 6.6.10 the first stage is simple linear regression and extraction of residuals. The procedure uses these residuals for interpolation rather than the original data points, but then it adds the regression surface data values back on completion. The interpolation process itself involves an iterative procedure that seeks to smooth the interpolated grid to a pre-specified parameter setting using a grid which is progressively made increasingly fine. The procedure is quite complex and is described in detail in Smith and Wessel (1990).

6.6.12 Moving average

This method of interpolation involves simple averaging using a moving window (usually a circle or ellipse). For each interpolated grid point a circle of specified radius (or an ellipse defined with semi-major axis, semi-minor axis and orientation angle) is placed with its centre at the grid point. All data points that lie within the window are averaged to produce the estimate at the grid point. If there are no data points within the window the selected point has no value assigned to it. The method is thus heavily dependent on the choice of window size and orientation, and the density of source data points. The resulting surface is simple and fast to generate, but may exhibit discontinuities.

6.6.13 Local polynomial

Local polynomial interpolation can be seen as a combination of (global) polynomial methods and the moving average procedure. As with global polynomials a least square polynomial fit to the data is applied, with options for Order 1, 2 or 3 equations. However, instead of fitting the polynomial to the entire dataset it is fitted to a local subset defined by a window, as in the moving average model. The size of this window needs to be large enough for a reasonable number of data points to be included in the process. One further adjustment is made to this procedure — a measure of distance-based weighting is included, so the least squares model is in fact a weighted least squares fit. The weights are computed using a power function of distance as a fraction of the window size. The simplest case is where the moving window is a circle, radius R. If the distance between grid point (x_i, y_i) and a data point (x, y) within the circle is denoted d_i, then the weight w_i is defined as:

$$w_i = \left(1 - \frac{d_i}{R}\right)^p$$

where p is a user definable power. The least squares procedure then involves minimising the expression:

$$\sum_{i=1}^{n} w_i (f(x_i, y_i) - z_i)^2$$

If *p*=0 all the weights are 1. The default value for *p* within Surfer is 2 and the maximum 20.

6.6.14 Topogrid/Topo to raster

This is a specific technique designed to facilitate conversion of vector surface data, in the form of contours, to hydrologically consistent DEM format. It is provided within ArcGIS and based directly on the work of Hutchinson (1988, 1989) and his ANUDEM (Australian National University DEM) program implementation. Surface interpolation uses a form of thin plate spline that is discretised to allow for discontinuities in the modelled surfaces (ridge lines, cliffs etc.). In order to ensure consistency in the drainage model it pre-fills any pits that it considers spurious (see further, Section 6.2.6 on pit filling and Section 6.4 on watersheds and drainage). Hutchinson describes ANUDEM as follows:

"ANUDEM ensures good shape and drainage structure in the calculated DEMs in five main ways by:

- Imposing a drainage enforcement condition on the fitted grid values that automatically removes spurious sinks or pits. This eliminates one of the main weaknesses of elevation grids produced by general purpose interpolation techniques. It greatly improves the utility of the DEM for hydrological applications. It can also aid in the efficient detection of data errors

- Incorporating surface drainage constraints directly from input streamline data

- Delineating ridges and streams automatically from input contour line data. This is achieved by inserting curvilinear ridge and streamlines associated with corners of contour lines that indicate where these lines cross the elevation contours

- Breaking the continuity of the DEM over data cliff lines

- Ensuring compatibility of lake boundaries with the elevations of connecting streamlines and neighbouring DEM points

The drainage enforcement algorithm is one of the principal innovations of ANUDEM. It has been found in practice to be a powerful condition that can significantly increase the accuracy, especially in terms of drainage properties, of digital elevation models interpolated from both sparse and dense elevation data... The essence of the drainage enforcement algorithm is to find for each sink point the lowest adjacent saddle point that leads to a lower data point, sink or edge and enforcing a descending chain condition from the sink, via the intervening saddle, to the lower data point, sink or edge (Hutchinson 1989). This action is not executed if a conflicting elevation data point has been allocated to the saddle. The action of the drainage enforcement algorithm is modified by the systematic application of two user supplied elevation tolerances."

For further details refer to the ANUDEM website (see Appendix: Web links for more details).

6.7 Geostatistical Interpolation Methods

6.7.1 Core concepts

Geostatistical interpolation differs from the procedures described in Section 6.6 in that it assumes the data point values represent a sample from some underlying true population. By analysing this sample it is often possible to derive a general model that describes how the sample values vary with distance (and optionally, direction). This model may be then used to interpolate, or *predict*, values at unsampled locations, in much the same way as with deterministic interpolation. If the samples meet various additional conditions it may also be possible to provide estimated confidence intervals for these predictions. Geostatistical interpolation methods attempt to address questions such as:

- how many points are need to compute the local average?

- what size, orientation and shape of neighbourhood should be chosen?

- what model and weights should be used to compute the local average?

- what errors (uncertainties) are associated with interpolated values?

A substantial number of specialist terms are used in the field of geostatistics and many of these are explained in subsection 6.7.1.1, before we proceed to describe specific geostatistical models. These include:

- Geostatistics

- Semivariance

- Sample size

- Support

- Declustering

- Variogram

- Stationarity

- Sill, range and nugget

- Transformation

- Anisotropy

- Indicator semivariance

- Cross-semivariance

- Comments on geostatistical software packages

- Semivariance modelling

- Fractal analysis

- Madograms and Rodograms, and

- Periodograms and Fourier analysis

There is an enormous volume of literature relating to geostatistics and geostatistical methods. Basic introductions in the GIS literature are provided in O'Sullivan and Unwin (2003) and Burrough and McDonnell (1998) amongst others. More advanced books include those by Goovaerts (1997, 1999), Deutsch and Journel (1992), Cressie (1991) and Isaaks and Srivastava (1989). The European Commission website devoted to Geostatistics (www.ai-geostats.org) is a useful starting point. This site lists articles, books, courses and conferences, software products and datasets available within the field of spatial statistics in general and geostatistics in particular. There is also some controversy about the usage of geostatistics, as there is with many attempts to apply statistical methods to spatial phenomena.

6.7.1.1 Geostatistics

We can define geostatistics as referring to "…models and methods for data observed at a discrete set of locations, such that the observed value, z_i, is either a direct measurement of, or is statistically related to, the value of an underlying continuous spatial phenomenon, $F(x,y)$, at the corresponding sampled location (x_i,y_i) within some spatial region A." This definition is based on that drawn up by Professor Diggle and his colleagues at the University of Lancaster, Spatial Statistics Group. It emphasises the fact that the term is particular to the analysis of continuous spatial phenomena, which is precisely the area that is of interest in 2D interpolation. It also highlights the statistical nature of the procedures involved, although rigid adherence to statistical requirements is not a pre-requisite for applying such techniques — rather it is a pre-requisite for interpretation of some of the results, especially those relating to statistical measures such as the estimation of prediction errors and confidence bands. As a more specific resource on geospatial analysis, in particular geostatistics and associated software packages, the European Commission's AI-GEOSTATS website (www.ai-geostats.org) is highly recommended.

For the purposes of the present Section we focus on the application of geostatistical methods to interpolation problems. As noted in Section 6.7.1, geostatistical interpolation differs from deterministic interpolation in that it assumes the source data points are a specific statistical sample or realisation from some true underlying surface function, and this sample must first be analysed in order to create a suitable model that will provide the best possible estimate of this underlying surface.

The initial analysis and modelling stage involves examining the dataset for spatial autocorrelation, essentially using similar techniques to those described in Section 5.5.2.2. Of particular relevance here is the Geary C statistics which (ignoring weights for the present) is given by:

$$C = \frac{1}{2} \frac{\sum_i \sum_j (z_i - z_j)^2}{\sum_i (z_i - \bar{z})^2}$$

and the correlogram form of the Moran I statistic:

$$I(h) = N(h) \frac{\sum_i \sum_j z_i z_j}{\sum_i z_i^2}$$

where, as in Section 5.5.2.2, $N(h)$ is the number of points in a distance band of width d or Δ and mean distance of h; z_i is the (standardised) value at point i; and z_j is the (standardised) value at a point j distance h from i (in practice within a distance band from h-$\Delta/2$ to h+$\Delta/2$). The summations apply for all point pairs in the selected band.

6.7.1.2 Geostatistical references

There is an enormous volume of literature relating to geostatistics and geostatistical methods. Basic introductions in the GIS literature are provided in O'Sullivan and Unwin (2003) and Burrough and McDonnell (1998) amongst others. More advanced books include those by Goovaerts (1997, 1999), Deutsch and Journel (1992), Cressie (1991) and Isaaks and Srivastava (1989). The European Commission website devoted to Geostatistics (www.ai-geostats.org) is a useful starting point. This site lists articles, books, courses and conferences, software products and datasets available within the field of spatial statistics in general and geostatistics in particular. There is also some controversy about the usage of geostatistics, as there is with many attempts to apply statistical methods to spatial phenomena.

6.7.1.3 Semivariance

The Geary and Moran ratios described in Section 6.7.1.1 are very similar to the functions used within geostatistics to understand the pattern of variation of observed surface values. These latter functions measure differences in observed values over a set of distance bands or *lags*. The function most commonly used is known as the (experimental) semivariance, and has the general form:

$$\hat{\gamma}(h) = \frac{1}{2N(h)} \sum_{d_{ij}=h-\Delta/2}^{d_{ij}=h+\Delta/2} (z_i - z_j)^2$$

where h is a fixed distance or lag, say 1000 metres, d_{ij} is the distance between points i and j, and Δ is the width of a band whose centre is at h, say 100m wide — hence in this case extending over the interval [950m,1050m]. The summation considers all pairs of observed data values whose spatial separation falls into the chosen band. There are $N(h)$ of these pairs, hence the expression measures the average squared dissimilarity between the data pairs in this band.

Important and useful relationships exist between the variogram, the variance (Var) and the covariance (Cov):

$2\gamma(h) = Var\{z(u+h) - z(u)\}, \forall u,$ and

$\gamma(h) = Cov(0) - Cov(h),$

where

$Cov(h) = E\{z(u+h)z(u)\} - E\{z(u)\}^2,$ and

$Cov(0) = Var(z(u))$

In these expressions the **u** vector is the set of locations at which the observations, $z(u)$, have been made, and **h** is a separation vector (i.e. with both distance and directional components). Cov(**h**) is the stationary variance and Cov(**u+h**) is the stationary covariance, for all **u** and **u+h**.

In practice, software implementations first compute the squared differences between all pairs of values in the dataset, and then allocate these to lag classes based on the distance (and optionally direction) separating the pair. This provides a set of semivariance values for distance lags, h, increasing in steps from 0 (or 0 plus some small increment) to a value somewhat less than the greatest distance between point pairs. When this set of values is plotted against the separation distance, h, it provides a graph known as a variogram (Figure 6-37).

Figure 6-37 Sample variogram

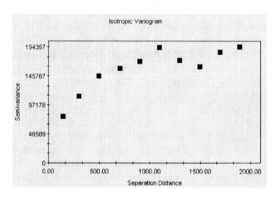

In this example the dataset consists of the 98 soil pollution data records provided in Burrough and McDonnell (1998, Appendix 3). The specific variable analysed is the level of zinc in parts per million (ppm) in the samples taken. Lags are shown at 200 metre intervals. The first black square shows the semivariance which has been calculated using 207 of the data pairs, i.e. those that fall into the distance band [0,200] metres. The total number of possible pairs in this dataset is 98*97/2=4753. This dataset will be used to illustrate several of the geostatistical procedures described in the following subsections. Note that the worked example in Burrough and McDonnell (1998, p140-141) using this dataset contains computational errors.

6.7.1.4 Sample size

With very dense data points many interpolation techniques will give similar results. As the density falls and becomes less regular selection of interpolation method becomes more critical and careful examination of the dataset and any known characteristics of the underlying surface becomes increasingly important. If the number of data points is too small (the sample size is too low) the method may either be unworkable (for example there are insufficient data pairs in a given distance band to provide meaningful information) or may produce unusable results. Some authors have suggested that ideally there should be at least 150 data points in order to obtain satisfactory semivariance estimates, although fewer points may be used with some success, especially where ancillary information about the data is available. Others suggest that the number of grid points interpolated should not be more than 20-50 times the number of data points. In practice it is the variability of the underlying surface and the importance of capturing a given level of detail that provides the best guidance. Conversely, prior knowledge of the patterns of variation of spatial data may be used to direct subsequent sampling schemes (e.g. point spacing on a regular grid) to have the best chance of ensuring that prediction errors lie within predetermined bounds.

6.7.1.5 Support

This is the term used to describe the length, area or volume applying to each of the (physical) samples taken, even though these may be nominally assigned to a point for computational purposes. Predicted values at other locations based on such data apply to similar supports at these locations. If a series of observations have been made in a small area or block, and then combined to produce an averaged value, again assigned to a point, the support would be comprised of this combined or "upscaled" data.

6.7.1.6 Declustering

Source data points are often preferentially sampled, such that some areas are more densely sampled than others. Where this occurs the data may not provide a truly random sample from the hypothetical population of all possible values, and hence the distribution of sampled values will not be Normal even after simple transforms are applied. Some packages (e.g. ArcGIS

Geostatistical Analyst, and those utilising GStat) offer procedures for declustering the source data points by assigning weights to these points. One simple procedure for weighting is to use the area of the Voronoi polygon surrounding the point to adjust for clustering (this topic is also discussed in Section 5.1.2.2).

6.7.1.7 Variogram

If the observed pattern of variation with distance (the variogram) can be modelled using a relatively simple equation (some function of distance), then values at unknown locations can be estimated by applying the model to obtain weights that may be used in an interpolation process, much the same as those described in Section 6.6. Indeed, there is a strong similarity between the methods used in radial basis interpolation (Section 6.6.4) and some forms of geostatistical interpolation. The modelling of variograms and similar graphs is described in Section 6.7.1.15. The use of such models in interpolation is described in Section 6.7.2. These methods are generally described as different forms of Kriging, so-named after the South African mining engineer, Krige, who introduced the general procedures used. Variograms can be re-scaled using the sample variance of the input data. This process is useful where several variograms on different variates are to be compared.

6.7.1.8 Stationarity

The term stationarity originates in the analysis of random processes, in particular in connection with time series. In this context a stationary random process is one for which all of its statistical properties (e.g. mean, variance, distribution, correlations etc) do not vary over time. In the context of spatial analysis a stationary process would be one that is stationary in space. Stationarity is often described in terms of first order, second order etc. First order stationary processes must have a constant mean, irrespective of the spatial configuration of sampled points. Second order stationarity implies that the autocorrelation function depends solely on the degree of separation of the observations in time or space. A non-stationary process would be one where the process varies depending on when in time or where in space it is examined. Several techniques, including a number of those in regression modelling (e.g. GWR) accept that many spatial processes will be non-stationary and seek to model this characteristic explicitly.

6.7.1.9 Sill, range and nugget

The increase in the semivariogram values with increasing lags seen in Figure 6-37 diminishes with distance and levels off, in this example at around 1100 metres. This distance is known as the *range* or active lag distance, and is the approximate distance at which spatial autocorrelation between data point pairs ceases or becomes much more variable. At this range a plateau or *sill* in the semivariance values has been reached (Figure 6-38). Such variograms are called *transitive*. *Non-transitive variograms* are ones in which the sill is not reached within the region of interest. When a curve is fitted to the set of points that lie within this range and sill the model used may have a zero or non-zero intercept with the y-axis. In Figure 6-38 a non-zero intercept is shown, which is known as the *nugget* or *nugget variance*, C_0, referring back to the original application of such methods to mineral exploration. The nugget is usually assumed to be non-spatial variation due to measurement error and variations in the data that relate to shorter ranges than the minimum sampled data spacing. The sill minus the nugget is sometimes known as the *partial sill* or *structural variance*, C. The values C and C_0 often appear as parameters in fitted models.

Figure 6-38 Sill, range and nugget

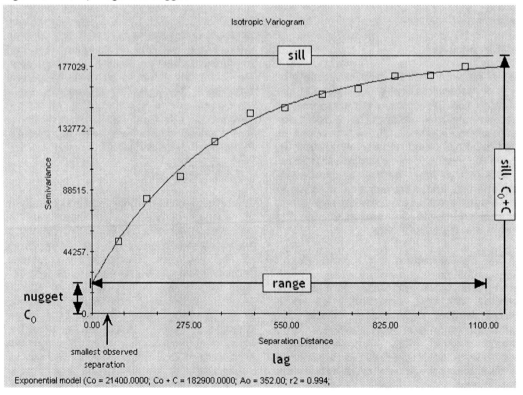

6.7.1.10 Transformation

Because the techniques involved make assumptions about the statistical nature of the observed data values it may be necessary to transform these values prior to analysis, in order to achieve a closer approximation to Normality. Frequently this is performed via a log or Box-Cox transform (see subsection 1.5.2.7 for a range of data transformation functions). In Figure 6-39 we show a series of transformations of 301 Radon level measurements made in part of South-West Ireland. These graphs and the associated test statistics were generated using the Minitab package. Three very large-valued outliers were removed from the original dataset prior to distribution analysis.

The raw data plotted in Figure 6-39A is clearly non-Normal, diverging substantially from the straight line which indicates a Normal distribution. In this case simple Log transformation of the data improves the fit to Normal but still diverges, even when corrections are made for background radiation (Figure 6-39C). In each of these three cases the Anderson-Darling test of fit to Normal fails, whereas for the Box-Cox transform (Figure 6-39D) with optimised parameter, k, the test passes and in this instance analysis proceeded using this specific transform. The Anderson-Darling test is a variation on the Kolmogorov-Smirnov test, again based on the cumulative distribution function, but is more sensitive to the tails of the distribution.

Figure 6-39 Data transformation for Normality

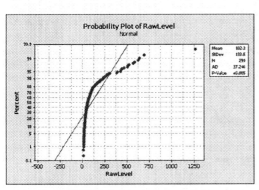

(a) Raw Radon Level

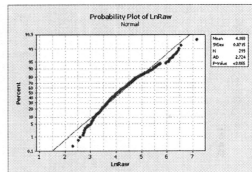

(b) Log$_e$ Raw Radon Level

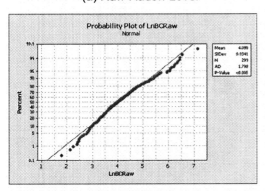

(c) Log$_e$ Background Corrected Radon Level

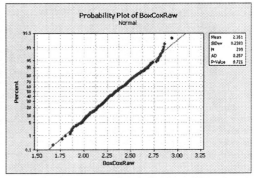

(d) Box-Cox Transformed Radon Level

Source: O'Sullivan F (2005)

Some software packages may automatically apply a transformation to the data before computing a variogram. Indicator semivariance analysis is a form of analysis on transformed variables, in this case thresholding to binary values. As with other variograms, indicator variograms may be calculated from the source data values transformed to binary, or re-scaled values using the variance of the sample data. Figure 6-40 illustrates both variants, using the same dataset as above. Typically the latter graph will show values in the range [0,1]. This scaling should be removed before proceeding with interpolation.

Figure 6-40 Indicator variograms

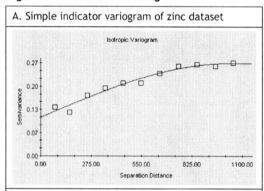

A. Simple indicator variogram of zinc dataset

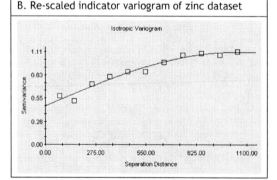

B. Re-scaled indicator variogram of zinc dataset

6.7.1.11 Anisotropy

In many instances the variogram is not of constant form in all directions. This feature is known as anisotropy, and the example shown

in Figure 6-37 is of an isotropic variogram, i.e. one in which any directional variations are ignored. An alternative approach to the calculation of the semivariance would be to divide the lag band into a series of discrete segments, say at 45 degree intervals, and then calculate separate variograms for each direction. Since the directions are not oriented (i.e. directions such as 0 and 180 degrees are treated as equivalent) there would be four such variograms in this case at 0, 45, 90 and 135 degrees. More generally the variogram can be thought of as a function of two variables or surface, $\gamma(\theta,h)$, that may or may not be radially symmetric.

The semivariance surface may be plotted as a 2D map or 3D surface representation (Figure 6-41). The centre of the map is the origin of the semivariogram, with a radially symmetric structure indicting that there is little or no anisotropy. If there is a strong directional bias, this can be regarded as the major axis of a directional ellipse, with a major axis providing the range in the primary direction, and the minor axis providing the range in the orthogonal direction. A single anisotropic model may be fitted to such datasets, or a series of separate models fitted to the data grouped into distinct directional bins.

Figure 6-41 Anisotropy 2D map, zinc data

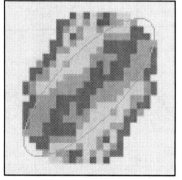

Not all software packages use radial sectors — some (including ArcGIS) use approximations to this form based on a grid structure.

6.7.1.12 Indicator semivariance

Some spatial variations of continuous datasets are less "smooth" than others. For example the concentrations of radioactive isotopes in soil samples is much lower than zinc or nickel, but may show marked peaks in some locations. Instead of using the z-values of the source data points directly in such cases, it may be more meaningful to record or transform the data as a binary set, with z-values above a pre-specified value as having a value of 1, and those below this threshold, t, having a value of 0 (or vice versa). The semivariogram formula is then applied, but this time to the $\{0,1\}$ values rather than the original values. These new values are called indicators, and are often denoted in the form $i(u,t)$. Indicator semivariance facilitates modelling of the probability that the variable of interest exceeds a given threshold value, t, at unknown points.

6.7.1.13 Cross-semivariance

Spatial datasets that contain measurements on two or more variables may be analysed in a similar manner to the procedures described for a single variate. Typically analysis is performed on pairs of variates, $\{z\}$ and $\{w\}$ say, where patterns of co-variation are of interest. Examination of such patterns may be of theoretical interest, for example in helping to understand underlying processes, or may assist in the prediction of one variate based on better information or more data that are available for another variate.

6.7.1.14 Comments on geostatistical software packages

The following (edited) observations on facilities provided in available geostatistical software (stand-alone or within/linked to GIS) are taken from AI-GEOSTATS. The AI-GEOSTATS original website (www.ai-geostats.org) has very recently undergone substantial re-development, so the comments have been updated and to some extent superseded with more up-to-date material.

Display of proportional/classed symbol maps: often the first step in the analysis of geostatistical data. Showing symbols with a size that is proportional or characteristic of their attributed value can help the data analyst to identify outliers and trends.

Moving windows statistics: geostatistics assumes second order stationarity in the data: the mean has to exist and is constant and independent of location within the region of stationarity. The covariance has to exist as well, and is only dependent on the distance between any two values, and not on the locations. One way to evaluate second order stationarity is to divide the studied area into cells and compare the mean and the covariance between all cells. The moving windows technique uses an overlap of the cells as if a window was moving over the whole dataset.

Sampling network analysis: techniques used to quantify the degree of irregularity of the sampling network. These include the analysis of: Thiessen polygons; the fractal dimension of the sampling network; and the use of the Morishita Index. Nearest-neighbours statistics and many other tests can also be used to describe the spatial distribution of the measurements.

Trend analysis: An underlying trend may appear in the data and needs to be taken into account when applying Ordinary Kriging (Universal Kriging is supposed to take the drift into account). The drift is frequently modelled with a simple linear or a quadratic model.

Declustering: When global estimations are desired one needs to obtain statistics that are representative of the whole area of interest. By attributing lower weights to clustered measurements one can reduce their influence and so limit the bias one would have obtained without a prior declustering. Two declustering techniques are frequently used: the cell and the polygonal declustering method.

Variogram analysis: The use of spatial autocorrelation being one of the main keys in a geostatistical case study, one would expect the GIS to provide at least the possibility to calculate the experimental semivariogram.

Interactive variography: The display of the experimental semivariogram is not enough to analyse the spatial autocorrelation. Prior to the modelling of the experimental semivariogram one needs to evaluate its robustness. The software Variowin is a tool for the interactive analysis of the spatial autocorrelation. Its success comes from the high level of interactivity between pairs of samples, h-scatterplots, variogram clouds and the experimental variogram. Such interactivity is necessary in order to identify and remove outliers as well as to evaluate the impact of the choice of the distance of the lags.

Whilst many software packages provide automatic fitting functions of the experimental semivariograms, most practitioners still prefer to adjust their model by hand. As an example, the additional knowledge of a component of the nugget effect (e.g. documented measurement errors) will help to define a more realistic value of the nugget than the one that would have been calculated with an automatic fitting. Therefore, the interactivity during the variogram modelling still remains an essential requirement of a good package.

In many case studies a spatial structuring of the studied phenomenon appears in a particular direction. Not taking this anisotropy into account may strongly affect conclusions derived from the analysis of the spatial correlation or from the estimations.

Models implemented: Very often, spatial interpolation functions are implemented like black boxes in GIS or contouring software. It is even truer with geostatistical techniques that require the users to understand properly the underlying theory. The quality of the documentation of the available geostatistical

functions should be a key for the GIS users willing to apply their tools to a geostatistical case study.

Additional interpolation/estimation functions are more than welcome since a geostatistical case study can be very time consuming, especially with very large datasets. On the other hand if one has the time to compare the efficiency of the various functions, one can expect to find that sometimes a non-geostatistical method has performed better, according to the user's criteria, than geostatistical techniques. Here again, clear documentation of the implemented functions is essential.

Search strategies: The search strategy of the neighbouring measurements to use during the estimation/interpolation step has clearly a strong impact on the estimates. One should ideally have the possibility to select: the number of measurements to use; the maximum distance at which one should search; and the measurements according to their locations. For example, one should have the possibility to select few points in each quarter of the circle used to select the neighbours. As for the analysis of the spatial correlation, one should have the possibility to use an ellipse of search rather than a circle

Masking: When the studied area has an irregular shape, or when estimates are not desired at certain locations, one should have the option to prevent certain regions from being interpolated by defining inclusive or exclusive polygons.

6.7.1.15 Semivariance modelling

Assuming a dataset has been examined, transformed where necessary and semivariances computed, modelling of the variogram may proceed. A large variety of alternative models have been proposed over recent years, and many of the more popular models are incorporated into GIS packages and geostatistical software. Table 6.3 shows a selection of the principal models provided within such packages and Figure 6-42 provides sample graphs for each of those listed. The majority of these models are included within ArcGIS Geostatistical Analyst (which also includes a number of other models), Vertical Mapper for MapInfo and Surfer, and in packages that are based on or utilise GStat, such as Idrisi, PCRaster and GRASS. Other products, such as GS+, GSLIB and Variowin (a free, interactive variogram analysis tool) may provide a smaller range of models but offer considerable flexibility in the modelling and display process. In Table 6.3 the distance variable, h, is pre-scaled by the effective range, a. In much of the literature and product documentation sets this scaling is not shown, so where we show h they may show h/a, or a similar ratio. Linear combinations of models are widely supported, but selecting and combining the component functions may be difficult.

Selecting and fitting variograms is something of an art rather than a science. Many packages provide a default model and attempt to find the best set of parameters to fit the dataset, whilst others apply this process for all models they support and select the one with the highest correlation coefficient or lowest residual sum of squares. Selecting only models that are asymptotic to the sill (i.e. to 1 in the diagrams shown) provides a useful first level of discrimination between functions. Both the selection of active lag distance to be considered, and the lag interval to be used, will affect the model fitted — the use of values that are too large in either case may result in over-large estimates of nugget variance. Examining the profiles of these functions it is clear that the main differences between them apply in the first quarter of the range, hence close examination of the data within this scale ($a/4$) is advisable. Anisotropy may be automatically modelled or require selection, with attributes such as size and type of sectors being specified. Optimised selection of model and/or parameters under anisotropy may well alter the major axis of anisotropy identified to some extent, depending on the model selected. Most packages do not allow interactive alteration of models and parameters (an exception being Variowin), but all will allow exploration by trial-and-error, re-calculating the model to user-defined specifications. One difficulty with these approaches is that it is very difficult to replicate results across packages, even when selected parameters are "forced" by the user.

Table 6.3 Variogram models (univariate, isotropic)

Model	Formula	Notes
Nugget effect	$\gamma(0) = C_0$	Simple constant. May be added to all models. Models with a nugget will not be exact
Linear	$\gamma(h) = C_1(h)$	No sill. Often used in combination with other functions. May be used as a ramp, with a constant sill value set at a range, a
Spherical Sph()	$\gamma(h) = C_1\left(\dfrac{3h}{2} - \dfrac{1}{2}h^3\right), h < 1$ $\gamma(h) = C_1, h \geq 1$	Useful when the nugget effect is important but small. Given as the default model in some packages.
Exponential Exp()	$\gamma(h) = C_1\left(1 - e^{-kh}\right)$	k is a constant, often $k=1$ or $k=3$. Useful when there is a larger nugget and slow rise to the sill — see for example, Figure 6-38
Gaussian	$\gamma(h) = C_1\left(1 - e^{-kh^2}\right)$	k is a constant, often $k=3$. Can be unstable without a nugget. Provides a more s-shaped curve
Quadratic	$\gamma(h) = C_1\left(2h - h^2\right), \gamma(h) = C_1, h \geq 1$	
Rational quadratic	$\gamma(h) = C_1\left(\dfrac{kh^2}{1 + kh^2}\right)$	k is a constant. ArcGIS uses $k=19$
Power	$\gamma(h) = C_1 h^n$	No sill. $0<n<2$ is a constant
Logarithmic	$\gamma(h) = C_1 \ln(h), h > 0$	No sill.
Cubic	$\gamma(h) = C_1(7h^2 - 8.75h^3 + 3.5h^5 - 0.75h^7)$	Compare to Gaussian — S-shaped curve with well-defined sill
Tetraspherical	$\gamma(h) = \dfrac{2C_1}{\pi}\left(\dfrac{\sin^{-1}(h) + h\sqrt{1 - h^2} +}{\dfrac{2h}{3}\left(1 - h^2\right)^{3/2}}\right), h \leq 1$ $\gamma(h) = 1, h > 1$	Similar to Circular with extra final term
Pentaspherical	$\gamma(h) = C_1(1.875h - 1.25h^3 + 0.375h^5)$	
Wave hole effect	$\gamma(h) = C_1\left(1 - \dfrac{\sin(kh)}{kh}\right), h > 0$	k is a constant, often 2π. Useful where periodic patterns in the data with distance are observed or expected
Circular	$\gamma(h) = \dfrac{2C_1}{\pi}\left(\sin^{-1}(h) + h\sqrt{1 - h^2}\right), h \leq 1$ $\gamma(h) = 1, h > 1$	

Figure 6-42 Variogram models — graphs

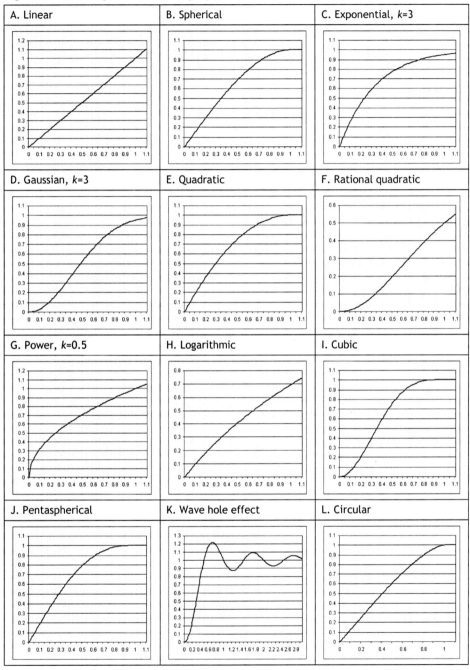

A. Linear

B. Spherical

C. Exponential, *k*=3

D. Gaussian, *k*=3

E. Quadratic

F. Rational quadratic

G. Power, *k*=0.5

H. Logarithmic

I. Cubic

J. Pentaspherical

K. Wave hole effect

L. Circular

6.7.1.16 Fractal analysis

Determining the fractal dimension of terrain surfaces is a contentious process. In practice such analysis involves obtaining an estimate that is highly dependent on the way in which the data have been captured and stored, and the modelling choices made by the user, rather than simply being a reflection of the complexity of original surface. For many datasets the fractal dimension as computed in one program will differ from that computed by other programs, and may only apply across a limited scale range.

Figure 6-43 shows the result of fractal analysis of one of our test surfaces, OS tile TQ81NE, using the Landserf package. The fractal dimension in this example is computed by selecting a fixed interval lag and computing the standard variogram for the surface based on this lag, assuming an isotropic process, and then plotting the log of the lag against the log of the variogram. The best fit line $y=ax+b$ is then used to compute the fractal or capacity dimension, D_C, using the expression:

$D_C=3-a/2$

Hence in this example $D_C=2.2$. The Pentland Hills test dataset (tile OS NT04) has a higher value of 2.4, although interpretation of the computed differences is not straightforward.

Figure 6-43 Fractal analysis of TQ81NE

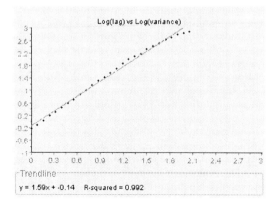

This approach yields a single measure of fractal dimension for the entire surface.

Idrisi computes fractal values for an entire grid using a 3x3 window or kernel, based on the slope values computed for that window. If s_{ij} is the slope value in degrees from 0 at location (i,j) then the fractal dimension, $D_{C,ij}$, is computed as:

$$D_{C,ij} = \frac{\log(2)}{\log(2) + \log\left(\sin\left(\frac{180 - s_{ij}}{2}\right)\right)}$$

where the logs can be in any base. This formula results in a linear (profile-like) fractal measure, providing an index of surface texture in the range [1,2].

6.7.1.17 Madograms and Rodograms

Madograms are computed in the same way as variograms, but the absolute values of lagged differences are calculated rather than the squared differences. Hence the equivalent expression to the semivariogram is:

$$\hat{\gamma}M(h) = \frac{1}{2N(h)} \sum_{d_{ij}=h-\Delta/2}^{d_{ij}=h+\Delta/2} \left|z_i - z_j\right|$$

The use of absolute values rather than squared differences reduces the impact of clustering and outliers, assisting in the determination of range and anisotropy for data with these characteristics. They may be seen as a supplement to variogram analysis and are not used subsequently in interpolation. Madograms should not be used for modelling nugget variance.

Rodograms are essentially the same as Madograms, and are used in a similar manner, but take the square root of the absolute values:

$$\hat{\gamma}R(h) = \frac{1}{2N(h)} \sum_{d_{ij}=h-\Delta/2}^{d_{ij}=h+\Delta/2} \left|z_i - z_j\right|^{1/2}$$

6.7.1.18 Periodograms and Fourier analysis

A surface represented as a grid, $z(x,y)$, may be analysed to identify possibly two-dimensional periodicities in the data using a combination of sin() and cos() functions. Surfer for example, analyses the grid looking for frequencies that are integer multiples of $2\pi/k$, where k represents the number of grid cells in the x or y directions. This form of analysis, whilst widespread in many disciplines, has only been applied in a few areas of spatial analysis, for example meteorological modelling.

6.7.2 Kriging interpolation

6.7.2.1 Core process

Burrough and McDonnell (1998, p.155) describe Kriging as a multi-stage process of interpolation that builds upon calculation and modelling of the experimental variogram. Slightly modifying their summary we have the following stages:

- examine the data for Normality and trends — transform where necessary. Note that Kriging only requires Normality in order to create confidence intervals (probability maps). Whatever the data distribution, Kriging will provide a best unbiased predictor of values at unsampled points

- compute the experimental variogram and fit a suitable model. If the nugget value is very high then interpolation is not sensible

- check the model by cross-validation, and/or by the inclusion of a separate validation file (e.g. additional measurements not included in the modelling process). Examine the detailed validation results for both size and sign of deviations, and any unexpected patterns that these show (e.g. spikes, spatial bias)

- decide on whether to apply Kriging or conditional simulation (see further Section 6.7.2.12. The latter is becoming increasingly widely used, but is not

supported in many geostatistical packages — GSLIB and GS+ Version 7 are exceptions)

- if using Kriging, then use the variogram model to interpolate sites onto a regular grid where the sites are either equal in size to the original sample (Point Kriging) or are larger blocks of land (Block Kriging). Block Kriging generates smoother surfaces — it estimates average values over a block. Kriging is not exact if a nugget is included

- display results as grid cell maps or contours, singly or draped over other data layers. Both predictions and standard deviations of predictions should be computed and mapped in most cases. If the random errors are Normally distributed and satisfy stationarity conditions then probability, quantile or confidence maps can be produced

The basic geostatistical interpolation procedure, assuming Ordinary Kriging (OK) for the time being, is essentially identical to that used in deterministic interpolation, notably interpolation using radial basis functions. The values at unsampled (grid) points are computed as a simple linear weighted average of neighbouring measured data points, where the (optimal) weights are determined from the fitted variogram rather than determined by the user. For a specific grid point, p, we have:

$$z_p = \sum_{i=1}^{n} \lambda_i z_i$$

with the condition that the weights must add up to 1:

$$\sum_{i=1}^{n} \lambda_i = 1$$

In Kriging these weights may be positive or negative. The condition that the weights must sum to 1 is equivalent to a process of re-estimating the mean value at each new location. Hence Ordinary Kriging is essentially the same as Simple Kriging (see 6.7.2.3) with a location-dependent mean.

Kriging assumes a general form of surface model which is a mix of:

- a deterministic model $m(x,y)$, sometimes called the *structural component* of z at locations (x,y) — in OK this is assumed to be an unknown mean value, i.e. a constant, within a given neighbourhood under consideration

- a regionalised statistical variation from $z(x,y)$, which we will denote $e_1(x,y)$ — in OK and other Kriging methods this regionalised variation is determined by modelling the semivariance and using the fitted function in the interpolation process

- a random noise (Normal error) component, with 0 mean, which we will denote $e_2(x,y)$ (actually two components, measurement error and random or "white" noise, which we cannot generally separate)

In equation form this model is of the form:

$$Z(x,y)=m(x,y)+e_1(x,y)+e_2(x,y)$$

or combining the two error terms and using Greek letter notation:

$$Z(x,y)=\mu(x,y)+\varepsilon(x,y)$$

If the vector **s** is used to represent the surface coordinates (x,y) then the standard model is often written as:

$$Z(s)=\mu(s)+\varepsilon(s)$$

If there is no overall trend or *drift* present then $m(x,y)$ or $\mu(s)$ is simply the overall (local) mean value of the sampled data. If a marked trend is observed this may be removed before modelling in most packages (usually a linear or quadratic regression surface) and then added back in at the end of the process. The procedure for Ordinary Kriging is essentially the same as we provided earlier for radial

basis interpolation (Section 6.6.4), but with the second step using the modelled semivariances rather than a user-selected function. As before, the steps are:

- Compute the $n \times n$ matrix, **D**, of inter-point distances between all (x,y) pairs in the source dataset (or a selected subset of these)

- Apply the chosen variogram model, $\phi()$, to each distance in **D**, to produce a new array **Φ**

- Augment **Φ** with a unit column vector and a unit row vector, plus a single entry 0 in position $(n+1),(n+1)$. Call this augmented array **A** — see below for an illustration

- Compute the column vector **r** of distances from a grid point, p, to each of the n source data points used to create **D**

- Apply the chosen variogram model, $\phi()$, to each distance in **r**, to produce a column vector $\varphi=\phi(r)$ of modelled semivariances augmented with a single 1 as the last entry. Call this augmented column vector **c** — see below for an illustration

- Compute the matrix product $b=A^{-1}c$. This provides the set of n weights, λ_i, to be used in the calculation of the estimated value at p, plus the Lagrangian value, m, which is utilised in the calculation of variances as described below

In matrix form the system of linear equations being solved is of the form:

$$\begin{bmatrix} \Phi & 1 \\ 1' & 0 \end{bmatrix} \begin{bmatrix} \lambda \\ m \end{bmatrix} = \begin{bmatrix} \varphi \\ 1 \end{bmatrix}$$

Or more simply **Ab=c**, hence $b=A^{-1}c$.

The estimated value for point p is computed using the n weights, λ_i, (but ignoring the Lagrangian, m) in the normal manner:

$$z_p = \sum_{i=1}^{n} \lambda_i z_i$$

or using matrix notation:

$$z_p = \boldsymbol{\lambda}' \mathbf{z}$$

The estimated variance at point p is then determined from the computed weights, times the modelled semivariance values, plus the Lagrangian, m:

$$\hat{\sigma}_p^2 = \sum_{i=1}^{n} \lambda_i c_i + m$$

or again using matrix notation:

$$\hat{\sigma}_p^2 = \boldsymbol{\lambda}' \boldsymbol{\varphi} + m = \mathbf{b}' \mathbf{c}$$

The variance estimates can be plotted directly and/or used to create confidence intervals for the modelled data surface under the (possibly risky) assumption that modelled random errors are distributed as Normal variates and are second-order stationary. Note that if the estimated points, p, lie on a fine grid, then average predicted values over discrete subsets of the grid (blocks) will be similar to those obtained by applying block Kriging to the same area, albeit with far more computation.

6.7.2.2 Goodness of fit

Having undertaken Kriging interpolation a variety of methods and procedures exist to enable the quality of the results, or goodness of fit, to be evaluated. These include:

- Residual plots (sometimes referred to as error plots) and standard deviation maps; maps of residuals from trend analyses

- Crossvalidation — this typically involves removing each observed data point one at a time (leaving the model otherwise unaltered) and then calculating the predicted values at these points. Best fit models should minimise residuals (often

referred to as errors) in these predictions: minimising maximum residuals/errors; minimising total sum of squared residuals/errors; removing extreme residual values (outliers/invalid results)…. Some packages allow the validation results to be exported for further examination and statistical analysis

- Jack-knifing — essentially a form of cross-validation achieved by statistically re-sampling the source dataset. Typically one or more subsets consisting of x% of data points are removed at random (without replacement) and statistics of interest calculated on these subsets, which are then compared to the global value. Note that the term *bootstrapping*, often used alongside jack-knifing, refers to procedures involving re-sampling with replacement

- Re-sampling the source material (field data) — an expensive or impractical option in many cases, but a realistic option in others

- Detailed modelling of related datasets (e.g. easy to measure variables)

- Detailed modelling incorporating related datasets: look at possible stratification of the source dataset (e.g. using underlying geological information); explicitly model non-stationarity, if appropriate; explicitly incorporate boundaries/faults or similar linear or areal elements that may substantive affect the results

- Comparison with independent data (e.g. satellite data or other aerial imagery)

- Examination the fit for possible artefacts and bull's-eyes (a form of circular artefact) — these may be generated by the interpolation process and/or be a feature inherited from the underlying dataset

6.7.2.3 Simple Kriging

Simple Kriging assumes that the data have a known, constant, mean value throughout the study area (i.e. a global mean value, in practice 0, is a suitable assumed value for modelling purposes) and exhibits second order stationarity. These assumptions are overly restrictive for most problems and hence this method is rarely used. De-trending and z-score (Normal) transformation may help to remove some of these problems. Ordinary Kriging (OK) and its variants have more relaxed assumptions than Simple Kriging (subsection 6.7.2.4). OK is the mostly widespread procedure implemented in GIS packages.

6.7.2.4 Ordinary Kriging

Ordinary Kriging (OK), plus variants of OK, is the most widespread procedure of this type implemented in GIS packages. OK has been described in more detail at the start of Section 6.7.2. As with other forms of Kriging, OK methods may use point or block computations, the latter resulting in a smoothed surface and inexact interpolation. As noted earlier, if the vector s is used to represent the surface coordinates (x,y) then the standard model is often written as:

$$Z(s)=\mu(s)+\varepsilon(s)$$

Figure 6-44 shows the result of applying Ordinary Kriging to the zinc dataset discussed earlier (Figure 6-37 and Figure 6-38; from Burrough and McDonnell (1998, Appendix 3)). Figure 6-44A provides a map of the predicted z-values, whilst Figure 6-44B shows the estimated standard deviations. In each of these the variogram model is based on the simple isotropic exponential formula shown earlier (Figure 6-38) and the analysis is based on untransformed data.

The predicted z-values appear as a smoothly varying map, whilst the predicted standard deviations appear somewhat as bull's-eyes around the original source data points. To the east of the region, where there are no source data points, uncertainty is much greater and the values of the results in this area are of little value. To the west of the region is a masked-off area, a facility which is explicitly supported within some packages. The fact that there are no data points in extensive parts of the sampled region presents problems, including issues of apparent anisotropy and questions as to the appropriateness of a broadly radial modelling approach. Substantial improvements to this approach (in terms of standard deviations) are reported by Burrough and McDonnell (1998, p.156) by the use of stratification (see further Section 6.7.2.10).

Figure 6-44 Ordinary Kriging of untransformed zinc dataset

A. z-values (predictions)

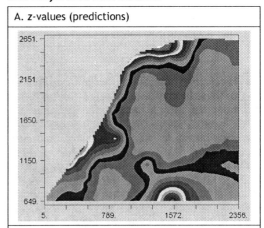

B. Standard deviations

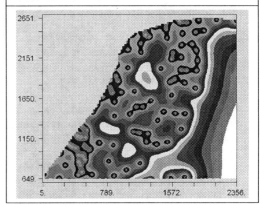

6.7.2.5 Universal Kriging

Also called 'Kriging with a trend', or KT Kriging, Universal Kriging (UK) is a procedure that uses a regression model as part of the Kriging process, typically modelling the unknown local mean values as having a local linear or quadratic trend. UK with no trend is the same as OK. Ordinary Kriging with a deterministic trend is often used in preference to UK as the modelling process is simpler and the results using the two approaches are often very similar.

6.7.2.6 Median-Polishing and Kriging

As an alternative to Universal Kriging, which involves identification and removal of trends in the data, some packages (such as Manifold) support a technique known as median-polishing. This involves finding the median of observed (sampled) data values in a neighbourhood (typically using a rectangular grid), generating row and column medians of these values in an iterative process ('polishing' the dataset). The median values are then subtracted from the original data points as local estimates of the mean value and Kriging of the residuals is carried out using either OK or UK. For a discussion and application of such methods see Berke (2000, 2004).

6.7.2.7 Indicator Kriging

Indicator Kriging is essentially the same as Ordinary Kriging, but using the binary data generated using the thresholding procedure described in Section 6.7.1.15 and illustrated in Figure 6-40. Indicator Kriging can be used with a nugget effect, and Co-Kriging (Section 6.7.2.11) of indicator data is also possible and supported in some packages. In the latter case, a second thresholding procedure is used and then both binary variables employed to predict values at unknown points for the first indicator variable. Note that the output does not provide a prediction of actual data values at grid points. The standard model for Indicator Kriging is often written as:

$$I(\mathbf{s}) = \mu + \varepsilon(\mathbf{s})$$

where μ is an unknown constant and $I(\mathbf{s})$ is a binary variable.

6.7.2.8 Probability Kriging

Like Indicator Kriging, Probability Kriging is a non-linear method employing indicator variables. It can be seen as a form of Co-Kriging in which the first variable is the indicator and the second variable is the original (un-transformed) data. As with Indicator Kriging the output does not provide a prediction of actual data values at grid points. With Probability Kriging the Indicator Kriging model is extended to the form:

$$I(\mathbf{s}) = \mu_1 + \varepsilon_1(\mathbf{s})$$

$$Z(\mathbf{s}) = \mu_2 + \varepsilon_2(\mathbf{s})$$

where μ_1 and μ_2 are unknown constants and $I(\mathbf{s})$ is a binary variable, as before. This procedure may be compared with that of Co-Kriging (see further, Section 6.7.2.11).

6.7.2.9 Disjunctive Kriging

Disjunctive Kriging is a non-linear procedure in which the original dataset is transformed using a series of additive functions, typically Hermite polynomials (see further, below). In this case the standard model is altered to the form:

$$F(Z(\mathbf{s})) = \mu(\mathbf{s}) + \varepsilon(\mathbf{s})$$

where $F()$ is the function in question. Indicator Kriging can be regarded as a special case of Disjunctive Kriging since the original data are transformed by the Indicator function. Disjunctive Kriging assumes pairwise bivariate Normality in the data, which is rarely verifiable in spatial datasets. A z-score (Normal) transformation of the data prior to Disjunctive Kriging is often desirable.

Hermite polynomials, $H_n(x)$, are a set of orthogonal polynomials that are most simply defined using the recurrence relation:

$H_{n+1}(x)=2xH_n(x)-2nH_{n-1}(x)$

with the first 3 terms being $H_0(x)=1$, $H_1(x)=2x$ and $H_2(x)=4x^2-2$. The range of x is $[-\infty,+\infty]$, whereas most orthogonal polynomials have a more restricted range.

6.7.2.10 Stratified Kriging

Variogram analysis and Kriging are often applied to a complete dataset as a single, global process. However, it may be desirable to sub-divide the study region into separate regions or strata, based on additional information. Stratification might be based on underlying geological formations, frequency of flooding, land and water regions, vegetation cover or other attributes that are of relevance to the problem at hand. In such cases the size and shape of sub-regions, and the effect of edges on the computed variograms, needs to be considered. Interpolated values using different models and parameters for each region will result in discontinuities at sub-region boundaries unless Kriging equations and/or the resulting surfaces are adjusted to support continuity. Currently, general purpose software does not support such procedures directly, but they may be implemented using standard facilities by manual stratification and re-combination of surfaces, albeit with problems of discontinuities to be resolved.

6.7.2.11 Co-Kriging

Many GIS packages support the analysis of a primary and secondary dataset, mapped across the same region, using a straightforward extension of standard Kriging techniques. With Co-Kriging the estimated value at an unsampled location is a linear weighted sum of all of the variables being examined (i.e. two or more). Co-Kriging extends the Ordinary Kriging model to the form:

$Z_1(\mathbf{s})=\mu_1(\mathbf{s})+\varepsilon_1(\mathbf{s})$

$Z_2(\mathbf{s})=\mu_2(\mathbf{s})+\varepsilon_2(\mathbf{s})$

where μ_1 and μ_2 are unknown mean values (constants) and ε_1 and ε_2 are random errors. Each of these sets of random errors may exhibit autocorrelation and cross-correlation between the datasets, which the procedure attempts to model. Co-Kriging uses this cross-correlation to improve the estimation of $Z_1(\mathbf{s})$. Co-Kriging can also be applied to models other than Ordinary Kriging (e.g. Indicator Co-Kriging, etc.).

6.7.2.12 Factorial Kriging

The term Factorial Kriging is used to refer to procedures in which the modelled variogram exhibits multi-scale variation. If these various scales of variation can be identified and extracted, Kriging may then be carried out using separate component parts of the variogram — for example, a spherical model component with range 2,000 metres, a similar model component with range 10,000 metres, and a separate drift component. Factorial Kriging is not widely supported in geostatistics packages at present — ISATIS, an expensive, but very powerful software package, is one of the few that does provide support for such analysis.

6.7.2.13 Conditional simulation

This is an interpolation procedure that can be seen as a development of Kriging. However, unlike Kriging it exactly reproduces the global characteristics of the source data (notably the source data points, the histogram of input data values and covariances) and it creates multiple maps which collectively provide an estimate of local and global uncertainty. Conditional simulation also reflects patterns of local variability more satisfactorily than Kriging, since the latter tends to smooth data locally, especially where source data points are sparse. Conditional simulation is an exact interpolator and cross-validation of the type used in other interpolation methods is not applicable, although a secondary (conditioning) dataset may be used.

The most widely available procedure for conditional simulation is one of three such methods provided in the GSLIB geostatistical package (the others being *simulated annealing* and *sequential indicator simulation*). The technique is known as *Gaussian Sequential Simulation* (GSS) and involves creating a surface using a random (or *stochastic*) process, applied many times to the same initial dataset. The specifics of implementations vary from package to package (further development of conditional simulation is provided in packages such as GMS, GS+ and ISATIS), but we shall first summarise the general procedure and then illustrate the results with the zinc dataset referred to in previous subsections (from Burrough and McDonnell (1998, Appendix 3)).

The inputs to conditional simulation are similar to those for Simple or Ordinary Kriging — a set of scattered data points for which one or more measured z-values have been recorded. The standard algorithm is then a three step process:

- **Step 1 initialisation:** analyse the input data for Normality, applying de-clustering and/or transformation if necessary; compute a best-fit experimental variogram model for the data; then define a fine grid (generally finer than is used for Kriging) over the entire study region. In some instances the definition of this grid is a multi-level process, starting with a relatively coarse grid and then progressively generating finer and finer grids. This has the advantage of improved representation of longer distance effects as well as short-range variability

- **Step 2 random walk:** pick a random location on the grid, X, and perform step 3, below. Then move to another (unvisited) grid location at random and continue until all grid nodes have been visited. If a multi-level grid approach is applied then repeat this process with the finer grid, using the results from the previous stage plus the input data points as the set of source data values, $z^*(x,y)$. The use of random walks or similar procedures minimises the risk of creating

artefacts that have been found to occur if systematic grid search is utilised

- **Step 3 local search and conditional estimation:** identify all $z^*(x,y)$ that lie within a pre-defined radius of X or locate the N nearest neighbouring $z^*(x,y)$ locations (e.g. N=16). Use the variogram model obtained in step 1 to provide a predicted (mean, m) value at X and estimated standard deviation (sd) at X, as per Ordinary Kriging, i.e. as a linear weighted combination of the N selected points (this is the conditional control applied); then, using the cumulative Normal distribution with mean m and standard deviation sd, select a random value (i.e. a random Normal variate in this case), and assign this as the estimate at X

These steps result in a single instance of a simulated grid at a given resolution (single level or multi-level process). The entire process is then repeated M times using a different random starting point and random number sequence. M may be quite large (e.g. M=1000) in order to obtain improved estimates of the surface mean and uncertainty (confidence bands), together with a smoother resulting surface. Inevitably the process is very computationally intensive, particularly if M is large, the grid resolution fine and/or multi-level processing is utilised.

Conditional simulation applied as an interpolation method on the zinc dataset yields results of the type shown in Figure 6-45. The results illustrated are based on 100 iterations. Although Burrough and McDonnell (1998) recommend 500+ iterations, increasing the iterations with this dataset and modelling system (GS+) results in very minor alteration of the z-values — it has been suggested that this particular implementation may over-smooth the results. Comparing this result with that obtained by Ordinary Kriging (Figure 6-44A) shows the resultant interpolations appear very similar, although the predicted data value range is greater in the conditional simulation model — Kriging tends to smooth and under-estimate the spatial variability of data as it is a form of weighted moving average. Predicted values in the unsampled areas are also often better behaved (smoother, less subject to artefacts) in the conditional simulation model.

Figure 6-45B shows the standard deviation map, which is substantially different from that obtained by Ordinary Kriging (Figure 6-44B) — both the pattern and the range are altered, with the range being very much reduced. This is particularly apparent in regions where the initial data points are sparse or absent. Increasing M results in minor alteration of the mapped patterns and value range, for the standard deviations in this example. A particular additional feature of conditional simulation lies in its generation of a set of individual realisations. These may be used, for example, to create a *distribution* of the areas above or below a given value, which is not the case with Kriging.

Figure 6-45 Conditional simulation of untransformed zinc test dataset

A. z-values, M=100 (predicted values)

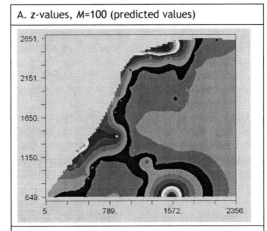

B. Standard deviations, M=100

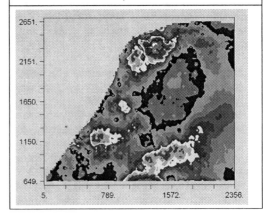

7 Network and location analysis

7.1 Introduction to network and location analysis

7.1.1 Overview of network and location analysis

This chapter briefly deals with two separate but related topics. The first involves questions regarding preferred routes, either traversing physical (transport) networks or by directly connecting pre-defined locations. The second topic covered involves the selection of locations and 'service areas' (principally within existing networks) that meet pre-specified criteria. We recommend the following books to accompany this chapter: Miller and Shaw (2001) "Geographic Information Systems for Transportation"; Ahuja *et al.* (1993) "Network flows: Theory, algorithms and applications"; Rodrigue *et al.* (2006) "The geography of transport systems"; and Daskin (1995) "Network and discrete location: Models, algorithms and applications". A software package, SITATION, to accompany Daskin's book, is available for free download.

Many GIS packages provide a limited set of functions designed to answer basic questions of this kind. Some packages provide more extensive tools using optional add-ins, such as ESRI's ArcGIS Network Analyst (now included in ArcGIS V9.1 as standard), ESRI's ArcLogistics option, Manifold's Business Tools option, and Intergraph's Incident Management System. Other packages, such as Caliper's TransCAD offering, aim to provide a fully integrated GIS-T (GIS for Transportation) suite with an extensive range of routing and modelling facilities. TransCAD provides facilities for: network analysis (shortest paths, travelling salesman problems and network partitioning); transit analysis (designing and modelling the use of public transit networks and associated fare structures); transportation planning and

travel demand modelling (trip modelling, modal split modelling and traffic assignment); vehicle routing and logistics (dispatching/collection problems, arc routing problems and flow-related problems); and territory management and site location modelling.

In many cases network analysis problems can be defined (*abstracted*) in terms that do not require a GIS for data manipulation, solution or presentation. For this reason algorithms for their exact or approximate solution are available as stand-alone code modules or as libraries providing solutions for generic problems, which then may be integrated into application suites with final output in map, list and/or table form. For example, problems involving the optimal selection of one or more locations in the plane (e.g. depots) given a set of point locations representing demand (e.g. customer sites) simply may require the relevant coordinate set (and associated demand levels) as input. The same observation applies to problems where a route traversing a set of points is required, but for which no explicit prior network has been defined. Likewise, assuming that a predefined network exists, many problems can be solved without detailed knowledge of the shape of links between network nodes — in general only their length or a related measure such as travel time in each direction is required, often provided as a shortest path or least-cost matrix. In this chapter, therefore, we discuss the solution to a range of network optimisation problems using both GIS and non-GIS toolsets.

7.1.2 Terminology

Table 7.1 provides a number of key definitions used within the discipline of network analysis. These terms are used within the various subsections of this chapter and widely within the cited references. The terms are all derived from abstract graph theory, itself part of the broader field of pure mathematics. Much work on graph theory bears little or no relation to the kind of spatial analysis discussed in this Guide, but nevertheless many of the results within this field are essential parts of the search for efficient (fast, high quality, compact, robust) solutions to many spatial problems. The disciplines that have focused the greatest attention on applications of graph theory in a geospatial context are those of operations research and combinatorial geometry. Within both areas an enormous range of results has been obtained and these provide the building blocks for the facilities seen increasingly in GIS and related software packages.

Table 7.1 Network analysis terminology

Term	Description
Vertex	A point location considered as a network node. Vertices are typically either distinct points or the end points or intersection points of lines or polylines. Note that intermediate points along a polyline are not generally vertices
Edge	A (directed or undirected) link between two vertices that are directly connected is called an edge. An undirected edge is determined by an unordered sequence of vertices, e.g. (3,8), which is the same as (8,3) whereas for a directed edge the order of the sequence matters, with (3,8) being from 3 to 8. An indirect link (via one or more other vertices) is not an edge. Edges are sometimes referred to as links or arcs
Degree (of a vertex)	In an undirected graph the number of edges meeting at a vertex. In a directed graph the degree usually refers to number of edges directed to a vertex (the *indegree*) minus the number of edges directed from a vertex (the *outdegree*)
Graph	A collection of vertices and edges constitutes a graph. Directed graphs are graphs that include one or more directed edges. If all edges are directed such graphs are known as *digraphs*. The mathematical study of the properties of graphs and paths through graphs is known as graph theory
Path	A (network) path is a sequence of connected edges between vertices
Connected graph	If at least one path exists between every vertex and every other vertex in a graph it is described as connected. A fully connected graph is one in which every vertex is directly connected to every other vertex. In such a graph, if there are *n* vertices there are $n(n-1)/2$ edges, assuming all edges are undirected
Planar graph	If a graph can be drawn in the plane (embedded) in such a way as to ensure edges only intersect at points that are vertices then the graph is described as planar
Network	A collection of vertices and edges together with associated attribute data that may be represented and analysed using graph theoretic methods. A network is often defined as a graph that has at least one real-valued attribute or weight (e.g. length) associated with every edge
Cycle	A path from a given vertex to itself that traverses other vertices is a cycle. A graph that has no cycles is called *acyclic*
Tree	An *n*-vertex acyclic network or subnetwork in which every vertex is connected, for which the number of edges is *n*-1. A unique path exists between every pair of vertices in a tree

7.1.3 Source data

A set of line segments or polylines, such as a road or street map, does not typically constitute a network. In order to conduct many forms of network analysis it is necessary to ensure that a set of distinct nodes or vertices (V) exist and between each vertex one or more links to other vertices are defined. These links are referred to as edges (E). At the simplest level edges may be directed (e.g. one way) or undirected (two-way) and may have one or more attributes associated with them (e.g. name, length, travel time/cost, mode of transport).

Alternatively, one can simply start with a point set in the plane representing network vertices and no edges. In this case a set of links between vertices is to be constructed, typically using straight line (Euclidean) segments that satisfy certain criteria. This can be considered as a form of network construction problem or simply a routing problem that ignores the details of any underlying fine network structure.

Many polyline spatial datasets that appear to satisfy the requirements for network analysis in fact do not. This is often because the required vertices do not exist as objects and/or the set of lines are not precisely linked — the endpoint of one polyline does not precisely coincide with the start or endpoint of other polylines in the set. Furthermore, where segments of polylines cross the nature of the intersection needs to be defined — for example, is the intersection to be regarded as a vertex or should we regard one line as passing over or under the other? The simplest arrangement is one that assumes all such intersections constitute valid network vertices, and that all lines whose endpoints are closer to each other than some specified tolerance value (typically a small Euclidean distance measure) are, in fact, coincident and constitute a network vertex.

With these assumptions made (or variants of these) many GIS packages will convert a set of line data into a network, i.e. effectively imposing a network structure or topology on the dataset. Edge and/or vertex labelling may form part of this process, enabling subsequent routing plans to be referenced by route or street name and flow direction. Having done so, and/or as part of this process, attributes may assigned to edges and vertices individually, across selected subsets or globally. Examples include: (i) definition of turn attributes at intersections (permitted, not-permitted by link and any penalty, such as additional time or cost associated with such a turn); (ii) definition of U-turn permissions; (iii) definition of weights (or impedances) to be associated with edges (typically transition

times, often defaulted to use edge length as a surrogate) — in some cases different definitions may be applied for each direction traversed along the edge, in others the link may be modelled as two separate edges, one for each direction of travel; (iv) definition of one-way edges and their direction; (v) specification of any permanent or temporary barriers, typically at vertices — links that are barred are often simply disabled or assigned a very high impedance value; and (vi) defining demand and capacity constraint levels (edge and/or vertex based).

Once such a network has been defined, attempts at optimal routing can be made — for example finding the shortest (or fastest) path between a given pair of vertices such as an accident location and an ambulance station or stand-by point. However, as noted earlier in this Guide, street network and similar physical infrastructures do not necessarily satisfy the requirements of a full metric space, and it may be the case that some points are simply not reachable from others, or that optimal routing can only be approximated and/or remain difficult or impossible to prove to be optimal. Again, the criterion one adopts in such cases is fitness for purpose — best known approximations, which in some instances are provably within a certain closeness to optimal solutions, are often the options that have to be taken. These can then be evaluated against operational criteria (e.g. maximum acceptable time for an ambulance to arrive at an incident for 99.8% of call-outs) and the problem or criteria re-evaluated if such criteria cannot be met. One of the great advantages of many of the toolsets available is the ability to run 'what if' scenarios, experimenting with alternative constraints, flows, changes in demand or supply, etc., and evaluating the varying costs or times resultant on these variations. This information provides a form of sensitivity analysis, which may (for example) enable depot location decisions to be made taking into account variations in availability of suitable sites, access routes and labour, forecast temporal variations etc., within areas of broadly similar generalised costs.

With pure point-based data no pre-existing network is assumed, and problems relate to determining the form of network (set of edges) or path that connects the point set in some particular manner. This assumes that either no underlying physical network exists and travel between vertices is possible and uniform across the plane or that an underlying transport infrastructure exists and is sufficiently fine and direct to allow its specific form to be ignored for the problem at hand. This latter assumption may or may not be entirely realistic, but has the advantage of enabling difficult problems to be addressed without the need for the additional burden of large numbers of network-based computations. Finally, depending on the problem being addressed, neither point (location) nor line (network) data are required — a pre-computed inter-point distance or cost matrix may be sufficient. However, this is not generally sufficient to re-construct a unique map of the locations, nor will real-world routes chosen be represented — only the sequence of links and associated distances/costs will be available.

7.1.4 Algorithms and computational complexity theory

Most of the key problems addressed in location and network analysis can be classed as "optimisation problems". As such, they are an important subset of a much broader set of problems. Understanding the computing resources required, in order to achieve solutions of standard optimisation and decision-making problems, is extremely important.

In the case of networks, which typically consist of a set of n vertices, V, and m edges, E, the amount of time taken to compute an exact solution as a function of the number of these elements is vital. Typically this information is provided in so-called "big O" notation and is based on the total number of steps involved in the operation and the way in which these steps are processed. This provides a very crude measure of the order (approximate amount) of time and possibly space (memory, disc) required to complete a run of a particular

algorithm, notably as the number of steps becomes large. The actual time required for any specific implementation will vary greatly, depending on the quality of the implementation and the size of the problem. As an example if the amount of time is stated as $O(n^k)$ where k is some Real number $k>0$, then the time required grows as a simple power function (or polynomial function) of the number of elements, n, in the problem. This is generally a worst-case scenario. Example polynomial orders include $O(n^2)$ and $O(n\log n)$. Examples of non-polynomial order include $O(n!)$ and $O(2^n)$. Note that for $n>3$, ignoring any additional constant multipliers, the following relation holds:

$$O(\log n)<O(n)<O(n\log n)<O(n^2)$$

i.e. typically some polynomial orders involve far fewer operations than others. The same is true for non-polynomial algorithms, where for example $O(2^n)<<O(n!)$ for $n>3$.

Assuming the problem in question can be solved exactly by a well-defined *deterministic* algorithm in polynomial time, it is said to be of type P. An example of a problem that can be solved in polynomial time is finding the shortest path from a single source point through a planar Euclidean network. Polynomial time algorithms are sometimes denoted using the abbreviation PTAS or PTAs.

It is possible to identify problems for which a given solution can be checked for optimality in polynomial time, but for which no deterministic algorithm can be devised. A non-deterministic algorithm (or computer) would be one that either: (i) guesses the correct solution and then must verify it to check it is correct; or (ii) utilises a computer that includes an infinite number of processors working in parallel to determine possible solutions that again must be verified. Non-deterministic Polynomial time or NP problems are the class of problems that can only be determined in this manner. It is clear that P⊆NP but as yet the hypothesis P=NP has not been proven or shown to be false (this hypothesis is the subject of one of the seven

$1Million Millennium Math Prizes offered by the Clay Mathematics Institute: see www.claymath.org/millennium/P_vs_NP/ for more details).

An example of an NP networking problem is that of determining whether a given non-directional network has a simple cycle that contains every vertex. It is clearly easy to check whether the cycle passes through every vertex and is not self-crossing, but it is very difficult to compute a specific instance of such a circuit as n increases. In fact this problem (known as the Hamiltonian Circuit Problem — HCP) is within the class NP. A specific, and even more challenging problem, related to the HCP, is the Travelling Salesman Problem (TSP). In this case one must find an HC for a network where each location may only be visited once and has minimum overall cost (or time/length).

A problem is described as being *NP-hard* if solving it would enable all problems in the class NP to be solved in polynomial time. Some NP-hard problems are in the class NP and some are not. Those that are in NP are called *NP-complete* (or *NP-full*). Proving that a specific problem is NP-complete involves first showing that it is a member of the set NP and then finding a problem that is known to be NP-complete and seeing if one can reduce (re-define/re-work) this problem such that it becomes identical to the problem being considered. For example, to show that the TSP is NP-complete we could proceed as follows: (i) since HCP is known to be NP-complete we could seek to show first that TSP is a member of the set NP; and then (ii) we could seek to show that the HCP is reducible in polynomial time to the TSP. A central result in complexity theory is the proof that every NP-complete problem is inter-changeable with every other one. NP-hard problems that are not NP-complete are considerably harder to solve, since even their solutions cannot be checked in polynomial time, unlike those of NP-complete problems.

7.2 Key problems in network and location analysis

7.2.1 Overview — network analysis problems

Table 7.2 provides a brief list of some of the key problems that are encompassed within our use of the term "network analysis". Solutions for many important practical problems are built upon these core optimisation problems and the various procedures (algorithms) applied to them. Examples include: the production of optimal routing and vehicle loading plans; multi-modal transport problems; identification of drive-time zones; network partitioning and territory definition; facility location on a network; and travel demand analysis. For selected problems exact solution methods exist, but in some cases these may require an excessive amount of time to complete. Commonly applied exact procedures include linear programming, in some instances augmented with cutting planes where an integer solution is required, branch and bound procedures, and dynamic programming. More details on many of these topics may be found in standard works on algorithms and complexity, for example Cormen *et al.* (2001) or Daskin (1995).

In many instances provably optimal solutions to key problems of certain sizes are not achievable and solutions based on heuristics that achieve near optimality in most cases are the norm. The majority of these solution procedures are essentially static, but they can be applied to more dynamic or real-time environments if they are fast enough. Many heuristics involve some form of local search or swapping (interchange) procedure. So-called greedy algorithms make choices based on the progress of steps so far and selecting the next step by choosing the local optimum in the hope of reaching the global optimum, rather than seeking a global optimum directly (e.g. by exhaustive search). A typical example would be attempting to find the shortest tour

of a set of cities (a travelling salesman problem) by selecting one city at random as the start point and then selecting the next location to visit by choosing the closest city. The resulting tour will visit all the cities but will almost always be suboptimal. Some greedy algorithms, such as Dijkstra's procedure for finding shortest paths through a network (see Section 7.3.4), always yield exact (optimal) solutions. Interchange heuristics involve swapping one or more components in a current solution, thus changing the order or arrangement of the solution. If the objective function — the value we are trying to optimise — improves, the new arrangement is retained and the process proceeds to further interchanges according to some rule. Tabu search methods are increasingly used in heuristic algorithms that adopt local searching. The term *tabu* (taboo) refers to the exclusion of certain options (e.g. adding or dropping a link in a network) based on options that have been recently searched. This has the effect of increasing the range of local searches and thus reducing the risk of solutions being trapped in local optima. There are several related methods such as simulated annealing, genetic algorithms and behavioural models (such as insect-based analogies) that have been shown to be effective in providing heuristics for some particularly difficult problems.

Many of the problems we describe can be modelled using one or more linear equations combined with a number of associated constraint expressions specified as equalities and/or inequalities. This form of standardised representation enables different problems to be compared within the discipline of the spatial sciences and across the broad discipline of optimisation as a whole. In addition it enables many problems to be expressed in a form that is suitable for submission to specialised optimisation packages such as LEDA, CPLEX, LP-solve and Xpress-MP. These provide source code sets and/or compiled libraries that offer optimal or heuristic solutions to a wide range of optimisation problems, including linear programming (LP) and mixed integer programming (MIP) tasks. However, many algorithms for solving networking problems do not use linear programming or related techniques as these may be less efficient (i.e. slower) than specialized procedures, or the problems are not amenable to formulation as an LP or MIP problem.

A good place to review such topics is the LP-solve website managed by Sourceforge:

http://lpsolve.sourceforge.net/5.5/

This site includes a useful Linear Programming FAQ page from which we have extracted the following explanatory quotation (with minor edits):

In the context of linear programming, the term "network" is most often associated with the minimum-cost network flow problem (MCFP — see Table 7.2). A network for this problem is viewed as a collection of nodes (vertices or locations) and arcs (or edges or links) connecting selected pairs of nodes. Arcs carry a physical or conceptual flow of some kind, and may be directed (one-way) or undirected (two-way). Some nodes may be sources (permitting flow to enter the network) or sinks (permitting flow to leave).

The network linear programming problem is to minimize the (linear) total cost of flows along all arcs of a network, subject to conservation of flow at each node, and upper and/or lower bounds on the flow along each arc. This is a special case of the general linear programming problem. The transportation problem is an even more special case in which the network is bipartite: all arcs run from nodes in one subset to the nodes in a disjoint subset. A variety of other well-known network problems, including shortest path problems, maximum flow problems, and certain assignment problems, can also be modelled and solved as network linear programs.

Network linear programs can be solved 10 to 100 times faster than general linear programs of the same size, by use of specialized optimisation algorithms. Some commercial LP solvers include a version of the network simplex method for this purpose. That method has the nice property that, if it is given integer flow data, it will return optimal flows that are integral [have integer values]. Integer network LPs can thus be solved efficiently without resort to complex integer programming software.

Unfortunately, many different network problems of practical interest do not have a formulation as a network LP. These include network LPs with additional linear "side constraints" (such as multicommodity flow problems) as well as problems of network routing and design that have completely different kinds of constraints. In principle, nearly all of these network problems can be modelled as integer programs. Some "easy" cases can be solved much more efficiently by specialized network algorithms, however, while other "hard" ones are so difficult that they require specialized methods that may or may not involve some integer programming.

A compendium of many problems of this type and the best known (approximation) algorithms for their solution can be found at:

http://www.nada.kth.se/~viggo/problemlist/

This website provides details for a very large range of problems for which checking that a result is optimal may be relatively simple, but for which finding such a solution may be extremely difficult (typically would take an extremely long time for any large problem). These are NP-hard or NP-complete problems in the terminology of computational complexity theory (see Section 7.1.4 for more details).

7.2.1.1 Key problems in network analysis

The survey by Mitchell (1998, 2000) of geometrical shortest path and network optimisation problems provides a further excellent source discussing such questions and solution procedures. Mitchell provides a useful summary of the range of questions that may be addressed by considering various associated parameters (Table 7.3).

Table 7.2 Some key problems in network analysis

Problem	Description
Hamiltonian circuit (HC)	If a cycle exists from a given vertex that passes through every other vertex exactly once it is called a Hamiltonian circuit. Testing for the existence of Hamiltonian circuits in a graph is known as the Hamiltonian circuit problem (HCP). NP-complete (see further, Section 7.1.4)
Eulerian circuit(EC)	A circuit in a directed graph that visits every arc exactly once. A condition that a graph contains an Eulerian circuit is that the number of arcs arriving at every included vertex, i, must be the same as the number of arcs leaving vertex i
Shortest path (SP)	A path between two vertices that minimises a pre-defined metric such as the total number of steps, total distance or time, is called a shortest path. Hence this term is relative to the metric applied and even then may not be unique for any given network. Determination of shortest paths is often described as shortest path analysis (SPA). This is perhaps the central computational problem in network analysis. There are many variants of this problem, including finding the 2nd, 3rd... n^{th} shortest path, finding the shortest path from a given node to all other nodes, and finding the longest path. Can be solved in linear time or better
Spanning tree (ST)	Given a fixed set of vertices, find a set of edges such that every vertex is connected and the network contains no cycles. Many spanning trees are possible for a given vertex set
Minimal spanning tree (MST)	Find a (Euclidean) spanning tree of minimum total length. Typically this will be unique, but uniqueness is not guaranteed. Solvable in near linear time
Steiner MST, Steiner tree	As per the MST but with additional nodes permitted that are not co-located with the original vertex set. In the (spatially) constant cost model, each additional point (known as a Steiner point) will be placed intermediate to three existing vertices and will provide a connection between these via three branches that are equally spaced (i.e. at 120 degrees) about the Steiner point. Steiner points are added to the MST, replacing MST links, if the total network length is reduced by their inclusion. NP-complete (for both Euclidean and Manhattan metrics)
Travelling salesman problem (TSP)	Given a set of vertices and symmetric or asymmetric distance matrix for each pair of vertices, find a Hamiltonian circuit of minimal length (cost). Typically the start location (vertex) is pre-specified and the vertices are not necessarily assumed to lie on a pre-existing network. If certain nodes must be visited before others, the task is known as a sequential ordering problem (SOP). NP-complete
Vehicle routing problem (VRP)	This class of problems relates to servicing customer demand (e.g. deliveries of fuel to retail garages) from a single depot, where each vehicle may have a known capacity (CVRP). If capacity is not restricted the problem in known simply as a vehicle routing problem (VRP). The number of vehicles and the number of tours of subsets of nodes are variables. The customer locations, depot location and customer demand levels are assumed to be known. The problem is to minimise the overall length of the tours, subject to the constraints. There are many variants of this problem, notably those in which there are pre-defined time windows for deliveries, problems involving pickups and deliveries, problems involving a series of depots, problems where demand is dynamically variable, problems in which link capacity constraints exist and hence may become congested, and problems where customer locations are generalised rather than fixed. Most such problems are classified as NP-hard
Transportation problem, Trans-	The general problem of completely servicing a set of target locations with given demand levels from a set of source locations with given supply levels such that total costs are minimised is known as the transportation problem. The unit cost of

Problem	Description
shipment problem	shipping from each supply point to each demand point is a key input to this problem specification. This problem is an example of a Minimum Cost Flow Problem (MCFP). A generalisation of the transportation problem is the trans-shipment problem. In the latter case flows from sources to targets can go via trans-shipment points, e.g. factories to warehouses to customers, rather than simply direct to customers
Arc routing problem (ARP)	Given a network (typically a street network or subset of a street network) find a route that completely traverses every edge, generally in both directions, that has the least cost (distance or time) subject to selected constraints (e.g. cost of turning). This problem applies to street cleaning, snow-ploughing, postal deliveries, meter reading, garbage collection etc. The capacitated version of the problem is known as CARP. NP-complete
Facility location: p-median/p-centre/ coverage	A collection of problems where the objective is to optimally locate one or more facilities within a network in order to satisfy customer requirements (demand, service level). The most commonly cited problem is minimisation of total (or average) travel cost/time to or from customers (the p-median problem). Minimisation of maximum distance or time is known as a p-centre problem. A related set of problems seeks to ensure that all customers can be served within a fixed upper time or cost, or at least, as many as possible are served within a fixed time or cost. These are known as coverage problems. They are discussed in Daskin (1995, Ch 4) but are not explored further in this Guide, mainly because their very restrictive constraints tend to generate solutions that are too costly or ineffective to be implemented in practice. Customer demand is often assumed to be located at vertices in which case p-median solutions for p facilities serving $n>p$ customer sites will always result in the facilities being located at network vertices (although this solution may not be unique). It is the network equivalent of the plane or free-space median location problem — a form of location-allocation task whereby facilities are located and customers are allocated to facilities. The p-median and p-centre problems are NP-hard

Table 7.3 Sample network analysis problem parameters

Parameter	Questions
Objective function	How do we measure the "length" of a path? Options include the Euclidean length, L_p length, link distance/time/cost etc.
Constraints on the path	Are we simply to get from point s to point t, or must we also visit other points or other regions along a path or cycle?
Input geometry	What types of obstacles or other entities are specified in the input map?
Dimension of the problem	Are we in 2D-space, 3D-space, or higher dimensions? Typically within GIS we only consider 2D, but transport networks may not be planar
Type of moving object	Are we moving a single point along the path, or is movement specified by some more complex geometry? In GIS off-road vehicular modelling is usually performed using Accumulated Cost Surface (ACS) or Distance Transform (DT) procedures applied to grid datasets rather than vector networks (see Section 4.4.2). Constraints on routes may also be applied to vehicles of particular sizes, types or weights (e.g. height restrictions)
Single shot vs. repetitive mode queries	Do we want to build an effective data structure for efficient queries? Many network problems involve very similar searches — for example determining an alternative route (2nd, 3rd best)

Static vs. dynamic environments	Do we allow obstacles to be inserted or deleted, or do we allow obstacles to be moving along known trajectories? Flow and event dynamics may also be important considerations
Exact vs. approximate algorithms	Are we content with an answer that is guaranteed to be within some small margin of optimal? Larger problems in many cases cannot be solved exactly in a finite amount of time. Ideally real-world problems should be solved to within a specified level of the optimum for suitably defined subset problems
Known vs. unknown map	Is the complete geometry of the map known in advance, or is it discovered on-line, using some kind of sensor? Typically the geometry is known (map-able) in advance, but flows or events may not be

After Mitchell J S B (1998, 2000)

7.2.1.2 Network analysis and logistics software

Whilst some GIS packages, such as TransCAD and the ESRI ArcLogistics Route option, include functionality designed to solve a wide range of network analysis and associated planning and modelling tasks, problems of this type are often addressed using a range of bespoke or specialised transportation and logistics software. Some of these, such as eRouteLogistics and RouteSmart are built on/integrate with GIS packages (ESRI software in these two examples), whilst many others are purpose-built components of logistics systems, Enterprise Resource Planning (ERP) software or navigation suites. Examples include REACT, from MJC^2 (www.mjc2.com), TruckStops from MicroAnalytics (www.bestroutes.com) and Optrak4 (www.optrak.co.uk).

Most of the latter class of software takes into account the many practical, real world constraints that must be incorporated in operational software. Examples (from REACT) include: vehicle routing taking one-way streets into account; trip routing taking restricted junctions into account; varying speeds by road type and time of day; trip routing of vehicles to avoid toll roads and toll bridges; trip routing vehicles taking account of congestion charges; delivery routing taking account of customer access constraints by time of day; night time/weekend lorry routing controls (e.g. the London 'brown' routes); weight restrictions (e.g. for truck routing); height restrictions (e.g. for truck routing); vehicle routing costs per mile/km; vehicle routing costs per hour; weight related vehicle routing costs; altitude related truck routing costs (e.g. trip routing to reduce hill climbing).

A recent review of packaged logistics software by OR/MS (the magazine of the Institute for Operations Research and Management Sciences in the USA) included a list of providers for networking/routing software, with routing functionality as shown in Table 7.4. Of these, ILOG Dispatcher is of wider interest as it is a component of the ILOG general purpose optimisation suite, which includes CPLEX, referred to later in this chapter.

Table 7.4 Routing functionality in selected logistics software packages

Product	Routing Functions				
	Node Routing	Arc Routing	Real-time Routing	Daily Routing	Route Planning & Analysis
A.MAZE	✓		✓	✓	✓
Clavis Route	✓	✓	✓	✓	✓
Compass			✓	✓	✓
Cube Route Platform	✓		✓	✓	✓
Direct Route	✓	✓	✓	✓	✓
DISC	✓	✓	✓	✓	✓
EDGAR Transportation Management	✓		✓	✓	✓
eRouteLogistics	✓	✓	✓	✓	✓
Global Road Data GIS	✓	✓	✓	✓	✓
ILOG Dispatcher	✓	✓	✓	✓	✓
Optrak4	✓		✓	✓	✓
Paragon Routing and Scheduling System	✓	✓	✓	✓	✓
Prophesy ShipperSeries Software	✓	✓	✓	✓	✓
REACT	✓	✓	✓	✓	✓
Roadnet Transportation Suite		✓	✓	✓	✓
Roadshow Enterprise			✓	✓	✓
RouteSmart	✓	✓		✓	✓
SHORTREC Software Suite	✓	✓	✓	✓	✓
STARS (Smart Truck Assignment & Routing System)	✓	✓	✓	✓	y
Trapeze MapNet	✓	✓	✓	✓	✓
TruckStops Routing & Scheduling	✓	✓	✓	✓	✓
VersaTrans Routing & Planning	✓		✓	✓	✓

Source: OR/MS. Blank entries indicate 'not available'

7.2.2 Overview of location analysis problems

Daskin (1995, Ch.1) provides a useful taxonomy of and commentary on location problems and models. We have adapted and extended his taxonomy (Table 7.5), in part using the work of Schietzelt and Densham (2003). Although Daskin's focus is on network and discrete location problems, rather than the broader sweep of network design and planar problems, this summary does offer an insight into many of the practical problems that may be encountered.

Table 7.5 Taxonomy of location analysis problems

Component	Description
Planar/network/discrete	**Planar** — demand occurs anywhere in the plane (possibly represented by a deterministic or probabilistic field). Facilities may be located anywhere in the plane; **Network** — demand and facilities can only be located at network nodes or on links, and travel is restricted to the network; **Discrete** — nodes are fixed but the cost of travel between nodes is not determined by an underlying network
Tree/graph	Some network problems are amenable to treatment as if the graph is a tree, which can make solution far simpler
Distance metric	We have previously noted (Section 4.4.1) that a variety of metrics can be used in spatial analysis and this applies equally to location problems, notably planar and discrete cases
Facilities	The number of facilities to locate can be pre-specified (e.g. as p) or generated as part of the optimisation process (as in the set coverage problem — see below). Problems involving the location of only one facility are often relatively straightforward to solve, and some procedures use this fact to develop heuristics that incrementally increase the number of facilities based on local optimisation of the next facility added
Static/dynamic	Most of the problems and techniques described in this chapter are essentially static. Such methods may be applied to some dynamic problems, but in many cases these procedures require extension or alternative approaches to deal with dynamic cases. Examples include: choosing when as well as where to locate the next 1,2... facilities; incorporating dynamic demand and possibly supply into the model; modifying vehicle locations as demand varies by time of day or week (e.g. relocating emergency vehicles on standby based on time of day)
Private/public	The principal issue here relates to the definition of the objective function — can this be done in purely monetary (or equivalent) terms or do social, environmental and other factors require evaluation? Generally all location selection processes are part of a broader decision-making environment (as identified in the PPDAC model, Section 3.2.3). In public facilities location it is also often possible to dictate allocations whereas for private facilities this is almost never possible, requiring the modelling of allocation (e.g. using spatial interaction models)
Single/multi-objective	In most instances single-objective solutions are sought (e.g. time, distance or cost minimisation). However, real-world situations are almost always multi-objective. One approach to this is to compute multiple solutions for variations in parameters that reflect the multiple objectives, and examine the nature and robustness of these solutions. This is not equivalent to true multi-objective analysis but may provide benchmarks and guidance for such problems as part of a broader analytical framework

Component	Description
Unique/diverse service	A unique service is one in which a single (principal) service is being modelled, such as ambulance provision. A diverse service might be a similar problem, but where different categories of ambulance service and associated vehicles and staffing were modelled — for example, full emergency call-out vehicles; individual paramedics on-call; and ambulances for non-emergency usage (e.g. transport of patients, medical supplies etc.)
Elastic/inelastic demand	Most models assume that demand and supply are independent. However, it must be recognised that improved supply frequently increases demand, i.e. that demand is elastic, and that supply may or may not be
Deterministic/adaptive/ stochastic	Many models are deterministic in terms of supply, demand and feasible solutions; others may assume demand is probabilistic and often dynamic, with solutions being sought that are optimal with respect to the inputs, but which are often adaptive reflecting changes in demand, supply and transport dynamics
Capacitated/uncapacitated	Basic models do not consider the capacity of facilities (e.g. warehouses, storage depots, hospitals, vehicles) or links (e.g. transport networks or pipelines) as a constraint. Tools are available that allow for the inclusion of such constraints (an example is provided in Section 7.3.5.1)
Nearest facility/ General demand allocation	Demand is often assumed to be allocated to the nearest facility. In capacitated models this may require that some demand is split between facilities (fractional allocation) — e.g. a retail store may need to be serviced by more than one warehouse, or patients may be sent to more distant hospitals in times of high demand or major emergencies
Hierarchical/single level	Some problems are intrinsically hierarchical, in that products or services are provided from larger centres (e.g. national) to smaller regional centres and on to district and local facilities (or vice versa, upwards flow rather than downwards distribution). This may apply for product or service offerings. In such problems the existence of facilities at one level (up or down) may be a pre-requisite for locating facilities at the next (e.g. large-scale regional hospitals are only provided where at least a certain number of local level medical facilities already exist). A similar example would be the design of a completely new multi-level health infrastructure for a region previously not served with these kinds of facility
Desirable/undesirable	Most location modelling relates to desirable facilities — the distance or cost of travel to or from facilities is to be minimised, according to some criterion. For some facilities, such as waste disposal sites, incinerators, nuclear power plants, etc. the facility location problem becomes more complex because there are often conflicting objectives. For example, the location of waste disposal sites need to be as near as possible to the waste creation points (towns say), but ideally as far away as possible from habitation

After Daskin (1995) and Schietzelt and Densham (2003)

7.3 Network construction, optimal routes and optimal tours

We now examine a number of the problems described in Table 7.2. Descriptions of each problem and examples and/or outline algorithms for their solution are provided. More detail can be found through the online and printed references provided.

This is a continually developing field, an area of research in which perhaps more papers have been written and algorithms published than almost any other, so comments made here on algorithms and solvability may (will) have altered by the time of publication.

7.3.1 Minimum spanning tree

Given a set of vertices (points, nodes) an enormous number of possible interconnections may be made to produce a network of direct or indirect connections between vertices. The set of connections that minimises total edge length whilst ensuring every point is reachable from every other point is known as a minimal spanning tree (MST). There is an exact solution to this problem. The algorithm involves a construction or growth process as follows: (i) connect every point to its nearest neighbour — typically this will result in a collection of unconnected sub-networks; (ii) connect each sub-network to its nearest neighbour sub-network; (iii) iterate step (ii) until every sub-network is inter-connected. This process is illustrated below, with the stages of nearest neighbour linkage identified by numerical sequence values.

Figure 7-1 Minimum Spanning Tree

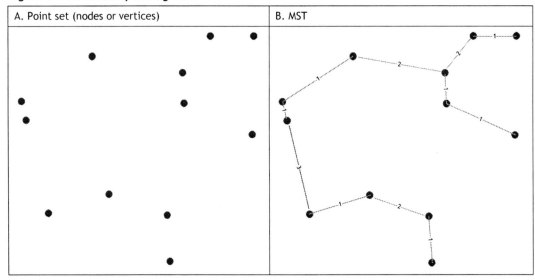

The MST for a set of vertices can be computed in O($n\log n$) time so is in the class P. An interesting feature of the solution is that it is a subset of various other more connected, forms of network, including the Delaunay triangulation — see Cheriton and Tarjan (1976) and Arora (1996) for more details. Some packages (e.g. the LEDA code suite from Algorithmic Solutions) first compute the Delaunay triangulation of the point set, then generate the Delaunay diagram if necessary (a subset of the triangulation in some instances) and run the minimum spanning tree algorithm on this graph, systematically deleting edges that do not belong to the MST. Note that whilst the MST provides the shortest possible connection between all vertices, it is very susceptible to disruption since deleting any link (e.g. link failure, road accident) will result in loss of overall connectivity. For real-world networks higher levels of vertex connectivity are almost always implemented.

7.3.2 Gabriel network

A form of network that is a super-set of the MST and has a variety of uses is known as the Gabriel network, named after its principal originator, K R Gabriel (see Gabriel and Sokal (1969) for more details). The Gabriel network for a point set, such as that shown in Figure 7-1A, is created by adding edges between pairs of points in the source set *iff* there are no other points from the set contained within a circle whose diameter passes through the two points (see Figure 7-2A — black circles). In this example the green (grey) circle shown encloses another point from the set so a link between the two points used to create this circle is not included in the final solution (the straight lines). The process continues until all point pairs have been examined in this manner and linked where the condition holds (Figure 7-2B).

Figure 7-2 Gabriel network construction

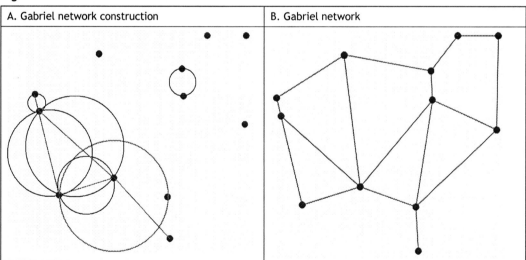

The Gabriel network provides a form of network that contains more links than a MST and hence provides greater connectivity especially between nearby points that are not actually nearest neighbours. It was introduced as one means of uniquely defining contiguity for a point set such that no other point could be regarded as lying 'between' connected pairs. It has been used in a variety of applications, most notably in genetic studies of populations (human and other). Examples include the research of Sokal *et al.* (1986) into the genetic makeup of the Yanomama Amerindian villages and the research by Arnaud *et al*. (1999) into land snail genetics. In each case the Gabriel network was judged to be a meaningful description of connectivity, with the option of binary or edge weighted measurement. The measure was then used to determine the spatial weights matrix in autocorrelation analyses.

The MST can be produced as a subset of the Gabriel network as follows: (i) construct an initial subset of the Gabriel network in which the additional constraint is applied that no other points may lie within the area of intersection defined by circles placed at each Gabriel network node with radius equal to the inter-node separation (see Figure 7-3B).

Figure 7-3 Relative neighbourhood network and related constructions

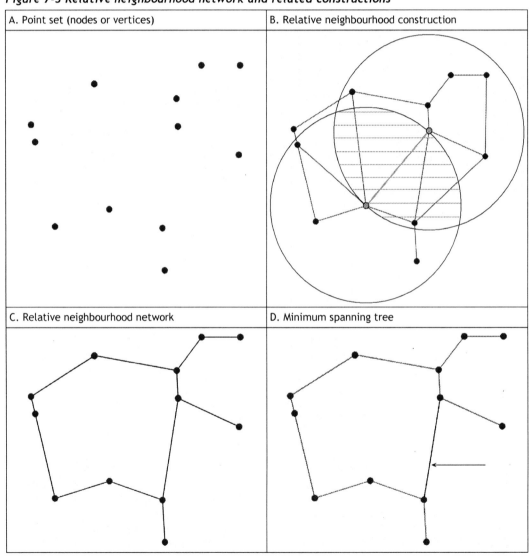

The resulting network is called a Relative Neighbourhood Network within the Manifold GIS package (see Figure 7-3C); (ii) remove the longest link in the relative network that does not break overall network connectivity; (iii) repeat step (ii) until no further reduction in overall length is achieved (see Figure 7-3D — the line identified by the arrow is the only edge requiring removal).

Although not specifically described in detail, this appears to be the general procedure adopted in the Manifold GIS for creating an MST. As opposed to the construction approached described earlier, this method commences with a promising solution that is a superset of the MST and reduces it until the MST is determined. A variant of the procedure is sometimes described as the k-MST problem, in which an MST is sought for a given subset, $k \leq n$ vertices, with the remaining vertices either connected via an existing set of edges or not connected.

The various networks described so far have involved construction using Euclidean straight lines and the Euclidean metric (L_2). Solutions to problems such as determining the MST and many of those that are discussed in the following subsections may also be sought using other metrics. Extensive research exists into solutions that apply for the L_1 metric, not merely because this may provide a more meaningful measure for street patterns in many cities, but also because in unrelated areas, such as electronic circuit design and utility infrastructure planning, such metrics are often more appropriate. Examples of solutions for alternative metrics will be provided later in this chapter.

7.3.3   Steiner trees

A feature of the MST solutions described is that they minimise network length for a spanning tree rather than minimising the length of a network of straight line segments connecting all vertices that permits the inclusion of intermediate points. Adding intermediate points can reduce the total network length, by a factor of up to $\sqrt{3}/2$ for the L_2 metric and 2/3 for the L_1 metric. For example, any three neighbouring vertices may be connected using an intermediate (or Steiner) point that will reduce their local MST, *iff* the angle between the two legs of the local MST is less than 120 degrees (Figure 7-4A) and the vertices are unweighted. The optimum location for an intermediate point in this case is such that each link from the chosen point to the three vertices lies at an angle of 120 degrees from the other two — i.e. they are equally spaced (Figure 7-4B). Whilst an MST may be systematically modified in this manner, accepting solutions that reduce total network length, this will not necessarily result in a minimal length Steiner tree. Algorithms that commence with the original point set and examine a much wider range of options may yield significantly improved solutions. For example, taking 4 points at a time may identify solutions in which two additional points are warranted, with these points being themselves inter-linked. In the most general case, with $m \leq n$-2 Steiner points and n vertices a very large number of alternative topologies are possible — over 60,000 with just n=7. The Steiner MST problem is NP-hard, although efficient approximation algorithms now exist. Such facilities are not provided in current GIS software — see for example, Robins and Zelikovsky (2000) for more details.

Figure 7-4 Steiner MST construction

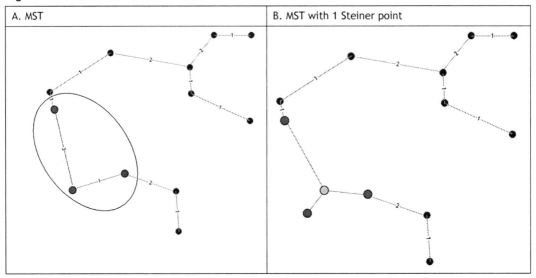

A. MST	B. MST with 1 Steiner point

Note that these networks and solution procedures assume that network flows and directions are not relevant, and that every vertex has equal weight. If vertices have unequal weights (e.g. number of beds in a hospital, demand for certain goods, weight of material input for some manufacturing operation) then the Steiner point solution will be altered. In fact such problems have been described briefly earlier in this Guide, in connection with locating centroids using the L_2 metric (Section 4.2.5.2). For n vertices each of which has an associated weight, w_i, the least distance network in the plane from a single additional point (the point of minimum aggregate travel, MAT, or Weber point) is defined by coordinates that are determined iteratively. The iteration equations for the MAT point, M6, are provided in Section 4.2.5.2. Typically the number of iterations required for convergence is very small (often 5 or fewer). This provides an example of a simple optimal location problem that is closely related to the Steiner network problem. If a pre-existing network is provided as part of the problem, and demand is located at the vertices, an optimal Weber point will always be found at a network vertex. An important generalisation of this problem is the case where p locations are sought (e.g. warehouses) to service demand at n vertices (e.g. retail outlets). In this case the network-based version is known as the p-median problem. Such problems are discussed in more detail in Section 7.4

7.3.4 Shortest (network) path problems

7.3.4.1 Overview of shortest path problems

Unlike some of the previous problems, the general shortest path (SP) problem requires a predefined network. The basic problem is then to determine one or more shortest (or least cost) routes between a source vertex and a target vertex where a set of edges are given. In several cases SP algorithms also provide shortest paths from a single source to all or many of the other vertices in a network as a side-effect of the procedure. Such algorithms are known as *single-source* SPAs or single-source queries. The set of shortest paths generated from a single source is known as a shortest path tree (SPT).

The determination of shortest paths can be specified as a linear programming problem, as follows:

Let s be the source vertex, t be the target vertex and let $c_{ij} > 0$ be the cost or distance associated with the link or edge (i,j). Then we seek to minimise z, where:

$$z = \sum_i \sum_j c_{ij} x_{ij},$$

subject to

$$\sum_j x_{ji} - \sum_k x_{ik} = m, \text{ where}$$

$m = 0 \text{ for } i \neq s,$

$m = 1 \text{ for } i = t$

$m = -1 \text{ for } i = s$

$x_{ij} \in \{0,1\}$

Although this formulation enables shortest path problems to be solved by standard LP software, and is guaranteed to yield all-integer solutions, in practice it is far less efficient than utilising specialised network aware algorithms.

Typical exact algorithms for computing shortest paths are those of Dijkstra (1959, see subsection 7.3.4.3) and Dantzig (1960, see subsection 7.3.4.2), both of which are single source SPAs. These 'greedy' algorithms applied to planar graphs with non-negative edge weights typically require $O(n^2)$ time, where n is the number of vertices. However, Henzinger, Klein and Rao (1997) have shown that a near linear time algorithm is possible, i.e. $O(n)$. The complexity of Dijkstra's original algorithm can be reduced to $O(m+n\log n)$ using Fibonacci heaps, where m represents the number of edges.

7.3.4.2 Dantzig algorithm

Figure 7-5 illustrates Dantzig's step-by-step algorithm for determining the shortest path from vertex 1, 2, 3 and 4 to all other vertices in a directed planar graph (a *digraph*) with

positive edge weights. The algorithm is a form of discrete dynamic programming. Figure 7-5A shows the graph and Figure 7-5B-D the steps involved in determining the shortest path from vertex 1 to 3, highlighted in red (bolder line/dark grey).

Figure 7-5 Dantzig shortest path algorithm

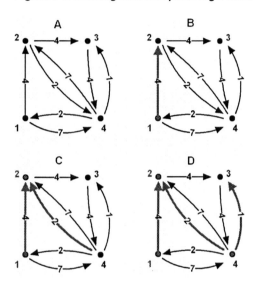

The basic steps in this procedure are as follows:

1: identify the shortest (least distance/cost/time) edge from vertex 1 — this is to vertex 2 (*cost*=4). Add vertex 2 and the edge or link from 1 to 2 to the tagged set. If a tie occurs, arbitrarily choose one of the edges

2: identify the shortest (least cumulative cost/time) edge from vertex 1 or from vertex 2 plus edge (1,2) distance — this is to vertex 4 from 2 (*cost*=6). Add vertex 4 and edge 2 to 4 to the tagged set

3: identify the shortest (least cumulative cost/time) edge from the tagged set — this is from vertex 1 to 2 to 4 to 3 (*cost*=7)

Stop — all vertices reached; repeat from vertex 2, 3 and 4 to compute all shortest paths from every vertex

This algorithm is very similar to the standard Dijkstra procedure, which describe in subsection 7.3.4.3:

7.3.4.3 Dijkstra algorithm

This algorithm works by storing the cost of the shortest path found so far between the start vertex, s, and each target vertex, t. We shall denote this distance $d(t)$.

The basic steps in this classic algorithm are as follows:

1: initialise all vertices such that $d(t)=\infty$ and $d(s)=0$

2: For each edge leading from s, add the edge length from s to the current value of the path length at s. If this new distance is less than the current value for $d(t)$ replace this with the lower value

3: choose the lowest value in the set $d(t)$ and move the current (active) vertex to this location

4: iterate steps 2 and 3 until the target vertex is reached or all vertices have been scanned

Note that the distance to every vertex reached is contained in the vector $d(t)$. If the entry is infinity then it means the vertex is unreachable or has not been scanned. Also note that unlike the Dantzig algorithm the shortest path(s) are not generated, only their lengths. To obtain these the algorithm must be altered to retain a list of shortest path vertices, e.g. by storing both the shortest path distance and the preceding vertex address of the shortest path in the vertex label.

It should also be noted that basic Dijkstra-type algorithms provide inconsistent run-times on real-world networks, principally because such networks have very limited vertex connectivity (their adjacency matrices are very sparse). SPAs need to be either very efficient implementations of these basic algorithms or should use heuristics, such as the A* algorithm, to obtain faster solutions (e.g. for real-time/interactive applications) — see further, subsection 7.3.4.4.

7.3.4.4 A* algorithm

Dijkstra's algorithm is a special case of a more general algorithm known as A*. Strictly speaking A* is a goal-directed best-first algorithm. It visits nodes in an order that may be preferable (faster) than the simple sweep of all nodes that the Dijkstra algorithm adopts. To select which node to visit next it typically computes the Euclidean distance of each vertex from the target and adds to this the distance currently recorded via the network to this vertex. Vertices are visited in overall distance priority order. Typically the A* algorithm will visit fewer nodes for a specified source-target (exact) route than other algorithms and will thus on average be faster — often a lot faster.

7.3.4.5 GIS implementations of SPA

Most GIS packages do not provide information on the algorithms used for this class or any similar network analysis problems, nor references to indicate the procedures used. Only the experience of using a particular implementation will confirm the nature of its behaviour, both in terms of computational complexity (time and space) and quality (optimality). Systematic testing of sample networks and problems based on a range of sizes (vertices) and plotting the results will provide a good indication as to the resource requirements and practicality of using any particular software. If a preferred package (e.g. the corporate standard) does not perform in the manner desired, then solving such problems externally using tools such as Concorde, Xpress-MP or LEDA, and then re-importing the results back into the GIS, may be the best option. These latter packages, and similar software, do provide details of the algorithms used, the time complexity of their solution procedures, and in many cases optional source code and linkable library code.

Figure 7-6 illustrates the process of carrying out simple variants of SPA using ArcGIS Network Analyst and a real street network. The source dataset for the network was a shapefile of the street layout for Salt Lake City

(SLC), Utah, USA. This is available as a downloadable file from the SLC authorities. Since the file is not in network-ready form it was processed using ArcCatalog to automatically create a network file that includes vertices at every intersection and end of line.

Figure 7-6 Salt Lake City — Sample networking problems and solutions

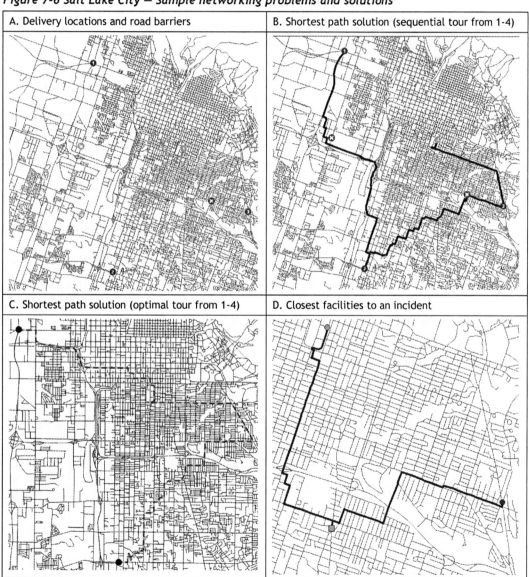

| A. Delivery locations and road barriers | B. Shortest path solution (sequential tour from 1-4) |
| C. Shortest path solution (optimal tour from 1-4) | D. Closest facilities to an incident |

Without further alteration (unrealistic in practice) this file was then used to compute the shortest path connecting four locations (Figure 7-6A). In this case the locations were selected interactively and parameters chosen for the computation. These included permitting U-turns and including two barriers (indicated by X symbols on the map).

The shortest path shown in Figure 7-6B is technically an open *tour*, or sequential ordering process, in that it is a connected series of shortest paths from 1-2, 2-3, and finally 3-4. It can also be seen as a shortest path from location 1 to location 4, with the constraint that the route must go via locations 2 and 3 — a very common requirement. A closed tour, returning to location 1, could have been generated by adding a shortest path from location 4 to 1. The tour described above is not optimal if the sequential condition is dropped — the tour 1-4-3-2 is shorter (Figure 7-6C). This latter tour was generated using the same dataset but this time using the Manifold package, Transform toolbar facilities. Note that there are some obvious differences in the solution map, other than the totally different routing. One of these is the lack of U-turns, an option which is not enabled in this instance. Another is the difference in default orientation of the display, which has no real significance. Manifold's solution is obtained by exhaustive search where there are 10 or fewer locations to be linked (i.e. an exact solution), and applies an heuristic algorithm (specific details are not provided) for 11+ sites. A discussion of optimal tours and associated algorithms in provided in the Section 7.3.5.

Figure 7-6D shows the result of addressing a related problem — that of identifying the closest facility or facilities to a specified location such as an incident. In this example the solution has again been generated using ArcGIS, which has identified two locations within a given distance or time from an incident location at the bottom of the map, together with optimal routes to or from the incident. Note that again U-turns have been permitted and that, depending on the network characteristics (turn permissions, one-way streets, traffic speeds etc.), the route *to* an incident may well differ from the route away *from* the incident. Often, of course, the routes to and from a location will be to and from different third locations (e.g. ambulance station and hospital).

7.3.5 Tours, travelling salesman problems and vehicle routing

The optimal tours described in Section 7.3.4 are subsets of a more general class of problem, the travelling salesman problem (TSP). The TSP involves making a tour of every vertex in a given set, returning to the starting point, such that the tour length is minimized. Applications include: salesmen visiting customers; rubbish trucks servicing business premises; delivery trucks servicing retail outlets; security staff patrolling premises and many more. There are many non-geographic applications, ranging from VLSI design to the analysis of DNA sequences. The majority of geographic instances involve modest numbers of locations to visit (e.g. stores, cities) but variants of the problem (e.g. the need to dispatch multiple vehicles from a central depot on tours to visit their own subset of stores) can increase the problem complexity substantially. Pure TSP problems (i.e. based on a point set with no pre-defined network) are not supported within GIS packages. Where such networks are defined, very few GIS packages provide TSP functionality — one exception being TransCAD.

Since every TSP tour involves visiting each location once, deleting one link from this tour will leave a spanning tree. This tree must be at least as long as the MST and will typically be longer (an upper bound for symmetric planar graphs is 1.5 times the MST length). The number, t, of possible tours with a symmetric (i.e. non-directed) graph consisting of n vertices is $t=(n-1)!/2$. This number grows extremely rapidly with n and cannot realistically be systematically evaluated for large n. For example, with $n=10$, $t=181,440$. Solving the problem exactly is thus an $O(n!)$ problem and has been shown to be NP-complete. An internationally collated set of test problems are provided in the dataset known as TSPLIB. These include test datasets for symmetric and asymmetric TSPs, HCPs, SOPs and CVRPs. See www.iwr.uni-heidelberg.de/groups/comopt/software/TSPLIB95/ for more details.

A formal statement of the basic TSP is:

$$1: \min\left\{\sum_{i=1}^{n}\sum_{j=1}^{n} c_{ij}x_{ij}\right\}, i \neq j$$

subject to:

$$2: \sum_{\substack{i=1 \\ i \neq j}}^{n} x_{ij} = 1, j = 1...n$$

$$3: \sum_{\substack{j=1 \\ j \neq i}}^{n} x_{ij} = 1, i = 1...n$$

$$4: x_{ij} \in \{0,1\}$$

where c_{ij} is the cost of travel from i to j (e.g. cost weighted distance), and $x_{ij}=1$ if a direct link exists from i to j in a tour, or 0 otherwise. Here n is the number of sites in the tour, where typically $i=1$ is the tour start point (e.g. a depot) and $i=2,3,...n$ are the sites to be visited (e.g. customers).

Solutions of many TSP problems have certain attributes that may be used to assist algorithm development. These include the fact that the TSP for a point set, S, must lie entirely within the convex hull of the set, and any vertices located on this hull must be traversed such that their clockwise or anti-clockwise sequential arrangement is preserved. Clearly also the TSP must not be self-intersecting. Heuristic solutions, notably the Lin-Kernighan or LK-heuristic, make use of any self-intersections that are found to generate an improved solution with the self-intersection removed by a form of symmetric resequencing or *flipping* of the order of visited nodes. The result is a very fast approximate algorithm that yields exact or near exact solutions for many problems. Exact algorithms also exist, some of which are fast. Concorde, cited earlier, uses a procedure based on linear

programming and cutting plane techniques to obtain exact solutions to relatively large problems (many 1000s of vertices). This code, which is very fast with moderate-sized problems, is amongst the best available. However, users should be aware that rounding issues exist in its treatment of decimal coordinates – to avoid problems it is advisable to multiply all values by one or more powers of 10 before undertaking analyses.

There are close relationships between the generation of the Delaunay triangulations, the generation of MSTs and the solution of Euclidean (plane) TSPs. To illustrate this linkage we will use a test dataset from TSPLIB consisting of $n=130$ vertices (CH130.TSP). Figure 7-7 illustrates several aspects of these relationships. Figure 7-7A shows the source point set and Figure 7-7B gives the Delaunay triangulation (DT) of this set. The total length of the edges in the triangulation is 30157 units. Figure 7-7C and Figure 7-7D show the minimum spanning tree (MST) for the point set, produced using the Concorde software and overlain on the DT. The MST consists of 129 edges with a total length of 5166 units. Figure 7-7E and Figure 7-7F show the same pair, but in this case showing the exact TSP solution obtained using the linear programming and cutting planes method that is implemented in Concorde. Processing on a modestly powered desktop PC was completed in a few seconds. The total number of edges in this case was 130 with a total length of 6110 units. Finally, Figure 7-8A shows the solution provided using the LK heuristic, which was generated almost instantly for this problem. Notice that the top right corner of this solution differs slightly from that of the exact solution in Figure 7-7F. This solution also has 130 edges but a slightly greater length of 6124 units. In all cases edge weights were computed using symmetric Euclidean distances. Interestingly, with a random pattern of n points whose MBR has area A, the expected length of the Euclidean TSP has been estimated as:

$$L = k\sqrt{nA}, k \approx 3/4$$

For the example illustrated in Figure 7-7A $n=130$ and $A=471{,}937$ units, giving an expected tour length if the point set were random of 5874 units. The optimal result is slightly greater than this, reflecting the fact that the point set is slightly more uniform than random. With the Manhattan or L_1 metric the expected tour length is greater, being the same general expression but with $k\approx19/20$.

Figure 7-7 MST, TSP and related problems

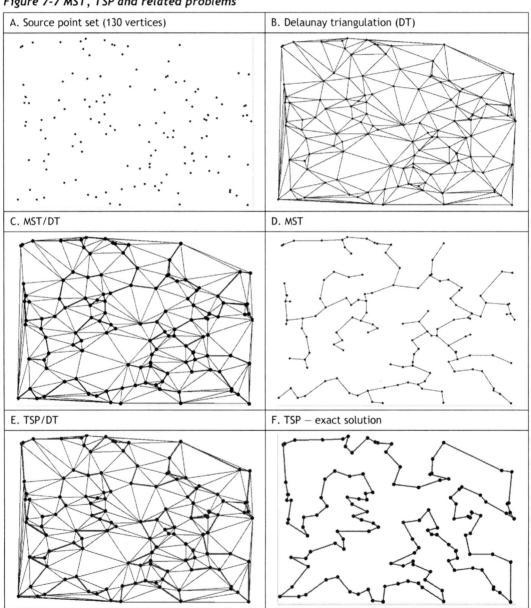

A. Source point set (130 vertices)

B. Delaunay triangulation (DT)

C. MST/DT

D. MST

E. TSP/DT

F. TSP — exact solution

Figure 7-8 Heuristic solution and dual circuit TSP examples

A. TSP – LK solution	B. Dual TSP
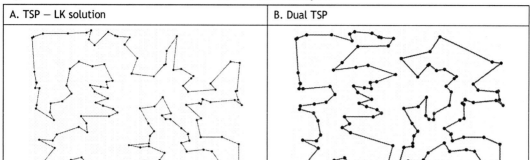	

7.3.5.1 Capacitated vehicle routing

Figure 7-8B provides one instance of a solution to a more general problem, that of determining the optimal set of $t>1$ tours for a given point set. This class of problems is known as the Vehicle Routing Problem (VRP) or Capacitated VRP (CVRP) if vehicle capacity constraints are included in the problem formulation. In this example the point set has simply been divided into left and right halves and separate optimal tours generated for each subset. The combined length of these tours is 6349 units, but it is perfectly possible for the tour length to be less than that for a single tour. Multiple tours will often utilise the same starting vertex (e.g. a single depot) but this does not substantially alter the approaches used. Separate tours can be created by breaking (dividing) the original optimal tour into discrete parts (route first-cluster second) or by separating the point sets into clusters and computing optimal or near optimal tours for each subset (cluster first-route second). There is no guarantee that the resulting multiple tours will be globally optimal, so repeated alternative subdivisions may be necessary to achieve improved solutions. Such a procedure is extremely processor intensive and hence use of algorithms such as the LK-heuristic would generally be preferred for the TSP generation step.

The two circuit TSP problem raises a number of practical issues:

- should the tours start at the same point (e.g. warehouse, bus depot...?)

- what if demand varies across the target points?

- the capacities of service vehicles may be limited and vary – what mix would be optimal?

- do tours/deliveries have to be made in certain time windows?

Questions such as these are addressed in several of the software packages referred to in the introduction to this chapter. Most seek optimal solutions for moderate sized problems, or provably good solutions that can be computed reasonably quickly – i.e. sub-optimal solutions, but much better than more arbitrary or user-defined plans. Individual packages vary in their approaches and often provide more than one option where complex real-world constraints and commercial criteria apply. This in turn requires careful formulation of the appropriate model, designing this using a standardized modelling structure or design tool, and then implementing the model using a high-level scripting language or code generator.

To illustrate this approach we shall use an example from the Xpress-MP optimisation product with its Mosel scripting language. In this case the problem involves delivery of liquid chemicals by tanker from a refinery to 6 customer sites. The demand, DEM, from the sites is known (these might be regular fixed or varying amounts, delivered on a scheduled basis). Total demand in this example is 59000 litres. A symmetric shortest distance matrix is provided between all locations — map support is not provided within Xpress-MP so distance matrix creation and final display would benefit from a direct link to a GIS package or simple import/export facilities. Pairs of sites vary in separation from 12 kms to 180 kms. The capacity, CAP, of the delivery tanker is fixed at 36000 litres and the task is to deliver to all customers with the shortest overall distance driven. Note that a single tour would not be able to service the customers, as the tanker capacity is not sufficient, so capacity-related constraints must be added to the model formulation provided at the start of this subsection (Equations 1: to 4: in Section 7.3.5). The objective function in this instance is specified by Equation 1:, whilst the range of i and j in Equations 2: and 3: is from 2...n as site 1 is the depot itself. We need to define additional expressions to account for customer demand and tanker capacity. Let DEM_i be the demand by customer i, and let q_i be the cumulative amount of chemical in litres delivered to all customers during a tour, up to and including customer i. If customer i is the first in a particular tour, then $DEM_i = q_i$, which can be written as the condition:

$$5 : q_i \leq CAP + (DEM_i - CAP)x_{1i}, \ \forall i \in \text{customers}$$

If j comes after i in a tour then q_i must be greater than the quantity delivered in the tour up to i plus the quantity to be delivered to j. This constraint can be written as:

$$6 : q_j \geq q_i + DEM_j - CAP + CAP * x_{ij} +$$
$$(CAP - DEM_j - DEM_i)x_{ji},$$
$$\forall i, j \in \text{customers}, i \neq j$$

Although this appears rather complicated, it picks up the various cases where i is not 1 and j may or may not follow or precede i. All that

remains is to add the basic quantity and capacity constraints:

$$7 : q_i \leq CAP, \ \forall i \in \text{customers}$$
$$8 : q_i \geq DEM_i, \ \forall i \in \text{customers}$$

The scripting language for Xpress-MP accepts input in almost the form shown in Equations 1: to 8:, together with the data and distance matrices described above, followed by a simple command to minimize Equation 1:. The result in this example is a solution consisting of two tours, one delivering 22000 litres to two customers, and the other delivering 37000 litres to the remaining 4 customers, with a combined tour of just under 500 kms (Figure 7-9).

Figure 7-9 Tanker delivery tours

Data-driven scripts that solve well-defined optimisation problems may be subjected to extensive tests, and then used as the back-ends to form-based or graphical/map interfaces in order to provide solutions to a wide range of real-world problems. These higher-level interfaces can provide location, route matrix, attribute and parameter inputs (data) and submit these to the pre-built scripts for execution. Outputs (such as lists of edges comprising routes) can then be read back into the visual interface software and presented as reports, tables and maps. GIS packages such as TransCAD and ArcGIS provide a range of scripting and programming interfaces to facilitate such integration, either with the GIS package or using GIS-package facilities within another application, such as a vehicle dispatch and management system for the emergency services.

7.4 Location and service area problems

7.4.1 Location problems

Optimally locating p facilities to serve customer demand at $n>p$ locations is a simplified description of many locational analysis problems (many of which may be formulated as p-median problems – see further, Section 7.4.2). If the existence of a network is ignored, and connectivity is assumed to be by direct connection in the plane from each customer to a single, closest, facility then a simple location-allocation problem is defined — selecting locations and then allocating customers or demand to these locations according to some rule (e.g. nearest facility). Frequently, however, a network (e.g. a road network) exists together with a matrix of shortest path/least cost distances between vertices, and this provides a key input into the optimisation process. Real world problems are, of course, far more complex, since there are many other variables to be considered (e.g. availability of suitable sites, cost of sites, size of facility, access to and from sites, regulatory issues, planning controls, availability of suitable labour, timing of developments, etc.) and all of these considerations exist within a dynamic environment that affects these and core variables such as customer demand patterns, materials supply and changes in the technological, commercial and political environment. Hence optimisation is only a part of a much larger decision-making process, whose importance will vary depending on the particular problem at hand. Obtaining near (provably) optimum solutions to many problems, and understanding their robustness to variation in parameters and locational changes is often at least as important as seeking absolute optima for problems that have been very narrowly defined. Such procedures may also provide a form of benchmarking, identifying the cost associated with the "best" solution, which may then be compared with other options within a broader decision-making framework.

Research into problems such as these has a long history, with substantial progress being made from the 1960s onwards. As the range of problems tackled became broader and solution procedures required increasing use of computational power, libraries of software were developed and made available by academics, which then (directly or indirectly) were incorporated into commercial software packages. One of the first of these was the set of FORTRAN programs published (as an edited monograph and in electronic form) by Rushton, Goodchild and Ostresh in 1973. The problems addressed in this software collection included single- and multi-facility location in the plane (Euclidean metric) and on a network, together with algorithms for allocation and spatial choice modelling. At this time the single facility optimisation procedures were exact, whilst multi-facility procedures utilized a variety of heuristics to obtain "good" solutions.

In the plane (i.e. with no network defined) the p-median problem is a form of point-set partitioning, similar to cluster analysis. Typically p supply or service points are sought to service n customer locations (vertices), such that each customer is allocated to a single supply point (often based on simple Euclidean proximity) and such that the total distance or cost of transport between the customers and the supply centres is minimised. If $p=1$ we have the simple computation of the Minimum Aggregate Travel (MAT) location, which may be determined using the iterative procedure described in Section 4.2.5.2. If $p>1$, however, the problem is more complicated and requires an extended solution procedure. Exact solution of the p-median problem in the plane is possible using branch and bound (tree-searching) algorithms. Ostresh, for example, provides such an algorithm in the FORTRAN suite referred to above (Rushton, Goodchild and Ostresh, eds., 1973). Exact solutions for larger problems and extended variants of the problem may be computationally demanding and several heuristic procedures have been devised that tests suggest will often provide optimal or near optimal solutions (with less than 3% deviation from the optimal objective function value) in many cases. Note that in planar or "free" space problems the form of the objective function can be viewed as a cost surface, whose shape may be quite complex with many small local optima. Furthermore, since sections of this surface may be relatively flat, a specific heuristic solution that is within 2% of the optimum objective function value could well be defined by a set of locations that are a considerable distance from the optimal sites. Evaluation of variation of the objective function in the neighbourhood of solution points one at a time (e.g. across a grid defined within the Voronoi region of this point), can provide an insight into the form of this surface.

A simple heuristic solution for the planar p-median problem, most commonly attributed to Cooper (1963, 1964), is as follows:

Step 1: randomly select p points in the Minimum Bounding Rectangle (MBR), or preferably within the convex hull, of the customer point set, V, as the initial locations for the median points

Step 2: allocate every point in V to its (Euclidean) closest median point. This partitions V into p subsets, V_p

Step 3: For each of the p subsets of V, compute the MAT point using the iterative equation referred to above (described in Section 4.2.5.2)

Step 4: iterate steps 2 and 3 until the change in the objective function falls below a preset tolerance level

Clearly the quality of the solution obtained by this heuristic is not provided. One approach to improving the solution found would be to repeat the process k times with a different set of random starting points and compare the final objective value with previous runs. A better approach is to make good initial choice of the median locations, for example by some form of simple cluster analysis (e.g. the density based procedure adopted by Crimestat

or a simple neighbour-based or quadrat-derived clustering). An extended version of this algorithm can be used to solve the same problem, but with capacity constraints on the supply centres. In this case allocation is more complicated and may result in some customers being served by more than one centre. Experience suggests that constrained problems of this kind often yield solutions that are closer on average to the true optimum than unconstrained problems.

More recent examples of the development of collections of location modelling utilities include the public-domain C++ library known as LOLA (Library of Location Algorithms), Daskin's SITATION software, and the S-Distance project developed at the University of Thessaly in Greece (a Visual Basic V6 implementation of many discrete and network location algorithms, including a wide range of heuristics, although not yet fully completed — see further, subsection 7.4.2.3). LOLA is available for various computing platforms in both source code and in pre-compiled form with optional basic graphical user interface (GUI) and link to ESRI's ArcView GIS (using ESRI's Avenue language). LOLA utilizes a number of third party code components, including LEDA and LP-solve which we have previously referred to (LOLA's V2.0 release in December 2000 pre-dates LEDA's move to a more commercial environment). LOLA addresses both median problems (minimisation of average or total travel costs) and so-called centre problems (minimisation of maximum travel costs, i.e. minimax problems).

As with the earlier FORTRAN-based project LOLA provides a mix of exact and heuristic solutions to problems in the plane or on networks, but in LOLA's case includes a much broader range of metrics (typically L_p metrics), optional restrictions on permitted solution areas (forbidden zones, barriers), and solutions for directed and undirected networks and on trees. For network (graph) problems most solutions are designed for 1-median or 1-centre problems, with solutions that may lie either at network vertices or anywhere on the network.

Figure 7-10 illustrates using LOLA to identify the best two cities in Italy from which to service demand from customer sites in a total of 39 cities. Each city included (network vertex) is taken as a customer site, with entries being of the form (x,y,z,n) where x and y are coordinates, z is the customer demand estimate and n is the city name — for example:

28.00 74.00 50000 [Milano]

Coupled with this information are details on the network structure and cost. In this case data are entered as adjacency lists, in the form <from,to,weight>.

Hence as Milano is location 1 and has direct connections to Piacenza, Novara and Verona (locations 34, 17 and 13) the adjacency list is of the form:

1 34 4
1 17 3
1 13 22

The weights in this case might be travel or transport costs or times. The geometry of the routes in this case is unimportant. The optimum solution consists of two city locations, shown here as Opt1 and Opt2 (Piacenza and Roma). This is a very simple example but serves to illustrate the ideas involved and the basic data required. In many instances the problem LOLA addresses is where to locate a new facility given that a number of such facilities already exist. New facilities typically involve an overall construction cost (cost of site plus construction etc.).

Figure 7-10 Optimum facility location on a network — LOLA solution

The majority of GIS packages do not provide such capabilities directly, although many offer basic operations, such as locating a weighted mean centre in the plane. TransCAD provides tools to add a number, m, of new facilities to an existing set (which could be empty) in a network environment, and thus has similar functionality to LOLA in this example, but here it is integrated within a GIS framework rather than linked to a separate product set. It provides tools to determine the best location to place one or more new facilities, or as many as might be required to meet a given service objective, subject to various optimisation objectives (minimisation of average costs, minimisation of maximum costs, maximisation of minimum costs). Custom-built software utilising general purpose optimisation libraries such as LEDA, CPLEX and Xpress-MP enables a wide range of problems of this type to be designed and solved in an optimal or near optimal manner. In many cases larger problems require the use of heuristic or approximation algorithms in order to obtain solutions within a reasonable time.

7.4.2 Larger p-median and p-centre problems

7.4.2.1 Simple heuristics

Exact solution is possible of the network version of the p-median problem. In this model there are n customer demand sites, each with demand weights w_i and p supply sites to be chosen from $m \le n$ candidates. These p "median" sites will lie at vertices of the network (in fact optimal solutions may exist anywhere on the network but a theorem due to Hakimi shows that at least one optimal solution will lie entirely at network nodes, which simplifies the overall problem in many cases). In principle, for small n, m and p, solutions can be found by enumerating the possible configurations, but this becomes infeasible very quickly as their values increase in size. Heuristic solutions may provide good or even optimal solutions in a reasonable time, but evaluating the quality of some of these approaches is not straightforward. Other

heuristics, by their very nature, do provide bounds on the quality of their approach to optimality.

A basic heuristic for the network-based p-median problem, due to Teitz and Bart (1968), is known as an "interchange" method and proceeds in the following manner. Let V be the set of m candidate vertices, then: (i) randomly select p vertices from V and call this set Q; (ii) for each vertex i in Q and each j not in Q (i.e. in the set V but not in Q) swap i and j and see if the value of the objective function is improved; if so keep this new solution as the new set Q; (iii) iterate step (ii) until no further improvements are found. Note that this algorithm has the potential to produce improved solutions by repeating step (i) with a different random set and/or by examining good alternative unoccupied locations (e.g. within sub-regions of the network) as preferred choices for the interchange process — see Densham and Rushton (1992) for more details.

7.4.2.2 Lagrangian relaxation

Because the p-median problem is of such importance and generality we will describe it in greater detail, together with a particularly effective solution procedure for larger problems known by the rather cumbersome name of "Lagrangian relaxation with subgradient optimisation" or LR procedures. Developments of this type of approach are described by Christofides and Beasley (1982) and Beasley (1985). Essentially, this procedure is a form of "branch and bound" algorithm, whereby upper and lower bounds are determined for the problem, optionally used in conjunction with a form of tree search (branching) to produce steadily improved results.

As with many optimisation problems a standard formulation of the p-median problem can be made as follows:

$$1: \min \sum_{i=1}^{n} \sum_{j=1}^{m} s_{ij} x_{ij}$$

subject to:

$$2: \sum_{j=1}^{m} x_{ij} = 1, \ i = 1,2..n,$$

i.e.

$$2': D=0 \text{ where } D = 1 - \sum_{j=1}^{m} x_{ij}$$

$$3: x_{jj} \geq x_{ij}, \forall i,j, i \neq j$$

$$4: \sum_{j=1}^{m} x_{jj} = p$$

In this formulation $x_{ij}=1$ if demand location i is allocated to supply location j, otherwise it is 0, and $x_{jj}=1$ if a facility is opened at location j otherwise it is 0. These requirements can be expressed as an additional constraint equation, requiring that $x_{jj} \in \{0,1\}$.

Equation 1: is the objective function, the total (or average) cost, s_{ij}, of servicing customer demand at locations $i=1,2..n$ from the supply sites, j. Here, $s_{ij}=w_i c_{ij}$ where w_i is a non-zero weight value indicating the demand at customer site i, and c_{ij} is the cost of supplying a unit of demand from supply location j to customer i. Typically c_{ij} is the shortest path distance across the network between i and j. This information may be computed in real time or supplied as a cost/distance matrix or list. Note that in general s_{ij} is not equal to s_{ji}. If Equation 1: is replaced with $\min\{\max\{s_{ij}x_{ij}\}\}$ then the formulation is known as the p-centre problem (which is also known to be NP-hard). In this latter case we are trying to minimize the maximum cost of servicing demand, for example to meet pre-defined service criteria.

Equation 2: states that every customer is served by exactly one site. An alternative formulation, 2', is shown as it is used later in this subsection.

Equation 3: states that if a facility is not located at site j then customer demand cannot be allocated to it.

Finally Equation 4: states that a total of p facilities need to be allocated.

The p-median problem on a network can be solved exactly for reasonably large problems (e.g. 1000 customer sites and 50 depots) using the LR algorithm. The solution procedure involves solving a simpler problem (a relaxation of the original problem) that does not satisfy all of the constraints (e.g. fails to satisfy Equation 2:) and then using this solution to obtain upper and lower bounds on the optimum solution. These bounds are then systematically narrowed (the subgradient optimisation phase) until, ideally, the solution upper bound (which does satisfy the constraints) meets the lower bound (exactly or to within a given tolerance), providing the optimal solution.

The simpler (relaxed) problem to solve actually looks more complicated. It involves adding a set of n variables (Lagrangian multipliers, $\lambda_i > 0$) into the original Equation 1:, "relaxing" the constraint in Equation 2:, as follows:

$$\max_{\lambda} \left\{ \min \left\{ \sum_{i} \sum_{j} s_{ij} x_{ij} + \sum_{i} \lambda_i \left(1 - \sum_{j} x_{ij} \right) \right\} \right\}$$

Observe that the last term in brackets in this expression is the constraint D, which now forms part of the objective function. The problem is 'relaxed' because the constraint in Equation 2: has been removed, so it is possible for a customer to be served by/assigned to more than one site — and indeed, when one comes to solve the relaxed problem this frequently occurs. Also note that the final summation can be expanded and combined with the initial summation, as follows:

$$\max_{\lambda} \left\{ \min \left\{ \sum_{i} \sum_{j} \left(s_{ij} - \lambda_i \right) x_{ij} + \sum_{i} \lambda_i \right\} \right\}$$

Minimisation of the main term in brackets above is thus a simple enumeration task using the first term of the expression, since the second summation is a constant. We select each of the potential sites, j, in turn by setting $x_{jj}=1$ and then compute the sum of the cost of servicing customer demand, C_j, using the selected facility minus the Lagrangian, λ_i, i.e.

$$C_j = \sum_{i=1}^{n}\left(s_{ij} - \lambda_i\right)$$

If this sum is greater than 0 it is set to 0 otherwise the value is stored and the computation proceeds for the next j. The p smallest (largest negative) values obtained in this way are taken as the locations of the initial set of p-median sites. The allocation of customer sites to these selected locations is made by setting $x_{ij}=1$ if $x_{jj}=1$ and $s_{ij}-\lambda_i<0$, otherwise $x_{ij}=0$. This allocation will not generally satisfy the original constraints, but the computed total cost obtained using Equation 1: will provide a lower bound, LB, to the true optimum solution.

We now obtain an upper bound, UB, by allocating the customer sites to their nearest supply location and again computing the total cost using Equation 1:.

To obtain an optimal or near-optimal solution the values of the initial Lagrangian multipliers need to be adjusted (reduced) in such a way as to eventually satisfy the original constraints. The updated values for these multipliers are obtained using a step adjustment to their values during the previous iteration:

$$\lambda_i^{(n+1)} = \max\left\{0, \lambda_i^{(n)} - t^{(n)}\left(\sum x_{ij}^{(n)} - 1\right)\right\}$$

The stepsize value at the nth iteration, $t^{(n)}$, is determined using an expression of the form:

$$t^{(n)} = \frac{a^{(n)}\left(UB^{(n)} - LB^{(n)}\right)}{\sum_i\left(\sum_j x_{ij}^{(n)} - 1\right)^2}$$

Selection of suitable values for $a^{(1)}$ and the $\lambda_i^{(1)}$ is important to ensure convergence. Christofides and Beasley (1982) and Beasley (1985) suggest using $\lambda_i^{(1)}=\min(s_{ij})$ across all j, for all $i\neq j$, and $a^{(1)}=2$, with procedures for reducing this in subsequent iterations (e.g. progressive halving) depending upon the convergence behaviour. Note that the term in the denominator of this expression is the sum of squared 'violations' of the constraint we have relaxed, so in this case it is based on the number of customers who have been assigned to more than one supply centre.

After each iteration new UB and LB value are obtained as before and the process continues until satisfactory or complete convergence has been achieved. An optional standard tree search procedure can be included at this stage if the process fails to converge. Daskin (1995, pp 225-228) provides a fuller explanation and simple numerical worked example of this procedure. For large problems the overhead of computing the initial upper bound has been found to be high, and as an alternative a simple heuristic such as the interchange method described earlier can be used. Indeed, as we have noted, the heuristic itself can sometimes produce optimal solutions directly.

Performance of the LR algorithm for a range of problems is described by Beasley (1985) and for heuristic implementations of essentially the same procedure, by Beasley (1993). A set of test data for the problems described by Beasley in these articles and a range of related problems (e.g. other problems involving depot and warehouse location, with or without capacity constraints) is available at his web site (see Appendix for details). TSPLIB can also be used as a test dataset for p-median software.

Problems that involve <500 vertices (customers) and modest values of p (e.g. $p<50$) can be solved directly and optimally by commercial mixed integer program (MIP) solvers such as CPLEX and LINGO. Larger problems remain difficult for such software and/or may require an excessive amount of time to complete — typically an hour or more of run time. Beltran *et al.* (2004) use the TSPLIB dataset in their semi-Lagrangian solution approach which is built on similar ideas to those of Beasley. Their tests were implemented in MATLab in combination with the CPLEX library V8.1. They use this approach to solve large *p*-median problems (3000+ customers and up to *p*=500 supply locations) to optimality or within 1% of optimality. More recently Avella *et al.* (2006) have reported results on their earlier study of their "Branch and Cut and Price" (BCP) algorithm. BCP can be used to solve large *p*-median and related problems in conjunction with CPLEX or similar library software. They have tested their methodology against the TSPLIB and OR-Library datasets, with very encouraging results, greatly outperforming CPLEX used on its own.

7.4.2.3 Comparison of alternative p-median heuristics

The S-Distance software project, mentioned earlier in this subsection and described further in Sirigos and Photis (2005), provides tools to compute solutions to quite large *p*-median and *p*-centre problems using a range of heuristics, together with solutions for selected coverage problems. The heuristics they implement include many of those described in Taillard (2003), Daskin (1995) and Dyer and Frieze (1985). Data may be read from various file formats, including OR-Library files and network node/link files in dbf format. These various methods may then be compared in order to obtain an estimate of their relative efficiency. For the purposes of this exercise we used an S-Distance test dataset comprising the street network in Tripolis, in the region of Arcadia, Southern Greece. The data consist of 1358 vertices and 2256 edges. Each vertex or node has an integer demand value associated with it, ranging from 1 to 137, plus 35 nodes with 0 demand. Each edge or street link has a "to-from" and/or "from-to" cost of travel assigned to it, with selected edges being designated as one-way. In this test we sought 5 locations to service the demand using several different heuristics. Figure 7-11A to Figure

7-11C show the results of running these procedures on the test dataset.

In these figures the red (dark grey) circles identify the solution points (median locations) and the darker lines show the network edges that each centre uses to service the demand. The figure in brackets in each case shows the objective function value, z, achieved, in millions of units (sum of demand times cost of travel). The basic greedy-random heuristic used by S-Distance is that defined by Dyer and Frieze (1985). This simply involves selecting the highest weighted demand point (or any at random if there are multiple locations with equal weights). The next location is then chosen by selecting the demand point that has the largest weighted distance from its nearest centre. The algorithm is of complexity O(*np*) and yields results that are within 2-3 times the optimum (and so are not very good!). By contrast, the CLS procedure often produces very good results (within 1% generally) but requires longer to run, with a complexity of at least O(*np*(*n*+1)). CLS is essentially a form of alternating location-allocation (ALT) procedure, like that of Cooper described earlier. Initial solutions are then perturbed according to a greedy interchange rule that attempts to swap each ALT solution point in turn with another candidate site, and examines whether the resultant solution improves the ALT local optimum.

The *greedy* heuristics shown in Figure 7-11A and B ran in almost no processor time, with the greedy-add algorithm producing a much improved z-value. The five locations it chose, however, are very different from those in Figure 7-11C, which was the solution found by three other algorithms: candidate list search (CLS), Lagrangian relaxation (upper bound solution, LR), and variable neighbourhood search (VNS). Of these VNS was fastest to run, requiring under a second of processor time, whilst CLS took around 3 seconds and LR over 3 minutes. However, the LR procedure provides both upper and lower bounds on the solution, which in this case showed that the lower bound was 1.177, but this solution is infeasible, so it is extremely likely that the solution shown is, in fact, the true optimum. The final image, Figure 7-11D, shows the allocation of customer demand to the individual centres. The size of the coloured circles reflects the demand weights assigned to the network vertices.

Figure 7-11 Comparison of heuristic p-median solutions, Tripolis, Greece

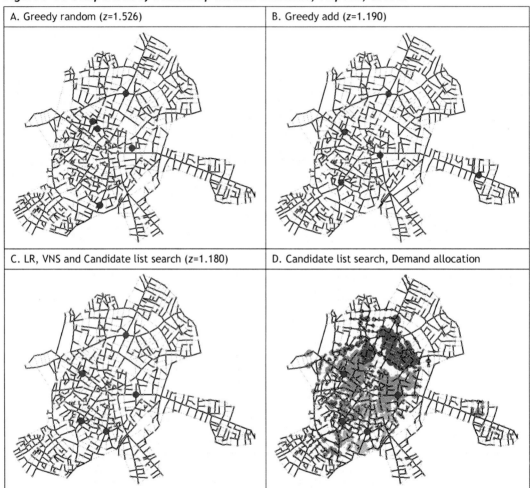

| A. Greedy random (z=1.526) | B. Greedy add (z=1.190) |
| C. LR, VNS and Candidate list search (z=1.180) | D. Candidate list search, Demand allocation |

Similar heuristics may be applied to the *p*-centre problem. The total cost of servicing demand in this case will always be greater than or equal to that for the *p*-median solution, but the maximum distance (cost of travel) for customers travelling to or from service centres will be minimised. The "Optimised service centre location" illustrated at the start of this Guide is the *p*-centre solution equivalent to that shown in Figure 7-11D. As can be seen, the selected centres are more evenly spread across the City as the algorithm seeks to avoid any very long links between customers and service centres. Particularly noticeable are the locations of the most easterly and south-westerly service centres in this solution. It is interesting to note in this case that this *p*-centre solution is very similar to the *p*-median solution illustrated in Figure 7-11B.

These network-based *p*-median and *p*-centre models can be compared with the discrete location planar equivalent, where demand is located at nodes as before, but facilities may be located anywhere in the plane and the objective function is for median location. In the example shown in Figure 7-12 the facilities located are approximately optimal with respect to demand using the Euclidean metric rather than network distance. The locations are similar to those identified by the *p*-centre optimisation process and significantly different from the best network-based *p*-median solutions.

Figure 7-12 Facility location in Triplois, Greece, planar model

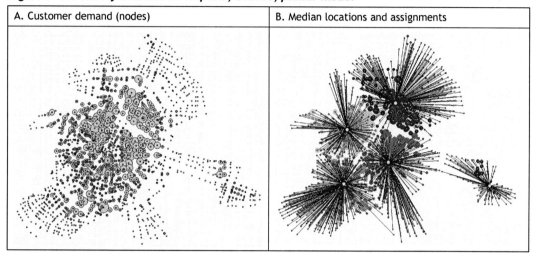

A. Customer demand (nodes)	B. Median locations and assignments

7.4.3 Service areas

The maps illustrating demand assignment in Figure 7-11D and Figure 7-12B provide examples of service area definition. Defining service areas or zones (often without regard to specific demand levels) is a widely supported facility within many GIS packages. These areas are typically discrete zones that are closer in distance, time or cost to specific points in the network than to any other specified point (rather like Voronoi regions in the plane).

Note that the specification of service areas in this case is not the same as solving coverage problems. The latter does not assume that the locations or numbers of facility sites are known — these are outcomes of the modelling process. In the pure or set coverage problem the number of facilities must be increased until demand from all eligible customers is met within the given service parameters. This will often result in unacceptably high numbers of facilities, requiring some form of relaxation of the constraints to be of use in real-world situations. An example of such relaxation, known as the maximum coverage problem, is to pre-specify the number of facilities (or vehicles etc.) to service demand, and then seek to maximise the customer coverage within the service constraints. Relaxing the service constraints (e.g. allowing the time to reach a patient with an ambulance not to be an absolute maximum, but an average for example) leads to *p*-centre and *p*-median type problem specifications.

An example of GIS-based service area functionality is provided in Figure 7-13A. This shows the locations of three ambulance stations situated in an urbanised region. In

Figure 7-13B all streets closest in network distance to each station are assigned as the primary service area for each station. As noted above, this form of simple network partitioning assumes pre-defined facility locations and takes no account of variations in demand (e.g. expected accident or illness rates) or in supply (e.g. number of vehicles available). The example illustrated was generated using the TransCAD software, but similar facilities are provided in many other GIS packages, such as the v.net.alloc and v.net.iso functions in GRASS, the "service areas" facility in ArcGIS Network Analyst, and the Allocation functionality provided in the TNTMips Network Analysis facility. The latter tool, for example, incorporates options to allocate service areas represented by network links based on flows to or from selected locations, with or without specifying from a range of parameters (such as demand levels, capacity constraints and travel impedance).

Figure 7-13 Service area definition

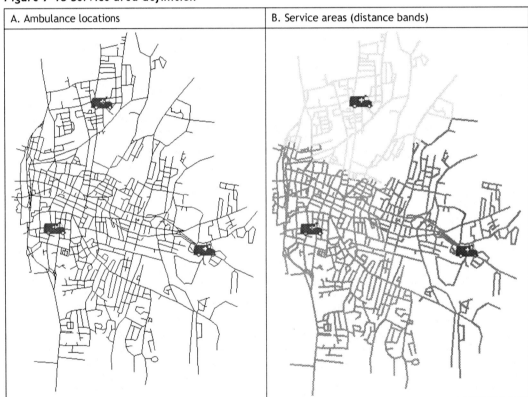

A. Ambulance locations	B. Service areas (distance bands)

7.4.3.1 Travel time zones

A related facility to service area definition, also provided in many GIS packages, is that of identifying travel time zones. These are typically generated as a polygon layer, overlaid on the network, indicating bands of travel times or distances. Figure 7-14 shows a set of bands that have been constructed around a point location on the Salt Lake City street network. These bands have been generated using distances in metres, with the edge of the band being the upper limit of the distance interval. Note that these zones include areas that have no roads, and the procedures for constructing such areas varies between packages, and may or may not allow zone construction procedures to be defined. Manifold implements three different options: (i) a convex hull; (ii) a buffer; and (iii) a zoning system, similar to the buffer operation, but in which off-road speeds (i.e. walking, off-road driving) are separately specified. Computation times can be very lengthy depending on the options selected and the complexity of the local network.

Figure 7-14 Travel time zones

Drive time zones — Bands: 500,1000,2000,5000 m

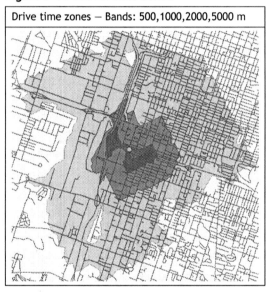

7.5 Arc Routing

The final topic in this chapter is known as Arc Routing. This is the term used to apply to a family of problems in which all links within a defined service area must be visited. It is thus a specialised routing problem. Common variants of the problem include: those in which lnot all links need to be traversed; cases where some or all links are directed; cases in which links have variable traversal costs depending on direction; cases where the links and/or vehicles have capacity constraints; and heirachical problems, in which one set of edges must be served before another. A thorough review and explanation of arc routing problems and their solution may be found in Moshe *et al.* (2000).

A simple example of an arc routing problem is the need to clear snow from all streets in a town in the most efficient manner. This problem and its solution are illustrated in Figure 7-15 and Figure 7-16A and B. On completion of the optimisation process the operator may be issued with a route plan as a list in addition/as an alternative to a map, as shown in Figure 7-15. In this example, three depots have pre-assigned street areas to clear, as shown in Figure 7-16A. A solution plan is then shown in Figure 7-16B for one of the snow ploughs. The plan involves a systematic coverage of all streets, on both sides, with no streets being covered twice (if possible) and as few turns as possible, especially where these incur penalties. Minimising coverage of streets that have been serviced already or have no demand for service is known as minimisation of deadheading.

Note that in this example U-turns are permitted, the solution involves no deadheading links (see Figure 7-15), and no explicit turn penalty has been specified. The solution illustrated was implemented using TransCAD — similar facilities are not provided in other GIS packages but are provided in a number of logistics software packages and in add-ons to widely used GIS packages — for example, IIT's eRouteLogistics add-on for ESRI

GIS products).

Figure 7-15 Routing directions

```
ROUTE ID 1          : depot/district 7310
Total Time          : 296.12
Deadhead Cost       : 0.00
# Deadhead Links : 0 (out of 1090 links)
# Left Turns        : 328
# Right Turns       : 265
# U-Turns           : 215
# Straight Turns    : 281
Turn Penalty Cost: 0.00

No.   Movement         Street_Name            Service
-----------------------------------------------------------
  1   Start West on    W 3RD ST                 Right
  2   Left on          HUMPHREY AVE             Right
  3   Left on          W 2ND ST                 Right
  4   U-turn on        W 2ND ST                 Right
  5   Right on         NEWMAN AVE               Right
  6   Left on          W 4TH ST                 Right
  7   Left on          GARRETSON AVE            Right
  8   U-turn on        GARRETSON AVE            Right
  9   Left on          W 4TH ST                 Right
 10   U-turn on        W 4TH ST                 Right
 11   Left on          GARRETSON AVE            Right
 12   Left on          W 5TH ST                 Right
 13   Left on          TRASK AVE                Right
```

Another variant of this problem includes the so-called CARP or Capacitated Arc Routing Problem. This class of problems assumes that capacity constraints apply, for example the capacity of a rubbish collection or gritting vehicle, or the carrying capacity of a postman. A range of exact and heuristic algorithms have been developed for both classes of problem, often based on similar techniques to those used in solving the TSP. CARP, like the TSP, is known to be NP-complete. Yet another, closely related class of problem involves capacity constraints on the links, typically described as capacity constrained flow problems. Example applications include traffic planning, pipeline routing and some forms of telecommunications planning.

A formal statement of the uncapacitated arc routing problem is as follows: Let E be a set of edges and V a set of vertices for the graph $G(E,V)$ and denote by (i,j) the edge from vertex i to vertex j, with edge length d_{ij}. Then we have:

$$1: \min\left\{ \sum_{(i,j)\in E} d_{ij}u_{ij} \right\}, \text{ subject to}$$

$$2: \sum_{(i,j)\in E} u_{ij} = \sum_{(j,i)\in E} u_{ji}$$

$$3: u_{ij} \geq 1 \; \forall (i,j) \in E$$

Figure 7-16 Arc routing

A. Snow plough depots and service areas

B. Arc routing plan — zone 7310

Here u_{ij} denotes the number of times that link (i,j) is used, with $u_{ij}=1$ when the problem is initialised, i.e. every edge must appear at least once in the solution. The objective function is simply minimisation of overall tour length subject to constraints that ensure each edge is traversed at least once.

A solution to this problem can be obtained by a two-stage process:

Step 1: check to see if every vertex satisfies the necessary condition for an Eulerian circuit — i.e. do the number of inflows at every vertex equal the number of outflows. If the network is not directed this condition requires that the *degree* of every vertex is even. If the network satisfies this condition proceed to step 2. If not, then systematically add (directed) edges until the vertices do all satisfy the condition.

Step 2: Solve the ECP. This turns out to be a relatively straightforward task, although the resulting circuit is not necessarily unique. In fact there are $O(k^n)$ Eulerian circuits in any graph, where n is the number of vertices and $k=K-1$, where K is the minimum degree of vertices in V. A simple algorithm for solving the ECP is due to a little-known Frenchman called M. Fleury dating from 1885, although a perhaps better algorithm, due to Hierholzer (1873) is still widely used today. In Fleury's algorithm the term *bridge* refers to an edge that, if removed, would result in the graph becoming disconnected, i.e. forming two unconnected subgraphs. The steps are as follows (for a connected Eulerian graph):

- select any vertex as the starting point and any edge leaving that vertex, but only use a bridge if no other options are available (this requires determining whether a link is, in fact, a bridge). Add the chosen edge to the current set

- move the current vertex to the end of the chosen edge and delete this edge from the graph. If this results in the previous vertex becoming isolated (having no links) then delete the vertex also

- repeat the above steps until all edges have been deleted. The current set should now contain an Eulerian circuit

The capacitated version of this class of problems, CARP, can be solved in a similar manner. However, as it is known to be NP-hard much attention has been focused on finding good heuristics. Amongst the most effective of these is the so-called tabu search algorithm introduced by Hertz *et al.* (2000). Their procedure involves finding an initial feasible solution and seeking to improve on this by a process of systematic dropping and adding of edges. Computational tests show that this heuristic can solve medium sized problems (50 vertices, 100 edges) to within 1% of optimality in reasonable computation times, but that such times can be highly variable depending on the particular network configuration (variations of an order of magnitude in time for problems of similar size). Problems with much larger numbers of edges (e.g. detailed street networks) were also tested and solutions were obtained within 5% of the optimum on average within acceptable processing times.

Afterword

Our intention in developing this Guide is broadly to support a number of international initiatives in spatial literacy and geographic information science — of which the UK 'Spatial Literacy in Teaching' and US National Research Council of the National Academies (2006) 'Beyond Mapping' initiatives are currently the most prominent.

The formats in which the Guide is available make it accessible to a wide range of readers in education and the professions, and are amenable to frequent updating. We very much view the future development of the Guide as a project for the geospatial community that is never likely to be 'complete' — a more accurate title might have been 'Geospatial analysis: *towards* a comprehensive guide to principles, techniques and software tools'! To this end, we very much welcome suggestions as to topics that are missing from this current version of the Guide, as well as corrections to any errors readers may find.

These, together with any other comments, should be emailed to the authors via the web site, www.spatialanalysisonline.com

References

Ahuja R K, Magnanti T L, Orlin J B (1993) Network flows: Theory, algorithms and applications. Prentice Hall, Englewood Cliffs, NJ, USA

Allan A L (2004) Maths for map makers. 2nd ed. Whittles Publishing, Dubeath, Scotland

Angel S, Hyman G M (1976) Urban fields. Pion, London

Anselin L (1988) Spatial econometrics: Methods and models. Kluwer Academic Publishers, Dordrecht, NL

Anselin L (2002) Under the hood: Issues in the specification and interpretation of spatial regression models. Agricultural Economics, 17(3), 247-67

Anselin L (1995) Local indicators of spatial association - LISA. Geog. Anal., 27, 93-115

Anselin L, Bera A (1998) Spatial dependence in linear regression models with an introduction to spatial econometrics. In Ullah A, Giles D (eds.), Handbook of Applied Economic Statistics. Marcel Dekker, New York, 237-89

Anselin L, Syabri I, Kho Y (2006) GeoDa: An Introduction to Spatial Data Analysis. Geog. Anal., 38, 5-22

Arnaud J-F, Madec L, Bellido A, Guiller A (1999) Microspatial genetic structure in the land snail Helix aspersa (Gastropoda: Helicidae). Heredity, 83, 110-19

Arora S (1996) Polynomial time approximation scheme for Euclidean TSP and other geometric problems. Proc. 37th Ann. IEEE Symp. on Foundations of Comput. Sci., IEEE Computer Society, 2-11

Avella P, Sassano A, Vasil'ev I (2005) Computational study of large-scale p-median problems. Math. Program., A, 32, 215-30

Bailey T C, Gatrell A C (1995) Interactive spatial data analysis. Longman, Harlow, UK

Ball G H, Hall D J (1965) A novel method of data analysis and pattern classification. Stanford Research Institute, Menlo Park, California

Baskent E Z, Jordan G A (1995) Characterizing spatial structure of forest landscapes. Can. J. For. Res. 25, 1830-49

Beasley J E (1985) A note on solving large p-median problems. European Journal of Operational Research, 21, 270-73

Beasley J E (1990) OR-Library: distributing test problems by electronic mail. Journal of the Operational Research Society 41(11), 1069-72

Beasley J E (1993) Lagrangian heuristics for location problems. European Journal of Operational Research, 65, 383-99

Beltran C, Tadonki C, Vial J-P (2004) Semi-Lagrangian relaxation. Optimization Online, at: www.optimization-online.org

Berke O (2000) Modified median polish kriging and its application to the Wolfcamp-acquifer data. Discussion Paper 48/00, Univ. of Dortmund, Dept of Statistics

Berke O (2004) Exploratory disease mapping: kriging the spatial risk function from regional count data. Intl. J of Health Geographics, 3,18, 1-11 (available from Biomed Central: www.ij-healthgeographics.com)

Bernhardsen T (2002) Geographic Information Systems — an introduction. 3rd edition, J Wiley, Hoboken, NY, USA

Berry B J L, Marble D F (1968) Spatial analysis — a reader in statistical geography. Prentice Hall, New Jersey

Berry J K (1993) Beyond Mapping: Concepts, algorithms and issues in GIS. GIS World Books, and also available online in HTML format

Besag J, Kooperberg C L (1995) On conditional and intrinsic autoregressions. Biometrika, 82, 733-46

Besag J, York J, Mollie A (1991) Bayesian image restoration, with two applications in spatial statistics. Annals of the Institute of Statistical Maths, 43, 1-59

Bluman A G (2003) Elementary statistics: A sept by step approach. 5th edition, McGraw-Hill, New York

Bowman A W, Azzalini A (1997) Applied smoothing techniques for data analysis: The kernel approach with S-Plus illustrations. Oxford Statistical Science Series, Oxford University Press, Oxford

Brewer C A, Pickle L (2002) Evaluation of methods for classifying epidemiological data on choropleth maps in series. Annals Assoc. Amer. Geogs., 92, 662-81

Brown D G (1998) Classification and boundary vagueness in mapping presettlement forest types. Int. J. of Geog. Information Science, 12, 105-29

Brunsdon C, Fotheringham A S, Charlton M (1998) Geographically Weighted Regression — modelling spatial non-stationarity. The Statistician, 47, 431-443

Burrough P A (1986) Principles of Geographical Information Systems for land resources assessment. Monographs on Soil and Resources Survey No. 12. Clarendon Press, Oxford

Burrough P A (1989) Fuzzy mathematical methods for soil survey and land evaluation. Journal of Soil Science 40, 477-92

Burrough P A, McDonnell R A (1998) Principles of Geographical Information Systems. Oxford University Press, Oxford

Carlson R E, Foley T A (1991) Radial Basis methods on track data. Proc. of the 13th IMACS World Congress on Computation, R. Vichnevetsky (ed.), Criterion Press, Dublin, 428- 29

Carlson R E, Foley T A (1992) Interpolation of track data with radial basis methods. Computers & Mathematics with Applications, 24, 27-34

Carr D B, Wallin J F, Carr D A (2000) Two new templates for epidemiology applications: Linked micromap plots and conditioned choropleth maps. Statistics in Medicine, 19, John Wiley & Sons, 2521-38

Carr D B, Chen J, Bell S, Pickle L, Zhang Y (2002) Interactive linked micromaps and dynamically conditioned choropleth maps. Proceedings of the Second National Conference on Digital Government Research, Digital Government Research Center, 61-7

Chatfield C (1975) The analysis of time series: theory and practice. Chapman and Hall, London

Cheriton D, Tarjan R E (1976) Finding minimum spanning trees. SIAM J. Comput., 5, 724-42

Clark P J, Evans F C (1954) Distance to nearest-neighbour as a measure of spatial relationships in populations. Ecology, 35, 445-53

Christofides N, Beasley J (1982) A tree search algorithm for the p-median problem. European Journal of Operational Research, 10, 196-204

Cliff A, Ord J (1973) Spatial autocorrelation. Pion, London

Cliff A, Ord J (1981) Spatial processes: models and applications. Pion, London (an updated and extended version of their 1973 book)

Cooper L (1963) Location-allocation problems. Operations research, 11, 331-43

Cooper L (1964) Heuristic methods for location-allocation problems. SIAM Review, 6, 37-53

Cormen T H, Leiserson C E, Rivest R L, Stein C (2001) Introduction to algorithms. 2nd ed., MIT Press, Cambridge, Mass.

Cressie N A C, Chan N H (1989) Spatial modelling of regional variables. J Amer. Stat. Assoc., 84, 393-401

Cressie N A C (1991) Statistics for spatial data. John Wiley, New York

Cromley R G (1996) A comparison of optimal classification strategies for choroplethic displays of spatially aggregated data. Int. J. of Geog. Information Science, 10, 101-109

Dacey M (1968) A review of measures of contiguity for two and k-colour maps, pp 479-95 in Berry B J L, Marble D F eds. (1968) Spatial analysis. Prentice Hall, New Jersey

Dale M R T (2000) Lacunarity analysis of spatial pattern: a comparison. Landscape Ecology 15, 467-8

Dale P (2005) Introduction to mathematical techniques used in GIS. CRC Press, Boca Raton, USA

Dantzig G B (1960) On the shortest route through a network. Management Science, 6, 187-90

Daskin M S (1995) Network and discrete location — models, algorithms and applications. Wiley, NJ, USA

Densham P J, Rushton G (1992) A more efficient heuristic for solving large p-median problems by heuristic methods. Papers in Regional Science, 71, 307-329

de Smith M J (2004) Distance transforms as a new tool in spatial analysis, urban planning and GIS. Environment & Planning B, 31, 85-104

de Smith M J (2006) Maths for the mystified. Troubador Publishing Ltd, Leicester, UK

de Smith M J (2006a) Determination of gradient and curvature constrained optimal paths. Computer-aided Civil and Infrastructure Engineering, 21, 17-31

Deutsch C V, Journel A G (1992, 1998) GSLIB geostatistical software library and user's guide. Oxford University Press, New York

Diggle P J (1983) Statistical analysis of spatial point patterns. Academic Press, London

Diggle P J (1990) A point process modelling approach to raised incidence of a rare phenomenon in the vicinity of a prespecified point. J. Royal Stat. Soc., A, 153, 349-62

Dijkstra E W (1959) A note on two problems in connexion with graphs. Numerische Mathematik, 1, 269-71

Douglas D H, Peucker T K (1973) Algorithms for the reduction of the number of points required to represent a digitised line or its caricature. The Canadian Cartographer, 10, 112-22

Douglas D H (1994) Least cost path in GIS using an accumulated cost surface and slope lines. Cartographica, 31, 37-51

Draper G, Vincent T, Kroll M E, Swanson J (2005) Childhood cancer in relation to distance from high voltage power lines in England and Wales: a case-control study. British Medical Journal, 330, 4 June 2005, 1-5

Dubois G, Saisana M (2002) Optimal spatial declustering weights — comparison of methods. Proc. Annual Conf. International Assoc. for Math. Geology, Berlin, 479-84

Dunn J C (1973) A fuzzy relative of the ISODATA process and its use in detecting compact well-separated clusters. Journal of Cybernetics, 3, 32-57

Dyer M E, Frieze A M (1985) A simple heuristic for the p-centre problem. Operations Research Letters, 3, 285-8

Eastman J R (1989) Pushbroom algorithms for calculating distances in raster grids. Proceedings, Autocarto 9, 288-97

Ebisch K (2002) A correction to the Douglas-Peuker line generalisation algorithm. Computers & Geosciences, 28, 995-7

Epanechnikov V A (1969) Nonparametric estimation of a multidimensional probability density. Theor. Probab. Appl., 14, 153-8

Fisher N I (1993) Statistical analysis of circular data. Cambridge University Press, Cambridge

Fisher W D (1958) On grouping for maximal homogeneity. Journal of the American Statistical Association, 53, 789-98

Foresman T W ed. (1998) The history of geographic information systems: Perspectives from the pioneers. Prentice Hall, Upper Saddle River, NJ, USA

Fotheringham A S, Brunsdon C, Charlton M (2002) Geographically Weighted Regression: The analysis of spatial varying relationships. Wiley, New York

Freeman M F, Tukey J (1950) Transformations related to the angular and the square root. Annals of Mathematical Statistics, 21, 607-11

Gabriel K R, Sokal R R (1969) A new statistical approach to geographic variation analysis. Systematic Zoology, 18, 259-78

Gardner R H, O'Neill R V (1991) Pattern, process, and predictability: the use of neutral models for landscape analysis. Pages 289-307 in Turner M G and Gardner R H, eds. (1991) Quantitative Methods in Landscape Ecology. Springer-Verlag, New York

Getis A D, Griffith D A (2002) Comparative spatial filtering in regression analysis. Geog. Anal., 34, 130-40

Getis A D, Ord J K (1992) The analysis of spatial association using distance statistics. Geog. Anal., 24, 189-206

Getis A D, Ord J K (1995) Local spatial autocorrelation statistics: Distributional issues and an application. Geog. Anal., 27, 287-306

Gold C M, Thibault D (2001) Map generalisation by skeleton retraction. Proceedings, International Cartographic Association, Beijing, China, 2072-81

Golledge R G, Rushton G (1972) Multidimensional scaling: Review and geographical applications. Commission on college geography, Tech. Paper 10, Washington DC, Assoc. of Amer. Geogs.

Goodchild M F, Mark D M (1987) The fractal nature of geographic phenomena. Annals of Association of American Geographers, 77, 265-78

Goodchild M F, Kemp K K eds. (1990) NCGIA Core Curriculum in GIS. National Centre for Geographic Information and Analysis, University of California, Santa Barbara, CA, USA (see Web links section for more details and contents)

Goodchild M F, Longley P A (1999) The future of GIS and spatial analysis. In Longley et al.(1999, op cit), 567-80

Goodchild M F, Parks B O, Steyaert L T eds. (1993) Environmental modelling with GIS. Oxford University Press, New York

Goodchild M F, Steyaert L T, Parks B O, Johnston C, Maidment D, Crane M, Glendenning S eds. (1996) Environmental modelling: Progress and research issues. GIS World Books, Fort Collins, CO, USA

Goovaerts P (1997) Geostatistics for natural resources evaluation. Oxford University Press, New York

Goovaerts P (1999) Geostatistics in soil science: state-of-the-art and perspectives. Geoderma, 89, 1-45

Gustafson E J (1998) Quantifying landscape spatial pattern: What is the state of the art. Ecosystems, 143-56

Gustafson E J, Parker G R (1992) Relationships between landcover proportion and indices of landscape spatial pattern. Landscape Ecology 7, 101-10

Haklay M (2004) Map Calculus in GIS: a proposal and demonstration. Int. J. Geog. Information Science, 18, 107-25

Haklay M (2006) Comparing map calculus and map algebra in dynamic GIS. Ch 5 in Drummond J, Billen R, Forrest D and Jo E. eds. (2006) Dynamic & mobile GIS: Investigating change in space and time, Taylor & Francis/CRC

Haggett P, Chorley R J (1969) Network analysis in geography. E Arnold, London

Haining R (2003) Spatial data analysis – theory and practice. Cambridge University Press, Cambridge, UK

Hanrahan P, Ebert D S, Musgrave F K, Peachey D, Perlin K, Worley S eds. (2003) Texturing and modelling: A procedural approach. 3rd ed., Morgan and Kaufmann, San Francisco

Harris R, Sleight P, Webber R (2005) Geodemographics, GIS and neighbourhood targeting. J Wiley & Sons

Harvey D (1969) Explanation in geography. E Arnold, London

Harvey M J, Barbour R W (1965) Polygon fitting of *Microtus ochrogaster* as determined by a modified minimum area method. J of Mammology, 46, 398-402

Hawley K, Moellering H (2005) A comparative analysis of areal interpolation methods: A preliminary report. Proc. Auto-Carto 2005, CaGIS

Henzinger M R, Klein P, Rao S (1997) Faster shortest-path algorithms for planar graphs. J. Comput. Syst. Sci., 55, 3-23

Hertz A, Laporte G, Mittaz M (2000) A tabu search heuristic for the capacited arc routing problem. Operations Research, 48, 129-35

Horne B K P (1981) Hill shading and the reflectance map. Proc. IEEE 69, 14-47

Hutchinson M F (1988) Calculation of hydrologically sound digital elevation models. Proceedings, Third International Symposium on Spatial Data Handling, Sydney, Columbus, International Geographical Union, 117-33

Hutchinson M F (1989) A new procedure for gridding elevation and stream line data with automatic removal of spurious pits. Journal of Hydrology 106, 211-32

Hutchinson M F (1996) A locally adaptive approach to the interpolation of digital elevation models. In Proceedings, *Third International Conference/Workshop on Integrating GIS and Environmental Modelling*, Santa Fe, NM. Available online from NCGIA

Iliffe J (1999) Datums and Map Projections for Remote Sensing, GIS and Surveying. Whittles Publishing, Dubeath, Scotland

Isaaks E H, Srivastava R M (1989) An introduction to applied geostatistics. Oxford University Press, New York

Jaeger J A G (2000) Landscape division, splitting index, and effective mesh size: new measures of landscape fragmentation. Landscape Ecology 15, 115-130

Jenks G F, Caspall F C (1971) Error on choroplethic maps: Definition, measurement, reduction. Annals of American Geographers, 61, 217-44

Jenson S, Domingue J (1988) Extracting topographic structure from digital elevation data for geographic information system analysis. Photogrammetric Engineering and Remote Sensing, 54, 1593-600

Jones K H (1998) A comparison of algorithms used to compute hill slope as a property of the DEM. Computers & Geosciences, 24, 315-23

Keitt T H, Urban D L, Milne B T (1997) Detecting critical scales in fragmented landscapes. Conservation Ecology, 1, 4

Kempf-Leonard K ed. (2005) Encyclopaedia of social measurement. Elsevier, San Diego, CA, USA, 921-6

Kristinsson H, Jonsson M T, Jonsdottir F (2005) Pipe route design using variable topography distance transforms. Proc. IDETC/CIE ASME 2005 International Design Engineering Technical Conference

LaGro J Jr (1991) Assessing patch shape in landscape mosaics. Photogrammetric Eng. Remote Sens. 57, 285-93

Li H, Reynolds J F (1993) A new contagion index to quantify spatial patterns of landscapes. Landscape Ecology, 8,155-62

Lichstein J W, Simons T R, Shriner S A, Franzreb K E (2002) Spatial autocorrelation and autoregressive models in Ecology. Ecological Monographs, 72, 445-63

Lindsay J B (2005) The Terrain Analysis System: A tool for hydro-geomorphic applications. Hydrological Processes, 19, 1123-30

Longley P A, Batty M, eds. (1997) Spatial analysis: Modelling in a GIS environment. J Wiley & Sons, New York

Longley P A, Batty M, eds. (2001) Advanced spatial analysis – the CASA book of GIS. ESRI Press, Redlands, CA, USA

Longley P A, Goodchild M F, Maguire D J, Rhind D W (1999) Geographic information systems: Principles, techniques, management and applications. 2nd ed., J Wiley, New York

Longley P A, Goodchild M F, Maguire D J, Rhind D W (2005) Geographic information systems: Principles, techniques, management and applications. Abridged edition, J Wiley, Hoboken

Longley P A, Goodchild M F, Maguire D J, Rhind D W (2005) Geographic information systems and science. 2nd ed., J Wiley, Chichester, UK

Longley P A, Harris R (1999) Towards a new digital data infrastructure for urban analysis and modelling. Environment and Planning B, 26, 855-78

Love R F, Morris J G (1972) Modelling intercity road distances by mathematical functions. Oper. Res. Quart., 23, 61-72

Love R F, Morris J G, Wesolowsky G O (1988) Facilities location, Models and methods. North-Holland, New York

Mackay R J, Oldford R W (2000) Scientific method, statistical method, and the speed of light. Working Paper 2000-02, Department of Statistics and Actuarial Science, University of Waterloo, Ontario, Canada

Maguire D, Batty M, Goodchild M F eds. (2005) GIS, Spatial Analysis and Modeling. ESRI Press, Redlands, CA, USA

Mandelbrot B B (1982) The fractal geometry of nature. W. H. Freeman and Co., New York

Mardia K V, Jupp P E (1999) Directional statistics. 2nd Ed., John Wiley, Chichester

Martin D (2000) Census 2001: making the best of zonal geographies. Paper presented at the "Census of population: 2000 and beyond, conference", University of Manchester, 22-23rd June 2000

Masser I, Campbell H, Craglia M eds. (1996) GIS Diffusion: The adoption and use of geographical information systems in local government in Europe. Taylor & Francis, Bristol, PA, USA

Matheron G (1973) The intrinsic random functions and their application. Advances in Applied Prob., 5, 439-68

McGarigal K, Marks B J (1995) FRAGSTATS: spatial pattern analysis program for quantifying landscape structure. Gen. Tech. Report PNW-GTR-351, USDA Forest Service, Pacific Northwest Research Station, Portland, OR, USA

McHarg I L (1969) Design with nature. Natural History press, Garden City, New York

McMaster R B, Shea K S (1992) Generalisation in digital cartography. Washington DC, Association of American Geographers

Miller H J, Shaw S-L (2001) Geographic information systems for transportation. Oxford University Press, Oxford

Milne BT (1991) Lessons from applying fractal models to landscape patterns. pp 199-235 in Turner, M.G., and R.H. Gardner, editors. Quantitative methods in landscape ecology. Springer-Verlag, New York

Mitasova H, Mitas L (1993) Interpolation by regularised spline with tension. Math. Geol., 25, 641-55

Mitchell A (1999) The ESRI guide to GIS analysis, Volume 1: Geographic patterns and relationships. ESRI Press, Redlands, CA, USA

Mitchell A (2005) The ESRI guide to GIS analysis, Volume 2: Spatial measurements and statistics. ESRI Press, Redlands, CA, USA

Mitchell B R (2006) Comparison of programs for fixed kernel homerange analysis. Jun2 2006 newsletter, The Wildlife Society, GIS Working Group. Paper retrieved from www.wildlife.org/wg/gis/newsletter/jun06/hr compar.htm 10th Jan 07.

Mitchell J S B (1998) Geometric shortest paths and network optimisation. Handbook of Computational Geometry, Elsevier Science, Sack J-R, Urrutia J, eds. (2000) Chapter 15, 633-701. Draft available as a PDF from http://www.ams.sunysb.edu/~jsbm/

Moore I D, Grayson R B, Landson A R (1991) Digital terrain modelling: A review of hydrological, geomorphological and biological applications. Hydrological processes, 5, 3-30

Moore I D, Lewis A, Gallant J C (1995) Terrain attributes: Estimation methods and scale effects, in Jakeman A J, Beck M B, McAleer M J (1995) Modelling change in environmental systems. John Wiley, New York

Moran P A P (1948) The interpretation of statistical maps. Journal of the Royal Statistical Society B, 10, 243-51

Moran P A P (1950) Notes on continuous stochastic phenomena. Biometrika, 37, 17-23

Moshe D ed. (2000) Arc routing: Theory, solutions and applications. Kluwer Academic Publishers, Amsterdam

Nakaya T, Fotheringham A S, Brunsdon C, Charlton M (2005) Geographically weighted Poisson regression for disease association mapping. Statist. in Med., 24, 2695-717

National Research Council of the National Academies (2006) Beyond Mapping: meeting national needs through enhanced geographic information science. The National Academies Press, Washington DC

Oden N (1995) Adjusting Moran's I for population density. Statistics in Medicine 14, 17-26

Okabe A, Boots B, Sugihara K, Chiu S N (2000) Spatial tessellations: Concepts and applications of Voronoi diagrams. 2nd ed., J Wiley, New Jersey

Openshaw S (1977) A geographical solution to scale and aggregation problems in region-building, partitioning and spatial modelling. Transactions of the Institute of British Geographers NS 2, 459-72

O'Sullivan D, Unwin D J (2003) Geographic information analysis. J Wiley, New Jersey

O'Sullivan F (2005) Factors affecting Radon levels in houses: A GIS study around Castleisland, County Kerry, South-West Ireland. Unpub. MSc Thesis, Dept of Geomatic Engineering, University College London

Pickover C A (1990) Computers, pattern, chaos and beauty: graphics from an unseen world. New York, St. Martin's

Plotnick R E, Gardner R H, O'Neill R V (1993) Lacunarity indices as measures of landscape texture. Landscape Ecology 8, 201-11.

Plotnick R E, Gardner R H, Hargrove W W, Pretegaard K, Perlmutter M (1996) Lacunarity analysis: a general technique for the analysis of spatial patterns. Phys. Rev. E53, 5461-8

Rana S ed. (2004a) Topological data structures for surfaces: An introduction to geographical information science. J Wiley & Sons

Rana S (2004b) Two approximate solutions to the Art Gallery problem. International Conference on Computer Graphics and Interactive techniques, Los Angeles (available from http://eprints.ucl.ac.uk/

Rana, S (2006) Isovist Analyst - An ArcView Extension for Planning Visual Surveillance. in: Proceedings ESRI User Conference 2006, San Diego, CA, USA

Rana S, Wood J (2000) Weighted and metric surface networks — new insights and an interactive application for their generalisation in TCL/TK. Working Paper 25, Centre for Advanced Spatial Analysis, University College London

Ratti C (2001) Urban analysis for environmental prediction. Unpub PhD thesis, Univ. of Cambridge, UK

Ratti C, Baker N, Steemers K (2004) Energy consumption and urban texture. Energy and Buildings, 37, 762-6

Ratti C, Richens S (2004) Raster analysis of urban texture. Environment and Planning B, Planning and Design, 31, 297-309

Rey S J, Janikas M V (2006) STARS: Space-time analysis of regional systems. Geog. Anal., 38, 67-86

Robins G, Zelikovsky A (2000) Improved Steiner tree approximation in graphs. Proc. 10th Ann. ACM-SIAM Symp. on Discrete Algorithms, ACM-SIAM, 770-9

Rodrigue J-P, Comtois C, Slack B (2006) The geography of transport systems. Routledge

Sawada M (1999) ROOKCASE: An Excel 97/2000 Visual Basic (VB) Add-in for exploring global and local spatial autocorrelation. Bull. Ecological Soc. America, 80, 231-4

Scott A J (1971) Combinatorial programming, spatial analysis and planning. Methuen, London

Shannon C, Weaver W (1949) The mathematical theory of communication. Univ. Illinois Press, Urbana, IL, USA

Schietzelt T, Densham P (2003) Location-allocation in GIS. In Longley & Batty (2003) *op cit*, 345-65

Silverman B W (1986) Density estimation for statistics and data analysis. Chapman and Hall, New York

Simpson E H (1949) Measurement of diversity. Nature, 163, 688

Sirigos S A, Photis Y N (2005) A prototype planning support system for scenario location analysis. Proceedings of Xth International Symposium on Locational Decisions, 2-8 June, Seville

Smith W H F, Wessel P (1990) Gridding with continuous curvature splines in tension. Geophysics, 55, 293-305

Sokal R R, Smouse P E, Neel J V (1986) The genetic structure of a tribal population, the Yanomama indians. XV. Patterns inferred by autocorrelation analysis. Genetics, 114, 259-87

Stouffler S A (1940) Intervening opportunities: A theory relating mobility and distance. Amer. Soc. Rev., 5, No. 6

Stratton R D (2004) Assessing the effectiveness of landscape fuel treatments on fire growth and behavior. Journal of Forestry, 102(7), 32-40

Taillard E D (2003) Heuristic methods for large centroid clustering problems. J. of Heuristics, 9, 51-74

Tarboton D G, Bras R L, Rodriguez-Iturbe I (1991) On the extraction of channel networks from digital elevation data. Hydrological Processes, 5, 81-100

Tarboton D G (1997) A new method for the determination of flow directions and upslope areas in grid digital elevation models. Water Resources Research, 33, 309-19

Teitz M B, Bart P (1968) Heuristic methods for estimating the generalised vertex median of a weighted graph. Operations Research, 16, 955-61

Tobler W, Wineberg S (1971) A Cappadocian speculation. Nature, 231(5297), 39-42

Tomlin C Dana (1990) Geographic Information systems and cartographic modelling. Prentice-Hall, New Jersey

Tomlinson R F (1967) An introduction to the geo-information system of the Canada Land Inventory. ARDA, Canada Land Inventory, Department of Forestry and Rural Development, Ottawa

Tomlinson R F (1970) Computer based geographical data handling methods. In New Possibilities and Techniques for Land Use and Related Surveys, ed. I. H. Cox. World Land Use Occasional Papers No. 8, Geographical Publications Ltd., Berkhamsted, Herts, England

Tou J T, Gonzales R C (1974) Pattern classification by distance functions. Addison-Wesley, Reading, MA

Tukey J W (1977) Explanatory data analysis. Addison-Wesley, Reading, MA

Turner M G, Gardner R H, eds. (1991) Quantitative methods in landscape ecology. Springer-Verlag, New York

Vaughan R (1987) Urban spatial traffic patterns. Pion, London

Vincenty T (1975) Direct and inverse solutions of geodesics on the ellipsoid with application of nested equations. Survey Review XXII, 176, 88-93. A spreadsheet implementation of this algorithm can be found at: www.ga.gov.au/geodesy/datums/calcs.jsp

Visvalingam M, Whyatt J D (1991) "The Douglas-Peucker algorithm for line simplification: re-evaluation through visualisation". Computer Graphics Forum 9, 213-28

Wall M M (2004) A close look at the spatial structure implied by the CAR and SAR models. J of Statistical Planning and Inference, 121, 311-24

Whyatt J D, Wade P R (1988) The Douglas-Peucker line simplification algorithm. Society of University Cartographers Bulletin 22, 17-25

Whitcomb R F, Lynch J F, Klimkiewwicz M K, Robbins C S, Whitcomb B L, Bystrak D (1981) Effects of forest fragmentation on avifauna of the eastern deciduous forest. pp. 125-205 in R L Burgess and D M Sharpe, editors. Forest island dynamics in man-dominated landscapes. Springer-Verlag, New York

White G C, Garrott R A (1990) Analysis of wildlife radio-tracking data. Academic Press, San Diego

Wilson A G (1967) A statistical theory of spatial distribution models. Transportation Research, 1, 253-69

Womble W H (1951) Differential systematics. Science, 114, 315-22

Worboys M, Duckham M (2004) GIS — A computing perspective. 2nd ed., CRC Press, Boca Raton, Florida

Yasui Y, Liu H, Beach J, Winget M (2000) An empirical evaluation of various priors in the empirical Bayes estimation of small area disease risks. Statistics in Medicine, 19, 2409-20

Zadeh L (1965) Fuzzy sets. Information and Control, 8, 338-22

Zeverbergen L W, Thorne C R (1987) Quantitative analysis of land surface topography. Earth Surface Processes and Landforms, 12, 47-56

Zhang J, Goodchild M F (2002) Uncertainty in geographic information. Taylor and Francis, London

Zhou Q, Liu X (2004) Analysis of errors of derived slope and aspect related to DEM data properties. Computers & Geosciences 30, 369-78

Web links

Principal software products cited

Package Name	Link
ArcGIS	http://www.esri.com/
ANUDEM	http://cres.anu.edu.au/outputs/anudem.php
CCmaps	http://www.galaxy.gmu.edu/~dcarr/ccmaps/
Concorde	http://www.tsp.gatech.edu/concorde/index.html
CPLEX (ILOG)	http://www.ilog.com/products/cplex/
Crimestat	http://www.icpsr.umich.edu/CRIMESTAT/
ENVI	http://www.ittvis.com/envi/index.asp
Fragstats	http://www.umass.edu/landeco/research/fragstats/fragstats.html
GAM	http://www.ccg.leeds.ac.uk/software/gam/
GeoDa	https://geoda.uiuc.edu/
Geomedia	http://www.intergraph.com/geomedia/
GRASS	http://grass.itc.it/
GS+	http://gammadesign.com/
GWR	http://www.nuim.ie/ncg/GWR/
Hawth's Tools	http://www.spatialecology.com/
Idrisi	http://www.clarklabs.org/
ILOG Dispatcher	http://www.ilog.com/products/dispatcher/
ISATIS	http://www.geovariances.com/
Landserf	http://www.landserf.org
LEDA	http://www.algorithmic-solutions.com/
LOLA	http://www.mathematik.uni-kl.de/~lola
Manifold	http://www.manifold.net/

Package Name	Link
MapCalc	http://www.farmgis.com/products/software/mapcalc/default.asp
MapInfo	http://www.mapinfo.com/ http://jratcliffe.net/hsd/ (MapInfo HotSpot Detective)
MATLab	http://www.mathworks.com/
MFWorks	http://www.keigansystems.com/software/mfworks/index.html
Oriana	http://www.kovcomp.com
PCRaster	http://pcraster.geog.uu.nl/
RiverTools	http://www.rivix.com/
Rookcase	http://www.lpc.uottawa.ca/data/scripts/
SANET	http://okabe.t.u-tokyo.ac.jp/okabelab/atsu/sanet/sanet-index2.html
S-Distance	http://www.prd.uth.gr/res_labs/spatial_analysis/software/SdHome_en.asp
SatScan	http://www.satscan.org/
SITATION	http://users.iems.northwestern.edu/~msdaskin/Mark%20S.%20Daskin%20Software.html
SPLANCS	R-Plus: http://rss.acs.unt.edu/Rdoc/library/splancs/html/00Index.html S-Plus: http://www.maths.lancs.ac.uk/~rowlings/Splancs/
STARS	http://regal.sdsu.edu/index.php/Main/STARS
Surfer	http://www.goldensoftware.com/products/surfer/surfer.shtml
TAS	http://www.sed.manchester.ac.uk/geography/research/tas/
TAUDEM	http://hydrology.neng.usu.edu/taudem/
Terraseer	http://www.terraseer.com/
TNTMips	http://www.microimages.com/
TransCAD	http://www.caliper.com/tcovu.htm
Vincenty (xls)	http://www.ga.gov.au/geodesy/datums/vincenty.xls
WinBUGS	http://www.mrc-bsu.cam.ac.uk/bugs/winbugs/geobugs.shtml
Xpress-MP	http://www.dashoptimization.com/
ZDES	http://www.geog.leeds.ac.uk/software/zdes/

Associations and academic bodies

Association of American Geographers (AAG)

http://www.aag.org/

Association of American Geographers (AAG), Spatial Analysis Special Interest Group

http://www.fsu.edu/~geog/sam/

Association of American Geographers (AAG), Geographic Information Systems and Science Special Interest Group

http://www.cas.sc.edu/gis/aaggis

Association for Geographic Information (AGI)

http://www.agi.org.uk/

Birkbeck College, University of London (GIS online Master's programme):

http://www.bbk.ac.uk/gisconline/

Centre for Advanced Spatial Analysis (CASA):

http://www.casa.ucl.ac.uk

Centre for Spatially Integrated Social Science (CSISS):

http://www.csiss.org/

Centre for Computational Geography:

http://www.ccg.leeds.ac.uk/

European Commission Joint Research Centre (CEC/JRC) – a web forum for Geostatistics and Spatial Statistics supported by the European Commission:

http://www.ai-geostats.org

EURO Working Group on Locational Analysis:

http://www.vub.ac.be/EWGLA/homepage.htm

Free GIS Organisation – portal and mailing list:

http://www.freegis.org/

Geoscience Australia:

http://www.ga.gov.au/

Geospatial Information and Technology Association (GITA):

http://www.gita.org/

GIS.COM – geospatial portal sponsored by ESRI:

http://www.gis.com

International Association of Crime Analysts (software list):

http://www.iaca.net/Software.asp

International Association for Mathematical Geology (IAMG)

http://www.iamg.org/

Market Research Society, Geodemographics Knowledgebase:

http://geodemographics.org.uk

NCGIA Core curriculum:

http://www.geog.ubc.ca/courses/klink/gis.notes/ncgia/toc.html

Open Geospatial Consortium (OGC):

http://www.opengeospatial.org

Open Source Technology Group (OSTG),

    Sourceforge:

    http://sourceforge.net

    LP-solve website managed at Sourceforge:

    http://lpsolve.sourceforge.net/5.5/

Optimisation compendium and best known solutions:

http://www.nada.kth.se/~viggo/problemlist/

Optimization Online:

www.optimization-online.org

Royal Geographical Society/Institute of British Geographers: GIS in Teaching

http://www.rgs.org/OurWork/Schools/Teachers/CurriculumDevelopments/GIS/

Spatial Analysis Laboratory, University of Thessaly, Greece:

http://www.prd.uth.gr/res_labs/spatial_analysis/

SPLINT: Spatial Literacy in Teaching

http://www.spatial-literacy.org

UK Census Dissemination Unit (CDU):

http://census.ac.uk/cdu/

UK Census Output area construction:

http://www.geog.soton.ac.uk/research/oa2001/

GB Ordnance Survey:

http://www.ordnancesurvey.co.uk/

University Consortium for Geographic Information Science (UCGIS):

http://www.ucgis.org

US National Institute of Justice (crime analysis software):

http://www.ojp.usdoj.gov/nij/maps/software.html

USGS:

http://www.usgs.gov/

USGS Spatial Data Transfer Standard:

http://mcmcweb.er.usgs.gov/sdts/

US National Research Council of the National Academies

http://www.nationalacademies.org/

US National Research Council of the National Academies: Geographical Sciences Committee

http://dels.nas.edu/besr/gsc.shtml

US National Research Council of the National Academies: Mapping Science Committee

http://dels.nas.edu/besr/msc.shtml

The Wildlife Society, GIS working Group:

http://www.wildlife.org/wg/gis/

Online technical dictionaries/definitions

ESRI GIS Dictionary:

http://support.esri.com/index.cfm?fa=knowledgebase.gisDictionary.gateway

Mathworld:

http://mathworld.wolfram.com/

University of California GIS Abbreviations Dictionary:

http://www.lib.berkeley.edu/EART/abbrev2.html#index

Spatial data, test data and spatial information sources

Selected data sources

GB Ordnance Survey:

http://www.ordnancesurvey.co.uk/

EDINA — Repository of UK Digital Mapping data and related resources (academic access):

http://www.edina.ac.uk/maps/ and

http://www.gogeo.ac.uk/ (the EDINA Geoportal)

MAGIC — Repository of UK Environmental and related data:

http://www.magic.gov.uk/

UK Geospatial information Gateway (Data and Standards documents):

http://www.gigateway.org.uk/

US Census Bureau:

http://www.census.gov/geo/www/index.html

US Geological Survey:

http://www.usgs.gov/

US National Spatial Data Infrastructure:

http://www.fgdc.gov/nsdi/nsdi.html

US one-stop geoportal:

http://gos2.geodata.gov/wps/portal/gos

ESRI Managed Geography Network:

http://www.geographynetwork.com/

Geo-community portal:

http://www.geocomm.com/

Test datasets

TSPLIB:

http://www.iwr.uni-heidelberg.de/groups/comopt/software/TSPLIB95/

OR-LIBRARY:

http://people.brunel.ac.uk/~mastjjb/jeb/info.html

Statistics links

Online statistical methods and definitions e-Handbook, the "NIST/SEMATECH e-Handbook of Statistical Methods":

http://www.itl.nist.gov/div898/handbook

Spatial Statistics Group, University of Lancaster, UK:

http://www.maths.lancs.ac.uk/department/research/statistics/spatial

Spatial statistics and environmental sciences (SSES) Program, Ohio State:

http://www.stat.ohio-state.edu/~sses/

Kelly Pace's spatial statistics page (MATLab toolkit and general spatial statistics site):

http://www.spatial-statistics.com/

Links page to spatial statistical material and sites:

http://www.statistical.org/

Trade sites

Geo:connexion magazine (monthly): http://www.geoconnexion.com/

GeoPlace: http://www.geoplace.com/

GEOInformatics: http://www.geoinformatics.com/

GeoSpatial Solutions: http://www.geospatial-solutions.com/

GIS Professional: http://www.gispro.com/

Directions Magazine: http://www.directionsmag.com/ and its sister publication

Location Intelligence: http://www.locationintelligence.net

GIS@development: http://www.gisdevelopment.net/

Index

A

Abundance · 195
  Proportional · 196
  Relative · 196
Accumulated Cost Surfaces · 142
Accuracy · 4, 7, 54, 59, 142, 192, 273, 284
ACS · 142-48, 153, 333
Adaptive Densification · 115, 254
Adjacency · 2, 15, 40, 155, 199, 219, 232
  Matrix · 219, 222
Affine Transformation · 19
AIC · 234, 239, 240
Akaike Information Criterion · 236-39
Alpha Hull · 108
Analysis of Covariance · 235
Analysis of Variance · 102, 235
Anderson-Darling Test · 309
Angular Transforms · 31
Anisotropic · 157
Anisotropy · 145, 157, 164, 295, 304, 310, 312,
  313, 316, 320
Arc · 15
Arc Routing · 335, 362
Arc Routing Problem · 333
Areal Interpolation · 92-94, 126
Artefact · 15
Aspect · 3, 15, 74, 161, 162, 164, 165, 253, 261,
  259-67, 280, 282
Atomistic Fallacy · 61
Attribute · 15
  Intrinsic · 15
Autocorrelation · 15, 213
Autocovariance · 214
Axial Data · 157
Azimuth · 15, 77, 146, 261
Azimuthal Projection · 15
AZP · 96

B

Back Transformation · 19
Bandwidth · 129, 131, 155, 209, 222, 239
Bayesian Information Criterion · 236
Bearing · 15, 77, 138, 163, 224
Bessel Function · 31, 160, 299
Bézier Curves · 78
Bi-linear Interpolation · 301
Binomial Distribution · 29, 193

Block Kriging · 317, 319
Blurring · 45, 271
Bonferroni Correction · 230
Bootstrapping · 189, 319
Box Plots · 101, 181, 182, 184
Box-Cox Transform · 30, 309
Breaklines · 63, 112, 256, 285, 294
Brownian Motion · 258
Brushing · 4, 54, 181, 187
B-splines · 78
Buffering
  Hybrid · 153
  Network · 153
  Raster · 153
  Vector · 151
Bull's-eye Effect · 394

C

CAR · 235, 242, 244, 245
CCmaps · 184, 185
Centering Transform · 232
Central Limit Theorem · 29, 30
Centres · 65, 79, 80, 82, 83, 86, 103, 120, 133,
  143, 144, 202, 211
Centroids · 71, 79, 82, 84, 87, 89, 93, 97, 103,
  109, 189, 202, 208, 218, 222, 236, 239, 248
Chebyshev Metric · 140
Chi-Square Tests · 48
Choropleth · 8, 15, 99, 100, 184, 236, 368
Circuity Index · 88
City Block Metric · 84
Classification · 63, 73, 100, 111, 165, 174, 175,
  185, 192, 194, 211, 263, 267
  Box · 101
  Census · 123
  Defined Interval · 100
  Equal Interval · 100
  Fuzzy · 110
  Fuzzy c-means · 105
  Fuzzy Membership · 110
  Image · 190
  ISODATA · 105
  Jenks · 101, 102
  K-Means · 105
  Manual · 100
  Maximum Likelihood · 105
  Multivariate · 105
  Natural Breaks · 101
  Percentile · 100

Quantile · 100
Slice · 100, 166
Standard Deviation · 101
Stepwise Linear · 105
Unique Values · 100
Classification Entropy · 111
Clip · 86
Closing · 171, 213
Clough-Tocher · 300
Cluster Analysis · 10, 126, 209-12
Cluster Hunting · 8, 189, 212
Clustering · 99, 102, 139, 155, 159, 179, 188, 190, 203, 210, 211, 227, 228, 243, 307, 316, 323
Cluster Analysis · 209
Heirarchical NN · 210
Kernel Density · 211
K-Means · 102, 139, 211
Coefficient of Determination · 237, 248
Coefficient of Eccentricity · 163
Coefficient of Relative Dispersion · 190
Coefficient of Variation · 27, 170
Cognostics · 185
Co-Kriging · 321, 322
Collapse · 79
Complete Spatial Randomness · See CSR
Conditional Choropleth
Example · 186
Conditional Choropleth mapping · 184
Conditional Simulation · 317, 322
Confidence Envelope · 63, 206
Conflation · 15
Confounding · 57
Confusion Index · 111
Confusion Matrix · 192
Connectivity · 40, 199, 201, 255
Constant of Path Maintenance · 133
Contagion · 196, 198, 201, 202, 217
Contiguity · 15, 59, 97, 99, 100, 170, 197, 200, 202, 217, 225, 226, 227, 228, 233, 243
Contingency Table · 191
Contouring · 107, 236, 284, 286, 291, 312
Contrast · 198, 201
Convex Hull · 70, 88, 106, 107, 108, 113, 117, 118, 180, 204, 236, 286, 296
Convolution · 42, 45
Linear · 168
Cookie Cutter · 86, 91
Core Area · 196, 197, 200
Correlated Walk Analysis · 163
Correlation Coefficient · 27, 213, 220, 232, 236, 238, 313
Correlogram · 44, 156, 164, 214, 222, 223, 224, 245, 247, 305
Cost Distance · 74, 116, 136, 142, 143, 147, 150, 153, 166

Costgrow · 145
Covariance · 27
Cross K Function · 208
Crossings Factor · 134
Cross-sections · 264
Cross-semivariance · 304, 311
Crosstabulation · 184, 190, 191, 192
Cross-validation · 239, 293, 295, 299, 317, 319, 322
CSR · 64, 203, 209, 210, 228
Curvature · 76, 137, 252, 253, 259, 264, 265, 267, 268, 269, 270, 274, 286, 287
Cross-sectional · 265, 268
Longitudinal · 268
Maximum · 269
Mean · 265, 269
Minimum · 267, 269, 287, 289, 302
Path · 270
Plan · 265, 268
Profile · 265, 267
Radius · 270
Tangential · 265, 268
Curve · 16
Cycle · 326

D

Data Mining · 181
Data Transform · 309
Data Transformation · 19
Datum · 16
Declustering · 175, 179, 233, 304, 307, 311
Delaunay Diagram · 115
Delaunay Triangulation · 113, 115, 117, 226, 287, 296, 300
DEM · 8, 16, 75, 76, 115, 145, 146, 148, 171, 180, 250, 254, 264, 273, 274, 275, 276, 280, 282, 291, 303
Density · 17, 39, 54, 63, 64, 77, 80, 94, 100, 106, 107, 109, 122, 124, 126-34, 179, 197, 198, 199, 200, 203, 208, 210, 211, 238, 246, 284, 302, 306
Line · 133
Patch · 199
Desire Lines · 89
Deterministic Interpolation · 293, 304, 305, 317
DGPS · 16, 136
Differential Global Positioning System · See DGPS
Digital Elevation Model · See DEM
Digital Terrain Model · 16
Digraph · 326, 343
Dilation · 19, 170, 171
Dimension · 28
Fractal · 28, 87, 196, 197, 200, 201, 311, 316

Toplogical · 28
D-infinity · 260, 280, 281
Directional Analysis · 157, 162, 164, 173
Directional Derivatives · 169, 259, 269
Discriminant Analysis · 102
Disjunctive Kriging · 321
Disperse · 146
Dispersion · 198
Dissolve · 89, 91, 92, 151, 152
Distance
    3D Euclidean · 139
    City Block · 140, 147
    Euclidean · 72
    Euclidean – Point to Line · 72
    Lp Metric · 140
    Manhattan · 140
    Metrics · 137
    nD Euclidean · 139
    Normalised Euclidean · 139
    Polar Coordinate Flat Earth · 136
    Rectilinear · 140
    Sphercial - Haversine · 72
    Spherical - Cosine · 72
    Taxicab · 140
Distance Decay · 94, 153, 155, 156, 226, 239, 245,
    246, 253, 257, 294
Distance Statistics · 202
Distance Transforms · 117, 146, 153
Diversity · 23, 28, 116, 190, 195, 196, 201
Drainage · 146, 164, 166, 260, 272, 280, 303
Drift · 311, 318, 322
Drive-Time Zones · 153, 330
DTM · 16

E

Ecological Fallacy · 61, 235
Ecological Regression · 235
EDA · 16, 181, 182
Edge · 200, 201
Edge Correction · 204, 207
Edge Effects · 37, 131, 196, 216, 246, 252, 262,
    287
    Filtering · 168
Edge Labelling · 327
EDM · 16
Effective Number of Parameters · 239
Eigenvalue · 32, 232
Eigenvector · 32
Electronic Distance Measurement · 16
Ellipsoid · 16, 72, 77, 136, 139
Entropy · 28, 111, 116, 188
Erosion · 170, 171
ESDA · 16, 181, 182, 187, 188, 229, 236, 244

ESTDA · 181
Eulerian Circuit · 332
Excess risk · 126, 246
Exponential Integral Function · 31, 299

F

Factor Analysis · 102, 104
Factorial Kriging · 322
Feature · 16, 71, 72, 108, 109, 118, 136, 152, 210,
    255, 267
    Amalgamation · 79
    Central · 82
    Central · 84
    Extraction · 115
    Generalisation · 79
Feature Extraction · 179, 255, 267
Feature, Central · 62
Field · 250
    Scalar · 17
Fill · 137, 272, 280
Filtering · 3, 70, 166, 248, 257, 269, 271
    Blurring · 169
    Edge · 169
    Embossing · 169
    Frequency Domain · 168
    Gaussian · 169
    High-pass · 167
    Low-pass · 167
    Non-linear · 170
    Roberts Method · 169
    Sharpening · 169
    Smoothing · 169
    Symmetric · 169
First-order Process · 45
Flow · 164, 166, 199, 280, 281, 282, 283, 334
Flow Accumulation · 282
Flow Analysis · 157
Focal · 119, 165, 166, 169, 170, 190, 198, 199,
    291
Focal Functions · 165
Fourier Analysis · 304, 317
Fractal · 28, 87, 106, 196, 197, 199, 200, 201,
    243, 257, 285, 304, 311, 316
Fractal Dimension · see Dimension:Fractal
Fractal Terrains · 258
Fragstats · 8, 106, 195, 200, 217
Freeman-Tukey · 246
    Angular Transform · 31
    Square Root Transform · 30
Friction · 145, 148, 164
    Angle · 146
Fuzzy Boundaries · 110

G

Gabriel Network · 339
GAM · 8, 131, 188
Gamma Function · 31
Gaussian · 155, 240, 271, 314
Gaussian Sequential Simulation · 323
Gaussian Smoothing · 291
Gaussian Surface · 278, 285
Geary C · 218, 227, 229
Generalised Least Squares · 233
Genetic Algorithms · 330
Geocomputation · 12
Geodemographics · 16, 96, 122
Geographically Weighted Regression · 8, 43, 83, 155, 238
Geoid · 16, 77, 138
Geometric Mean · See Mean:Geometric
Geoportal · 50, 66
Geostatistical Analyst · 182, 187, 295, 299, 307, 313
Geostatistical Interpolation · 304
Geostatistics · 10, 16, 157, 174, 225, 304, 305, 311, 322
Ghost Lines · 15, 264
GIS-T · 16, 138, 150, 325
Global Functions · 166
Global Outliers · 182
Global Positioning System · See GPS
GLS · 233
GPS · 16, 35, 47, 50, 52
Gradient · 3, 17, 64, 74, 111, 137, 148, 161, 164, 165, 166, 169, 256, 261, 259-61, 264, 266, 269, 281
Grading · 148
Graph · 17, 326
    Connected · 326
    Degree · 326
    Edge · 326
    Planar · 17, 326
Great Circle · 72, 76, 77, 136, 138, 266
Greedy Algorithms · 330
Grid
    Smoothing · 271
Gridding · 10, 73, 145, 284, 293
Ground Truth · 175, 180, 192
Groundwater Modelling System · 300
GWR · See Geographically Weighted Regression

H

Hamiltonian Circuit · 332
Harmonic Mean · See Mean:Harmonic

Hedonic Regression · 231, 235
Heteroskedastic · 233
Heteroskedasticity · 232, 237
Heuristics · 17, 354
Hinge · 101, 183
Histograms · 174, 181, 184
Home range · 107
Horn's Method · 252, 260
Hot Spot · 130, 133, 228
Hydrology · 2, 3, 9, 69, 134, 157, 164, 166, 266, 272, 280

I

IAR · 246
ICAR · 246
Identity · 91
Identity Matrix · 32
IDW · 154, 286, 294, 295, 297, 300
Indegree · 326
Indicator Kriging · 321
Indicator Semivariance · 304, 310, 311
Information Statistic
    Diversity · 28
    Entropy · 28
Interpolation · 16, 44, 79, 92, 93, 109, 113, 120, 126, 133, 144, 154, 165, 166, 173, 174, 190, 232, 250, 251, 256, 271
    Bull's-eye Effect · 300, 319, 320
Inter-quartile Range · 25, 100, 101, 116, 170, 182, 188
Intersection · 70, 84, 89, 92, 108, 117, 118, 129, 152, 178, 189, 210, 238, 266, 285, 291, 293, 294, 296, 298
Intersection Density · 133
Interspersion · 198, 201
Invariance · 17, 41, 142
Inverse Distance · 154, 225, 288, 294, 295, 298, 300
ISODATA · 104, 105
Isolation · 58, 196, 197, 198, 201
Isotropic · 157
Isovist · 274, 279
Iterative Thresholding · 291

J

Jack-knifing · 189, 293, 319
JOIN · 118, 199, 216
Join Count · 214, 215

K

Kappa Index · 192
Kernel · 17, 106, 128, 131, 133, 167, 286, 316
  Gaussian · 167, 212
Kernel Density · 106, 126-34, 211, 238
Kernel Density Function
  Exponential · 131
  Gaussian · 131, 134
  Quartic · 131
  Triangular · 131
Kernel Smoothing · 271
Kolmogorov-Smirnov Test · 194, 205, 309
Kriging · 45, 286, 288, 298, 307, 311, 317, 319, 320, 321, 322, 323
Kurtosis · 22, 27

L

Lacunarity · 199, 374
Lagrangian · 243, 298, 318, 356
Lagrangian Multipliers · 355
Lagrangian Relaxation · 354
Landscape Metrics · 87, 194, 195
Laplacian · 169
ldd network · 280, 282
Least Cost Paths · 142, 144
Least Squares · 231, 253, 256, 286, 302
  Iteratively Reweighted · 240
  Weighted · 233, 239, 302
Level Sets · 54, 273, 291
Line Frequency · 133
Line of Sight · 274, 276, 277
Line Smoothing · 77
Linear Programming · 8, 331
LineString · 16, 18
Linking · 4, 54, 159, 181, 188, 255
LISA · 181, 226, 227, 228
Local Drainage Direction · 164, 272, 280
Local Functions · 165
Local Indicators of Spatial Association · See LISA
Local Outliers · 182, 188
Local Polynomial · 287, 289, 298, 300, 301, 302
Locational Analysis · 3, 140, 351
Logistics · 8
Logit · 30, 234
Lower Quartile · 25, 222
Lp-Metric · 139

M

Madograms · 225, 316

Majority · 23, 165, 170
Map Algebra · 3, 9, 119, 134, 145, 165, 171, 285, 291
Map Calculus · 257
Map Hot Spot · 9
Map Transformation · 19
Markov Chain Monte Carlo (MCMC) · 245
Masking · 63, 112, 175, 190, 312
Matrix
  Determinant · 32, 75, 300
  Inverse · 32
  Positive Definite · 32
  Positive Semi-definite · 32
  Singular · 32
  Symmetric · 32
  Trace · 32, 239
  Transpose · 32
MAUP · 95
Maximum · 23
MBR · 17, 80, 83, 84, 85, 88, 108, 180, 187, 189, 204, 207, 294, 297, 299, 347, 352
MCFP · 331, 332
Mean
  Arithmetic · 23
  Directional · 159
  Geometric · 24, 84, 202
  Harmonic · 24, 84
  Olympic · 24
  Power · 24
  Trim · 24
  Weighted · 23, 83, 87, 133, 169, 185, 190
Mean Absolute Deviation · 26
Mean Deviation · 26
Measurement Scales
  Cyclic · 36, 157, 158
  Interval · 35
  Nominal · 35
  Ordinal · 35
  Ratio · 35
Medial Axis · 86, 279
Median · 24, 62, 83, 88, 101, 124, 170, 179, 183, 188, 197, 202, 246
Median Polish Kriging · 321
Membership Function · 73, 110, 111
MER · 17
Merging · 79, 92, 94, 151, 166, 281
Metadata · 50, 66
Metrics · 73, 79, 84, 87, 135, 142, 147, 190, 202, 209, 217
  Distance · 137-42
  Landscape · 194-201
Mid-range · 25, 165
Minimax · 84, 115
Minimax Metric · 140
Minimum · 23

Minimum Bounding Rectangles · 108
Minimum Cost Flow · 332
Minimum Curvature · 265, 286, 289, 302
Minimum Spanning Tree · 338
Minority · 23, 165
MIP · 8, 330, 356
Mixed Integer Programming · 8, 330, 356
Mixed Regressive Spatial Autoregressive (MRSA)
   Model · 242
Mode · 24
Modifiable Areal Unit Problem · *See* MAUP
Moments
   Statistical · 22
Monte Carlo · 7, 229
Moran I · 218, 226, 229, 237, 247, 248
Morishita Index · 311
Morphometric Analysis · 265
Mosaic · 195, 196, 249, 252, 274
Moving Average · 286, 287, 290, 302, 324
MST · 332, 338, 339, 341, 342, 346, 348
Multi-collinearity Condition Number · 232
Multidimensional Scaling (MDS) · 42, 102
Multi-fractal · 257
Multivariate Classification · 102

N

Natural Boundaries · 112
Natural Breaks · 116
Natural Neighbour · 286, 288, 296, 297
NCPH · 107
Nearest Neighbour · 180, 202, 210, 286, 288, 298,
   311, 323
   Kth Order · 204
Neighbourhood · 42
Network · 326
Network Distance · 88, 136, 140, 150, 155
NNh · 210, 211
   Risk Adjusted · 211
Node Routing · 335
Non-Convex Polygonal Hull · 107, *See* NCPH
Non-linear Filtering · 170
Non-Planar · 17
Non-Stationarity · 241, 246, 294, 319
Non-Stationary · 214
Normal Distribution · 27, 29, 128, 129, 160, 229
Normal Transform · *See* Z-Transform
Normalisation · 63, 100, 122, 128, 184
NP-complete · 329, 331, 332, 333, 346, 362
NP-full · 329
NP-hard · 279, 329, 331, 332, 333, 341, 355, 364
Nugget · 286, 308, 312, 313, 314, 316, 317, 321
Nugget Variance · *See* Nugget

O

Occupancy · 124, 125, 126, 216
OGC · 8, 70
OLS · 231, 232, 234, 239, 240, 243, 247, 248, 302
Opening · 171
Ordinary Kriging · 298, 311, 317, 320, 321, 322,
   323
Ordinary Least Squares · *See* OLS
Orientation · 73
Outdegree · 326
Outliers · 24, 25, 62, 80, 84, 101, 107, 170, 181,
   182, 187, 214, 233, 240, 311, 312, 316, 319
Over-determined · 231
Overlay · 3, 7, 8, 41, 86, 89, 91, 92, 107, 123,
   152, 171, 178, 179, 283, 285
Over-specified · 232

P

Parsimony Objective · 63, 232, 234
Path · 146, 270, 326
Pattern · 64
pCentre · 354
pCentre Problem · 333
PCP · 208
Percentile · 100, 124
Periodograms · 304, 317
Piecewise Function · 256
PIP · 85
Pits · 146, 166, 255, 267, 271, 272, 280, 282, 303
Pixel · 17, 120, 122, 166, 170, 250, 281, 298
Plan Convexity · 266
Planar · 17, 21, 74, 89, 117, 139, 173, 267, 269,
   300
   Enforcement · 17
Planimetric Area · 74
pMedian · 354
pMedian Problem · 333
Point in Polygon · 85
Point Kriging · 317
Point Sampling · 175
Point Weeding · 77, 78, 108
Poisson Cluster Process · *See* PCP
Poisson Distribution · 27, 29, 193
Polygon · 2, 15, 16, 17, 22, 69, 71, 73, 74, 77, 79,
   82, 85, 87, 89, 92, 106, 107, 110, 115, 116,
   117, 118, 119, 121, 151, 152, 174, 176, 178,
   204, 207, 218, 226, 273, 286, 296, 301, 307,
   371
Polygon Decomposition · 86
Polygon Fitting · 107

Polyline · 15, 18, 71, 72, 78, 84, 88, 136, 158, 264, 291
Polynomial Regression · 287, 290, 302
Power
   Statistical · 62
Power Transforms · 30
PPDAC · 57, 58, 59, 62, 65, 244
Precision Farming · 175
Predictor Variables · 231, 235, 239
Principal Components Analysis · 102, 104, 232
Probability Density · 47
Probability Kriging · 321
Product Moment Correlation · 22, 238
Profile Curvature · 253
Profiles · 136, 264, 269, 270, 277, 286, 313
Profiling · 264, 287, 301
Proximity · 41, 57, 91, 115, 166, 179, 197, 201, 214, 237, 275
   Bands · 214
PTAS · 328
Pushbroom Algorithm · 145
Pycnophylactic · 92

Q

Quadrat · 178, 190, 191, 193
Quadrat Sampling · *See* Sampling:Quadrat
Quantile · 100, 125, 317
Quintic functions · 256

R

Radial Basis · 156, 286, 289, 298, 299, 307, 317
Radial Basis Function
   Completely Regularised Spline · 299
   Inverse Multiquadric · 299
   Multilog · 299
   Multiquadric · 298, 299
   Spline
      Cubic · 299
      Tension · 299
   Thin Plate Spline · 298, 299
Radio Transmission Analysis · 276
Random Walk · 163, 323
Range · 25
Raster Data Model · 18, 251
Rate Smoothing · 125
Raw Rates · 124
Re-districting · 92, 94
Reflexive · 205

Regression · 8, 43, 78, 133, 174, 179, 188, 190, 191, 229, 231, 233, 235, 236, 237, 238, 239, 241, 247, 251, 287, 290, 302, 318, 321
   CAR · 235
   Conditional Autoregressive · 235
   Ecological · 235
   Hedonic · 231, 235
   Linear · 162, 235, 242
   Logistic · 235, 240
   Multiple · 235
   Multivariate · 235
   Poisson · 235, 240
   SAR · 243
   SAR · 235, 242
   Simultaneous Autoregressive · 235
Regression Analysis · *See* Regression
RELATE · 118
Relative Neighbourhood Network · 341
Representative Fraction · 39
Representative Point · 37, 79
Resampling · 18, 70, 73, 89, 119, 120, 171, 175, 190, 226, 232, 233, 284, 285, 298, 301
Re-sampling · 293
Resolution · 74
Response Variables · 231
Resultant Vector · 158
Rhumb Line · 138
Richness · 190, 195, 196, 201
Ridged Multi-fractal · 257
Ripley's K Function · 206, 207, 208, 225
Risk · 57, 61, 62, 124, 126, 208, 210, 222, 234, 246
Robust · 62, 246
Rodograms · 304, 316
Rookcase · 9, 216, 220, 222, 229
Root Mean Square · 25, 26
Root Mean Squared Error · 26
Rose Diagram · 159, 262
Route Factor · 140
Routing · 3, 7, 11, 54, 107, 137, 138, 150, 335, 362
Row Standardisation · 227
Rubber Sheeting · 18

S

Sample Size · 304
Sampling · 59, 64, 140, 173, 175, 176, 179, 284, 307
   Equalised Random · 175
   Gibbs · 245
   Grid · 177
   Line · 269
   NonFree · 216

Point · 178
Quadrat · 178, 193
Random · 175, 178
Replacement · 189
Resolution · 218
Stratified Random · 175
SubSampling · 180
Sampling Error · 62
Sampling Network · 311
SAR · 235, 242, 243, 244, 248
Scatter Plots · 181, 182
Schwartz Criterion · 236
S-Distance · 9, 352, 357
Second Order Stationarity · 311, 320
Second-order Process · 45
Seed Points · 211
Semivariance · 181, 187, 225, 305, 307, 308, 310, 311, 318, 319
Semivariance Modelling · 304, 312
Serial Correlation Coefficient · 213
Service Areas · 359
Shape · 16, 22, 74, 80, 82, 87, 92, 95, 97, 107, 108, 112, 116, 130, 134, 136, 140, 142, 160, 165, 171, 176, 178, 181, 183, 196, 200, 202, 207, 222, 267, 268, 285, 294, 304, 312, 322
Patch · 197
Shape Distribution · 88
Sharpening · 45, 166, 167
Shepard · 286, 289, 300
Shortest Path · 11, 147, 150, 332, 342, 345
A* Algorithm · 344
Dantzig Algorithm · 343
Dijkstra Algorithm · 344
Shortest Path Tree · 342
Sigmoidal · 110, 111
Signatures · 104
Sill · 304, 308, 313, 314
Simple Kriging · 318, 320
Simulated Annealing · 323, 330
Skeletonised · 86
Skewness · 22, 27, 183
Slice · 100, 136, 264, 266
Slivers · 91
Slope · 18, 75, 93, 111, 113, 115, 135, 145, 148, 161, 162, 253, 259-61, 259-67, 267, 268, 270, 280, 281, 282, 316, 369
Slope Distance · 135
Smoothing · 45, 77, 78, 86, 116, 125, 128, 131, 133, 155, 188, 189, 256, 260, 270, 271, 272, 286, 289, 291, 295, 298, 299, 300
Space-time Analysis · 9
Spanning Tree · 332, 338
Spatial Autocorrelation · 9, 16, 44, 99, 175, 179, 190, 191, 213, 214, 220, 221, 223, 225, 229, 232, 233, 235, 237, 238, 243, 248, 305, 308, 312
Spatial Dependence · 43, 44
Spatial Differencing · 247
Spatial Filter · 166, 169
Spatial Filtering · 166, 170, 247
Spatial Heterogeneity · 43
Spatial Overlay · 86, 91, 190
Spatial Probability · 47
Spatial Query · 18, 118
Spatial Sampling · 173, 175
Spatial Support · 36
Spatial Weights · 226, 245
Spatial Weights Matrix · 41, 155, 219, 225, 226, 228, 233, 238, 242, 243, 340
Spatially Extensive · 122, 126
Spatially Intensive · 100, 122, 124
Spatio-temporal Analysis · 2, 181, 208
Spherical Excess · 77
Spheroid · 16, 18, 20, 136
Spline · 31, 78, 270, 271, 287, 289, 291, 298, 299, 300, 301, 302, 303
Bicubic · 291
Regularised · 299
Smoothing · 78
Thin Plate · 299
Split · 89, 92
SQL · 18, 23, 91, 118, 119
Standard Deviation · 22, 26, 88, 100, 101, 105, 116, 123, 128, 131, 159, 162, 165, 170, 181, 188, 202, 204, 211, 213, 216, 224, 230, 257, 317, 320, 323
Standard Deviation Map · 319, 324
Standard Deviational Ellipse · 88, 162, 163, 202, 211, 224
Standard Distance · 80, 162, 190, 202, 204
Standard Error · 26, 30, 203, 210, 232, 239
Standardisation · 122, 123, 133, 227, 237
Stationarity · 304, 307, 311, 317, 368
Stationary · 214, 306, 319
Statistical Inference · 48
Steiner MST · 332, 342
Steiner Point · 86, 115, 332
Steiner Trees · 341
Stratified Kriging · 322
Stream Basins · 280, 283
Stream Networks · 282
Strength of Evidence Objective · 232
Structural Variance · 308
Study Error · 62
Sum · 23
Support · 304, 307
Surface · 18
Surface Area · 74, 75, 76, 272
SURPOP · 93

Symmetric Difference · 70, 92

T

Tabu · 330
Tangential Curvature · 253
Terrestrial Distance · 136, 138
Tessellation · 19, 113
Tesseral · 19
Thematic Map · 15, 99, 236
Thiessen · 115, 311
Time Series · 100, 213, 214, 220, 225, 242, 247
TIN · 19, 69, 74, 115, 251, 254, 256, 273, 284, 291, 300
Tobler's First Law · 44, 45, 46, 47, 48, 293
Tomlin · 9, 121, 165
    Map Analysis Package · 166
Topogrid · 303
Topological Overlay · 89, 91
Topology · 19, 40, 85, 91, 150, 226
Touching · 91, 199
Transformation · 304
Transformed Vegetation Index · *See* TVI
Transportation Problem · 332
Travel Time Zones · 360
Travelling Salesman Problem · *See* TSP
Tree · 326
Trend · 214, 244, 257, 318, 321
Trend Analysis · 188, 232, 302, 311, 319
Trend Surface · 236, 287
Trend, Linear · 236, 243
Triangulated Irregular Network · 19, 69, 109, 115
Trim-range · 25, 123
TSA · 213, 222
TSP · 8, 329, 332, 346, 348, 349, 362
    Dual Circuit · 349
TSPLIB · 347, 356, 383
TVI · 120, 121

U

Uncertainty · 47
Uniform Distribution · 29
Union · 70, 89
Univariate Classification · 99
Universal Kriging · 311, 321
Upper Quartile · 25, 101, 222
U-turns · 346

V

Variance · 26
    Instability · 124
Variance Inflation · 232
Variance Mean Ratio · 27
Variety · 23
Variogram · 246, 286, 288, 304, 305-15, 316, 317, 318, 320, 322, 323
    Graphs · 315
    Non-transitive · 308
    Transitive · 308
Variogram Model
    Circular · 314
    Cubic · 314
    Exponential · 314
    Linear · 314
    Logarithmic · 314
    Pentaspherical · 314
    Power · 314
    Quadratic · 314
    Spherical · 314
    Tetraspherical · 314
    Wave Hole · 314
Variography · 312
Vector · 19
Vehicle Routing Problem · 332, 349
Vertex · 326
Viewshed · 274, 275, 277, 278
Vincenty · 10, 139
Visibility · 3, 113, 164, 274, 277, 279
Volumetric Analysis · 72, 272
Voronoi · 92, 97, 107, 115-17, 141, 166, 179, 187, 188, 226, 236, 250, 286, 296, 307, 352, 359

W

Watersheds · 280, 283
Weighted Mean · *See* Mean:Weighted
WGS84 · 20, 72, 136
Whiskers · 183
Wind Rose · 161
Winding Number · 85
Wombling · 111
World Geodetic System · 20, 34

X

XOR · 89, 119

Z

Zonal · 106, 119, 165, 173, 184, 190, 222, 247

Zonal Functions · 166
Zone Membership · 106, 110
Zone of Indifference Weighting · 226, 228
Z-transform · 30, 101, 203, 216, 229

Index cross references

Complete Spatial Randomness • *See* CSR

Bull's-eye effect• *See* Interpolation:Bull's-eye effect

Digital Elevation Model • *See* DEM

Differential Global Positioning System • See DGPS

Fractal Dimension • see Dimension:Fractal

GWR • *See* Geographically Weighted Regression

Geometric Mean • *See* Mean:Geometric

Global Positioning System • *See* GPS

Harmonic Mean • *See* Mean:Harmonic

Local Indicators of Spatial Association • *See* LISA

Non-Convex Polygonal Hull • *See* NCPH

Normal Transform • *See* Z-Transform

Modifiable Areal Unit Problem • *See* MAUP

Nugget Variance • *See* Nugget

Ordinary Least Squares • *See* OLS

Poisson Cluster Process • *See* PCP

Quadrat Sampling • *See* Sampling:Quadrat

Regression Analysis • *See* Regression

Transformed Vegetation Index • *See* TVI

Travelling Salesman Problem • *See* TSP

Weighted Mean • *See* Mean:Weighted

Geospatial Analysis

Michael J de Smith, University College London, UK

Michael F Goodchild, University of California, Santa Barbara, USA

Paul A Longley, University College London, UK

This Guide addresses the full spectrum of spatial analysis and associated techniques that are provided within currently available and widely used Geographic Information Systems (GIS). It is both broad, in terms of concepts and techniques, and representative in terms of the software that people actually use.

Topics covered include:

- the principal concepts of geospatial analysis, their origins and methodological context
- core topics in geospatial analysis, including distance and directional analysis, geometrical processing, map algebra, and grid models
- basic methods of exploratory spatial data analysis (ESDA) and spatial statistics, including spatial autocorrelation and spatial regression
- surface analysis, including surface form analysis, gridding and interpolation methods
- network and locational analysis, including shortest path calculation, travelling salesman problems, facility location and arc routing

The Guide has been designed to be of use all those involved in geospatial analysis, from undergraduates and postgraduates studying GIS and spatial analysis to professional analysts, software engineers and GIS practitioners. It is much more than a cookbook of algorithms and techniques — it provides a thorough explanation of the key concepts and techniques of geospatial analysis using applications developed using widely available software packages. It builds upon the spatial analysis topics included in the US National Research Council 'Beyond Mapping' agenda, the NCGIA Core Curriculum and UCGIS Body of Knowledge, and is a useful accompaniment to courses built around them.

Software products referenced in this edition include:

ArcGIS ◆ ANUDEM ◆ Boundaryseer ◆ CCMaps ◆ Cellular Expert ◆ Clusterseer ◆ ComSiteDesign ◆ Concorde ◆ CPLEX ◆ Crimestat ◆ ENVI ◆ Fragstats ◆ GAM ◆ GeoBUGS ◆ GeoDa ◆ Geomed ◆ GMS ◆ Grapher ◆ GRASS ◆ GS+ GSLIB ◆ GStat ◆ GWR ◆ HertzMapper ◆ Hawth'sTools ◆ Idrisi ◆ Imagine ◆ ISATIS ◆ Land Desktop ◆ Landserf ◆ LEDA ◆ LOLA ◆ Manifold ◆ MapCalc ◆ MapInfo ◆ Maptitude ◆ MATLab ◆ MFWorks ◆ NETACT ◆ Oriana ◆ PCRaster ◆ RiverTools ◆ Rookcase ◆ SANET ◆ S-Distance ◆ SITATION ◆ SpaceStat ◆ SPLANCS ◆ SPSS ◆ STARS ◆ STATA ◆ STIS ◆ Surfer ◆ SURPOP ◆ TAP ◆ TAS ◆ TNTMips ◆ TAUDEM ◆TransCAD ◆ Variowin ◆ WinBUGS ◆ Xpress-MP ◆ ZDES

www.spatialanalysisonline.com

Printed in the United Kingdom
by Lightning Source UK Ltd.
121881UK00001B/9-16/A